Commercial Poultry Raising

A Practical and Complete Reference Work For The Amateur, Fancier or General Farmer, Especially Adapted To The Commercial Poultryman

by H. Armstrong Roberts

with an introduction by Jackson Chambers

This work contains material that was originally published in 1918.

This publication is within the Public Domain.

*This edition is reprinted for educational purposes
and in accordance with all applicable Federal Laws.*

Introduction Copyright 2018 by Jackson Chambers

Self Reliance Books

Get more historic titles on animal and stock breeding, gardening and old
fashioned skills by visiting us at:

http://selfreliancebooks.blogspot.com/

Introduction

I am pleased to present yet another title on Poultry.

The work is in the Public Domain and is re-printed here in accordance with Federal Laws.

As with all reprinted books of this age that are intended to perfectly reproduce the original edition, considerable pains and effort had to be undertaken to correct fading and sometimes outright damage to existing proofs of this title. At times, this task is quite monumental, requiring an almost total "rebuilding" of some pages from digital proofs of multiple copies. Despite this, imperfections still sometimes exist in the final proof and may detract from the visual appearance of the text.

I hope you enjoy reading this book as much as I enjoyed making it available to readers again.

Jackson Chambers

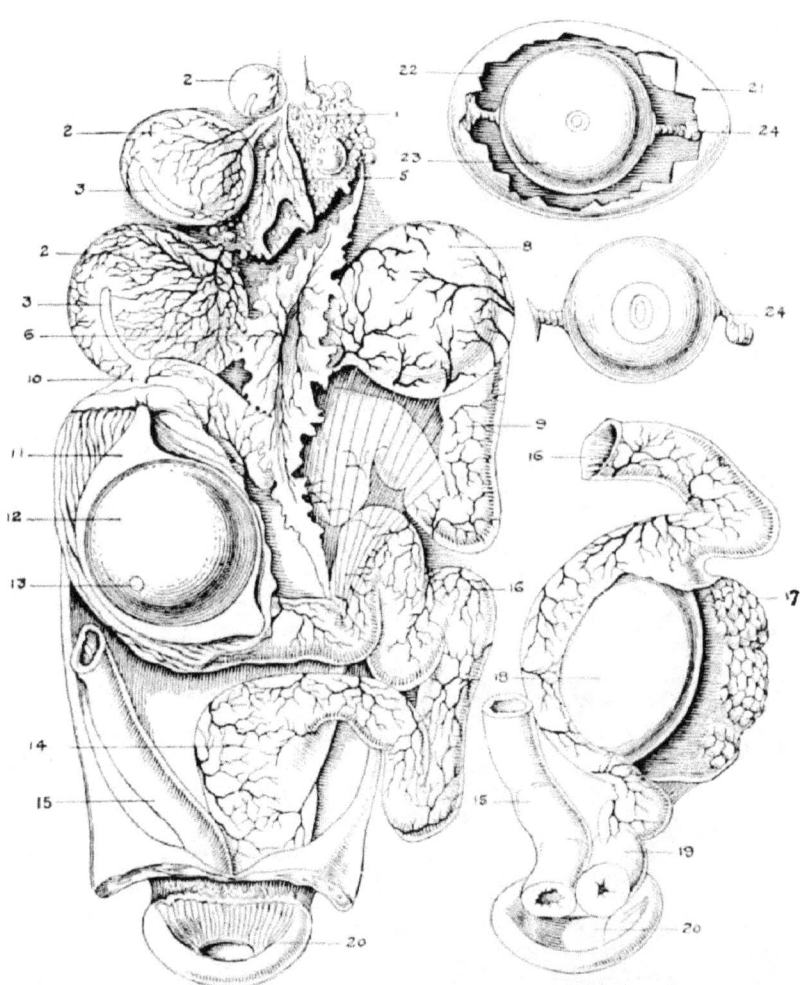

Diagram reproductive organs of laying hen.

1, Ovary, with minute ovules; 2, yolk sacs; 3, suture lines; 5, empty yolk sac; 6, funnel opening into oviduct; 8, yolk in oviduct; 9, albumen-secreting region; 10, the same; 11, albumen being secreted; 12, yolk passing through oviduct; 13, germinal disc; 14, uterus; 15, large intestine; 16, isthmus; 17, glands of uterus; 18, complete egg; 19, vagina; 20, cloaca; 21, egg shell; 22, albumen; 23, yolk; 24, chalaza.

To

𝕸. 𝕬. 𝕽.

WHOSE ENCOURAGEMENT WAS A
CONSTANT SOURCE OF INSPIRATION THIS
WORK IS AFFECTIONATELY
INSCRIBED

INTRODUCTION

There are numerous poultry books on the market. Some are valuable works, except that they are out of date. Others have been written by persons who knew more about theory than of practice. Still others were printed because the author had a system to advocate, a farm to advertise, or an axe to grind. Obviously, in reading these so-called systems the beginner generally obtains erroneous ideas, which sooner or later bring about costly mistakes, often failure.

In the chapters that follow the author has no hobbies to ride, no theories to advance, nothing to offer, in fact, except the practical information which he has acquired from many years of actual experience raising fowls on a commercial scale, and from the associations of other poultrymen with whom he has come in contact in a business or friendly way. His effort has been to give facts and state principles clearly, so as to establish a solid foundation for the study of poultry culture. The real study, of course, commences with the actual work with the fowls, and cannot be acquired from the printed page alone.

No phase of agriculture or animal husbandry has made such enormous progress in the past thirty years as poultry culture. Consequently, literature written on this subject a quarter of a century ago, or even a decade ago, is now mostly obsolete. Not that fowls have changed their habits to such an extent, but because we have learned more about their habits, and how to derive the greatest benefits from them at the least possible cost.

Poultry raising on a commercial scale could not be attempted on the methods practised by our grandfathers. It was not until the invention and perfection of appliances for the artificial rearing of little chicks in large numbers that poultry keeping really passed the "pin money" stage. Even then, it was not until we learned something about the scientific manner of feeding, breed- ing and housing the fowls that our efforts were assured any degree of success from the standpoint of dollars and cents. We have long known how to keep fowls; but it has been compara- tively recent that we found out how to make the fowls *keep us*.

There is money to be made from poultry. There is a living to be made from it, and a good living. Some claim there are for- tunes to be had from hens. The author begs leave to differ with this last statement, or rather to qualify it. Fortunes might be made from chickens if it were possible to look after them, personally, in large enough numbers; but this is quite out of the question. Very large flocks necessitate the employment of help, and it is the uncertainty or incompetence of this help that makes success with tens of thousands of birds more or less dependent upon chance.

The author has purposely avoided all reference to "big stories of big profits," likewise the fads and fancies of poultry keeping, which have brought disaster to so many beginners. He has aimed to show that hard work is necessary, that the caretaker must be always on the job, that some of the tasks are not as pleasant as they might be, that common sense is required above theoretical training, that disappointments and mistakes are to be expected, and that it is no get-rich-quick scheme, but a safe, sane, practical business enterprise, and as such it must be con- ducted. Where figures are quoted, especially in respect to profits, if anything the author has been too conservative. These figures, however, are based on prices prevailing before our entrance into the European War. Success does not fall into the lap of the poultryman; he must go out and dig for it. And if he digs hard enough, he is sure to be rewarded.

As previously inferred, the author has read most of the litera- ture written about poultry, and having found it either obsolete, incomplete or utterly fallacious, he has endeavored to prepare a work that is the most up-to-date, comprehensive, practical guide-book of its kind. Economy and efficiency are the under-

INTRODUCTION

lying motives of every chapter. They are what he conceives to be the cornerstones of success with poultry.

Acknowledgment is herewith made to The Country Gentleman and Public Ledger, in which publications most of these chapters have appeared, for the rights to publish them in book form. Appreciation is also acknowledged to those who kindly submitted photographs for reproduction as illustrations.

H. ARMSTRONG ROBERTS

Philadelphia, 1918

CONTENTS

ILLUSTRATIONS

ILLUSTRATIONS

ILLUSTRATIONS

ILLUSTRATIONS

COMMERCIAL POULTRY RAISING

CHAPTER I

AMERICAN POULTRY INDUSTRY

The term poultry, as it is commonly understood, applies collectively to those species of domestic birds which are kept for the purpose of furnishing eggs and meat for human consumption. Game birds are used on the table, but so long as they remain in a wild state they cannot be classed as poultry. On the other hand, we are disposed to consider certain species of birds as poultry, such as pigeons which are bred for ornament or as carriers, whereas they do not rightfully belong under this head. Peafowls are considered as poultry, and while formerly they were bred for the table, they are now raised almost exclusively for ornament. Swans are in pretty much the same position as peafowls.

Main Divisions.—Technically there are three main divisions of poultry: (1) GALLINACEA, or comb bearers, which include chickens, turkeys, guinea fowls, pheasants and quail. All resemble each other in general structure and habits, and all are distinguished from other birds in that the flesh on the breast and wings is lighter in color than on the rest of the body. (2) NATATORES, or swimmers, include ducks, geese and swans, and are characterized by their web feet and long, thick bills. (3) COLUMBIDAE, or doves, is the other order; pigeons are its only representatives in the poultry world.

Chickens comprise the bulk of the poultry industry, especially in this country, and their relation to the animal kingdom is as follows: They belong to the series, METAZOA, because they consist of animals with cellular tissues and true eggs. They are of the branch, VERTEBRATA, inasmuch as they are animals having an internal skeleton and backbone. They are in the division known

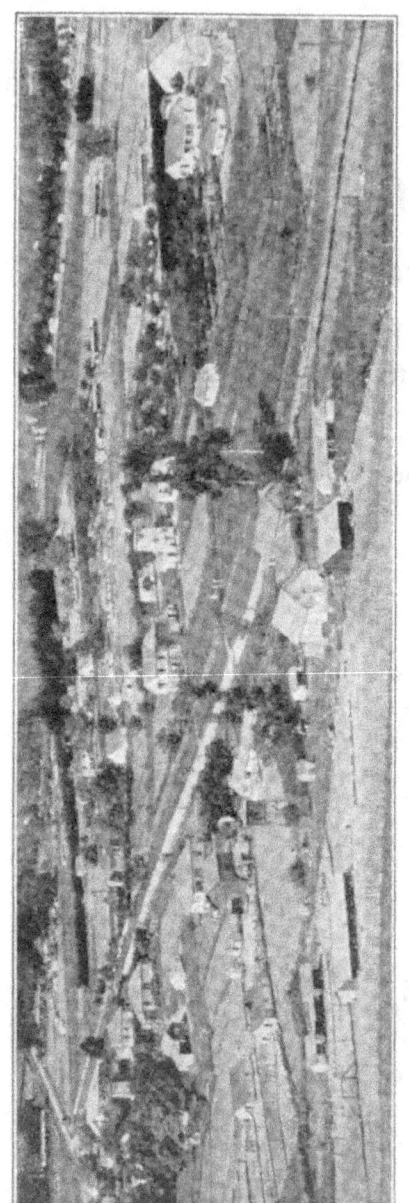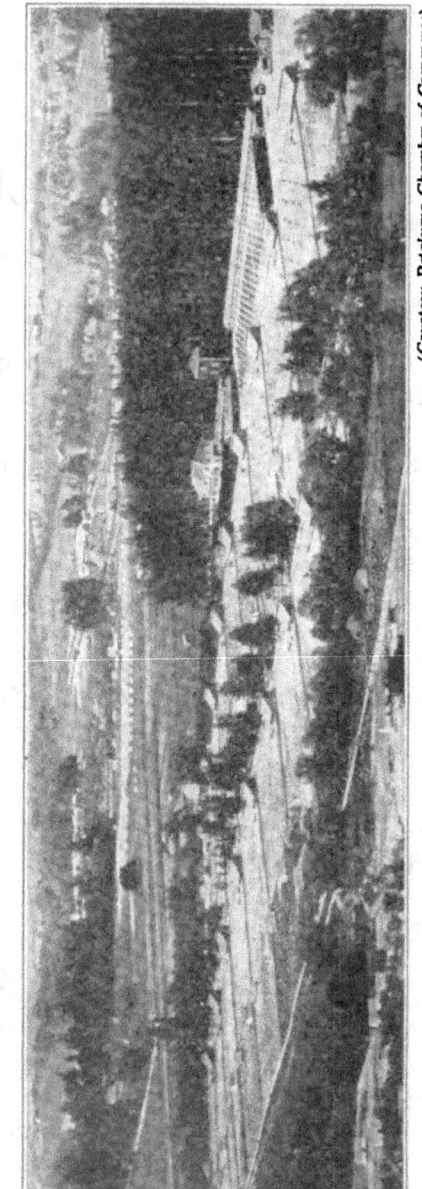

Fig. 1.—Petaluma District, California, where a million White Leghorns are kept.

2

as CRANIOTA, because they have a skull, heart and brain. They are classed as AVES, because they are true birds, feathered, and have four limbs, one pair for progression on land and one pair for flight; no teeth. They are sub-divided into the order, RASORES, because they are terrestrial in their habits, having stout legs suitable for scratching, and strong arched beaks suitable for seed eating. GALLUS is a true representative of this order, and is the common ancestor of all our domestic fowls; it was a jungle fowl native to southwestern Asia and Oceania.

To-day there are over a hundred different varieties of chickens. By variety we mean species of certain well defined characteristics, which are officially recognized, as by the American Standard of Perfection.

Scope of Poultry Industry.—The importance of the poultry industry, and the relative importance of chickens to the industry, may be gathered from the United States census report for 1910, as shown in Table I.

TABLE I.—SPECIES OF POULTRY IN UNITED STATES ACCORDING TO 1910 CENSUS

	FARMS REPORTING		NUMBER OF POULTRY	ESTIMATED VALUE
	Number	Per Cent of Farms		
Total Poultry......	5,585,012	88.1	295,876,176	$153,394,142
Chickens..........	5,577,218	88.0	280,340,643	$140,192,912
Turkeys...........	852,679	13.4	3,688,688	6,605,640
Ducks............	503,673	7.9	2,904,359	1,566,176
Geese............	661,189	10.4	4,431,623	3,192,861
Guinea fowls.......	339,922	5.4	1,765,033	613,282
Pigeons...........	99,409	1.6	2,730,996	162,372
All others........	2,005	.001	14,834	460,899

Unfortunately, Table I fails to take into account the numbers of poultry under three months of age, or those which are raised and kept in backyards of towns and villages all over the country. If these were added they would constitute a big increase over the

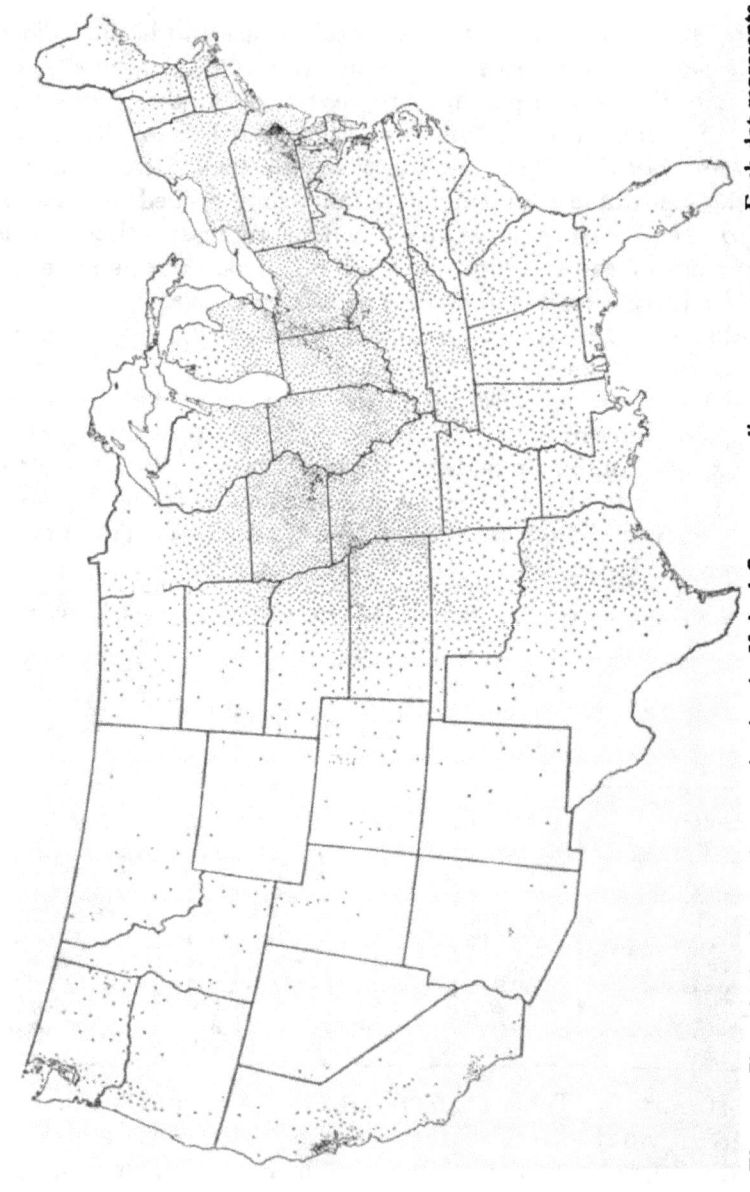

Fig. 2.—Chart indicating poultry production in United States according to 1910 census. Each dot represents 50,000 head of poultry.

figures in the census report, and also raise the ratio of chickens to other species.

Regarding the magnitude of the industry in still another way, in 1911 the Secretary of Agriculture placed the national income from poultry products at $750,000,000 for the year. This figure equaled the combined value of all the gold, silver, iron and coal mined that year, and exceeded the value of the wheat crop for the same period. To-day, the income from poultry products is estimated at one billion dollars annually, or one-twelfth the value of all agricultural products, or one-fiftieth of our total wealth on all manufactures and industries of every description.

Manifestly, the poultry industry occupies a very important part in our development, a very much more important part than most of us have any idea.

Eggs Leading Product.—According to the 1910 census the income from poultry products for the average farm was $104.98, or about two dollars per bird. Eggs

(*Courtesy Wisconsin Experiment Station*)

Fig. 3.—A hen with a brood of sturdy chicks is one of the farmer's best assets.

are the leading poultry product, and constitute about sixty-five per cent of the total value of the poultry as a whole. That most of these eggs are hens' eggs goes without saying. True, the eggs from ducks and guinea fowls find their way into the trade channels, but in such small numbers as to be a negligible factor. Turkey eggs and goose eggs are used almost exclusively for hatching purposes.

Demand Greater Than Supply.—The output of eggs is increasing each year, yet there are no gluts for the reason that the demand is always greater than the supply. According to the Year Book of the Department of Agriculture for 1910 the average price paid to farmers in 1899 was 11.5 cents a dozen. In 1909 the average price was 19.7 cents per dozen, an increase of almost a hundred per cent in ten years. In 1917 the average price was about 25 cents a dozen.

Table II shows a comparison of the prices paid for poultry and eggs in New York City for a number of years. The prices are those paid by wholesalers, all grades considered, except those of the strictly fancy marks from nearby points, which are in small volume, and handled mostly by express shipments.

TABLE II.—WHOLESALE PRICES OF EGGS AND POULTRY IN NEW YORK CITY

| Year | Prices Paid in Cents | | |
	Eggs Per Dozen	Live Poultry Per Pound	Dressed Poultry Per Pound
1896	14.0	9.0	10.0
1897	14.5	8.5	9.0
1898	15.5	8.8	9.1
1899	17.0	9.8	11.0
1900	16.0	9.3	10.0
1901	18.0	9.5	11.0
1902	20.0	11.5	14.5
1903	18.5	12.5	15.5
1904	20.0	12.5	15.0
1905	20.0	13.0	16.0
1906	19.0	12.8	15.0
1907	18.5	13.8	15.5
1908	19.0	13.5	15.8
1909	23.5	16.0	18.3
1910	23.0	17.0	19.8
1911	19.3	15.0	17.8
1912	22.8	15.5	18.3

For 1917 the average price on eggs was about 30 cents, for live poultry 22 cents, and for dressed poultry 26 cents.

Export Trade.—Aside from the increase in our home consumption of poultry products, our export trade is steadily growing. In 1912 we imported eggs to the extent of $147,173. For the same period we exported eggs to the extent of $3,400,000. Table III shows how these exports have grown, and to which continental divisions they were sent:

TABLE III.—EXPORTS OF EGGS FROM U. S.

YEAR	1908	1909	1910	1911	1912
Europe	$44,995	$2,265	$574	$639	$15,613
North America	1,490,279	1,195,446	1,256,998	1,782,141	3,375,413
South America	2,692	1,342	1,921	3,218	3,135
Asia	597	135	337	43	60
Oceania	1,451	334	666	978	1,731

The figures in Table III apply to eggs in the shell. There were additional exports, amounting to about thirty thousand dollars a year, for canned eggs, yolks, and so on.

Distribution of Poultry.—Geographically the poultry production of this country may be divided into three principal sections: (1) the northeastern states, including New England and the Middle Atlantic States, (2) the states bordering on the Pacific coast, and (3) the states lying in the Mississippi Valley. Each principal section is more or less distinctive for its kind of products. In the northeastern territory the poultry industry is highly specialized. A great many large farms devoted to nothing but fowls are found there, and their outputs are consumed in nearby markets in the large cities. In the Vineland, New Jersey, tract there are upwards of a quarter of a million White Leghorn hens, bred for high egg production.

The Pacific territory is pretty much the same as the northeastern section, only on a smaller scale. Conditions are pretty much identical, both as to methods of production and marketing. The Petaluma district corresponds to the Vineland tract, and it is the largest single poultry producing section in the world. See Fig. 1. It has close to a million White Leghorn hens.

Mississippi Valley Poultry.—Things are very different in the Mississippi Valley, which comprises the states of Minnesota, Wisconsin, Illinois, Michigan, Indiana, Ohio, Nebraska, Iowa, Kansas, Missouri, Kentucky, Tennessee, Oklahoma, Arkansas, and Texas. In this vast territory the great bulk of our poultry is produced, yet for all, there are very few farms which might be

Fig. 4.—This picture suggests what can be done in the way of combining poultry with other farm crops.

8

termed strictly poultry farms, or where the poultry reared is considered anything more than a by-product. The farms are devoted to other forms of livestock or to other agricultural pursuits, chiefly grain, and what chickens are kept represent small flocks, mostly of indiscriminate origin.

TABLE IV.—VALUE OF POULTRY IN TEN LEADING STATES, ACCORDING TO CENSUS OF 1910

RANK	STATE	TOTAL VALUE
I	Iowa	$12,270,000
2	Missouri	11,871,000
3	Illinois	11,697,000
4	Ohio	9,533,000
5	New York	7,879,000
6	Indiana	7,762,000
7	Pennsylvania	7,674,000
8	Kansas	7,377,000
9	Michigan	5,611,000
10	Texas	4,807,000

Collectively these flocks produce over fifty per cent of the total valuation of our poultry industry, the greater part of which is marketed elsewhere. The quality of the products produced in the Mississippi Valley is inferior to the grades produced in the other sections. This is especially true of eggs, and is due to the fact that the average farmer does not give his hens the correct care, neither does he take sufficient pains in marketing the eggs promptly. The bulk of the poultry and eggs produced in this section is sold to local buyers, thence to the city packing houses and wholesalers.

Fig. 5.—Model backyard poultry plant.

10

CHAPTER II

ANALYZING THE DIFFERENT BREEDS

Best Breed.—No inquiry comes to hand more frequently than: "Which is the best breed of fowls?" "Which variety of chickens is the most profitable?"

These and similar questions are entirely pertinent ones, of course, yet none are so difficult to answer, except at great length, accompanied by much explanatory matter. I have always held that there is no one BEST breed of chickens. And I am still of the same opinion.

To be sure, there are best breeds for a particular purpose, and there may be superior breeds for a given locality, also, there are varieties better suited to a special taste for ornamental fowls; but, these qualifications make the subject more or less specific, which is the only way that it can be discussed intelligently. Let us consider it from this impartial standpoint.

Standard Varieties.—It may surprise the layman to learn that there are upward of 110 standard and a large number of non-standard varieties of chickens. See Table V. By non-standard breeds is meant those that are well established, but not as yet admitted to the American Standard of Perfection, which is the authority on poultry in this country, just as the American Kennel Club is the authority on dogs. The American Standard of Perfection is a very illuminating work, by the way, gotten up by the American Poultry Association, and should find a place on every poultryman's bookshelf. No fancier should attempt to raise fowls for exhibition purposes without this authority, and even for commercial ends it will be found decidedly helpful.

11

TABLE V.—STANDARD CLASSES, BREEDS, AND VARIETIES OF POULTRY

ACCORDING TO AMERICAN STANDARD OF PERFECTION FOR 1910

CLASS	BREED	VARIETY
American	Plymouth Rock	Barred White Buff Silver Pencilled Partridge Columbian
	Wyandotte	Silver Golden White Buff Black Partridge Silver Pencilled Columbian
	Java	Black Mottled
	Dominique	Rose Comb
	Rhode Island Red	Single Comb Rose Comb
	Buckeye	Pea Comb
Asiatic	Brahma	Light Dark
	Cochin	Buff Partridge White Black
	Langshan	Black White
Mediterranean	Leghorn	Single Comb Brown Rose Comb Brown Single Comb White Rose Comb White Single Comb Buff Rose Comb Buff Single Comb Black Silver Single Comb Red Pyle
	Minorca	Single Comb Black Rose Comb Black Single Comb White Single Comb Buff
	Spanish	White-faced Black
	Andalusians	Blue
	Ancona	Mottled

TABLE V.—STANDARD CLASSES, BREEDS, AND VARIETIES OF POULTRY—
(*Continued*)

CLASS	BREED	VARIETY
English	Dorkings	White Silver Gray Colored
	Red Caps	Rose Comb
	Orpingtons	Single Comb Buff Single Comb Black Single Comb White
Polish	Polish	White-crested Black Bearded Golden Bearded Silver Bearded White Buff Laced Non-bearded Golden Non-bearded Silver Non-bearded White
Hamburg	Hamburg	Golden Spangled Silver Spangled Golden Pencilled Silver Pencilled White Black
French	Houdan	Mottled
	Crevecœurs	Black
	La Fleche	Black
Game and Game Bantam	Game	Black-breasted Red Brown Red Golden Duckwing Silver Duckwing Birchen Red Pyle White Black
	Game Bantam	Black-breasted Red Brown Red Golden Duckwing Silver Duckwing Birchen Red Pyle White Black

TABLE V.—STANDARD CLASSES, BREEDS, AND VARIETIES OF POULTRY—
(*Continued*)

CLASS	BREED	VARIETY
Oriental	Cornish	Dark
		White
		White-laced Red
	Sumatra	Black
	Malay	Black-breasted Red
	Malay Bantam	Black-breasted Red
Ornamental Bantam	Sebright	Golden
		Silver
	Rose Comb	White
		Black
	Booted	White
	Brahma	Light
		Dark
	Cochin	Buff
		Partridge
		White
		Black
	Japanese	Black Tailed
		White
		Black
	Polish	Bearded White
		Buff Laced
		Non-bearded
Miscellaneous	Silkies	White
	Sultans	White
	Frizzles	Any Color
Turkey	Turkey	Bronze
		Narragansett
		White
		Black
		Buff
		Slate
		Bourbon

TABLE V.—STANDARD CLASSES, BREEDS, AND VARIETIES OF POULTRY—
(*Continued*)

CLASS	BREED	VARIETY
Duck............	Pekin...........	White
	Aylesbury.......	White
	Rouen..........	Colored
	Cayuga.........	Black
	Call...........	Gray / White
	East India......	Black
	Crested........	White
	Muscovy........	Colored / White
	Indian Runner...	Fawn and White / White
	Swedish........	Blue
Goose............	Toulouse..........Gray	
	Embden..........White	
	African ....Gray	
	Chinese.........	Brown / White
	Wild or Canadian ...Gray	
	Egyptian ....Colored	

The standard varieties are generally divided into the following classes:

1. American Class: Plymouth Rocks, Wyandottes, Rhode Island Reds, Javas, Dominiques, Buckeyes.
2. Asiatic Class: Brahmas, Cochins and Langshans.
3. Mediterranean Class: Leghorns, Minorcas, Spanish, Andalusians, Campines and Anconas.
4. English Class: Dorkings, Orpingtons and Redcaps.
5. Polish Class: Polish fowls of which there are eight varieties.
6. Dutch Class: Hamburgs, of which there are six varieties.
7. French Class: Houdans, Crevecoeurs and La Fleche.
8. Games and Game Bantams: Eight varieties of games, and the same number and kind of game bantams.

9. Oriental Class: Cornish, Sumatras, Malays and Malay Bantams.
10. Ornamental Bantam Class: Sebrights, Rose-Comb, Booted, Brahmas, Cochins, Japanese and Polish.
11. Miscellaneous: Silkies, Sultans and Frizzles.

For practical purposes the above eleven classes are grouped into four general classes, about as follows:

1. Egg breeds, commonly called the non-sitting breeds.
2. Meat or table breeds.
3. General purpose breeds, sometimes spoken of as dual-purpose fowls.
4. Ornamental breeds, which, as the term implies, includes such ornamental varieties as the Polish, Games, and Bantams, and are raised almost exclusively by fanciers for exhibition purposes.

General Discussion.—It will appear obvious that the question of selecting a particular variety of poultry is one having considerable scope and many sides. It is manifestly impossible to take care of the subject in a single chapter. Therefore, in this chapter we will devote our attention to a general discussion of the problem.

No Iron-Clad Rules.—The writer has always worked on the assumption, if such it be, that there are no iron-clad rules for poultry keeping. It has been said that a man must be half hen with feathers growing on his back to make a success with chickens. We interpret this to mean that a man (or woman, we use the masculine form merely to simplify expression) must be fond of the work, and understand the nature of his *charge*, otherwise he can not attain that insight and competence, in a sense comradeship, which is the foundation of all success in handling animals. In other words, he must always have the interests of his birds at heart. Such a man can make a success of any breed of poultry.

Hens are considered downright stupid by some. However true this may be, there is considerable human nature in a chicken, more character, perhaps, than one would suppose. Failure to appreciate this is probably one of the reasons why many breeders are unsuccessful. There are temperaments in chickens, just as there are temperaments in the human kind. The point is to seek the fowl having some basis for mutual exchange—a similarity or congeniality between the keeper and his flock. For this reason I do not care to say offhand that a certain breed is best. I prefer to leave this selection to the individual. The wisdom of this stand will become apparent in the succeeding paragraphs.

Some prospective poultrymen contemplate a start in the business with a fixed interest or fondness for a particular variety firmly established in their minds. Others enter the field with a

Fig. 6.—Light Brahmas.

definite end, but no particular interest in any one breed. Still others engage in the work without any special interest in any breed or any particular phase of the work, simply a desire to raise chickens. There is yet another class of prospective poultrymen who, being possessed of certain real estate, from which it is impracticable to move, we will say, should select breeds that are best adapted to their particular localities.

Have a Definite Aim.—At first glance you may conclude that

2

this analysis is hair-splitting. Not so. It is common sense in the extreme. One of the fundamentals of good business is to apply one's talents to the best advantage, where they are the most congenial, the most productive, and where they can be made to specialize. Aimlessness is almost fatal to success in any line of endeavor. It is like the proverbial rolling stone. Fix your mind on a specific end, and then select the conditions best suited to accomplish that end.

Fig. 7.—White Wyandottes.

The person who starts out with a fondness for a particular breed of chickens should capitalize on that fondness if it is at all feasible. For example, we will say that a person is partial to Light Brahmas. See Fig. 6. That person should make his specialty meat, and not eggs. If, on the other hand, he favors Silver Campines to the exclusion of all other breeds, he should make his specialty eggs, and not meat. The Light Brahma is one of the

best meat breeds, either for soft roasters or capons; the Silver
Campine is essentially an egg producer, and is not desirable for
table poultry.   When I say "not desirable for table poultry" I
mean that it is not in popular demand, and not particularly profitable as such.   The Campine is small, inclined to be dry and
stringy, and the color of its carcass does not meet with market
requirements.   They find a sale, of course, but the prices received
are below those gotten for the Brahmas.

Fig. 8.—White Plymouth Rocks.

Breeds for Particular Purposes.—The person who starts out
with a particular specialty firmly fixed in his mind should make
everything subordinate to this end.   If he has his mind made up
on a broiler plant, well and good, he will do well to choose the
Wyandotte, all other things being equal.   See Fig. 7.   He might
also select the Plymouth Rock or the Rhode Island Red, or numerous others, but he would be heading wrong to pick out, let

TABLE VI.—WEIGHTS OF DIFFERENT BREEDS AS REQUIRED BY THE
AMERICAN STANDARD OF PERFECTION OF 1910

CLASS	COCKS	HENS	COCKERELS	PULLETS
	Pounds	*Pounds*	*Pounds*	*Pounds*
American class:				
Plymouth Rocks............	9½	7½	8	6
Wyandottes................	8½	6½	7½	5½
Javas.....................	9½	7½	8	6½
Dominiques...............	7	5	6	4
Rhode Island Reds..........	8½	6½	7½	5
Buckeyes..................	9	6	8	5
Asiatic class:				
Brahmas..................	12	9½	10	8
Cochins...................	11	9½	9	7
Langshans................	9½	7½	8	6½
Mediterranean class:				
Minorcas..................	9	7½	7½	6½
Spanish...................	8	6½	6½	5½
Blue Andalusians...........	6	5	5	4
English class:				
Dorkings..................	7½	6	6½	5
Redcaps...................	7½	6	6	5
Orpingtons................	10	8	8½	7
French class:				
Houdans..................	7½	6½	6½	5½
Crevecœurs................	8	7	7	6
La Fleche.................	8½	7½	7½	6½
Oriental class:				
Cornish	9	7	8	6
Malays...................	9	7	7	5
	Ounces	*Ounces*	*Ounces*	*Ounces*
Malay Bantams....	26	24	24	22
Game Bantam class:				
Game Bantams.............	22	20	20	18
Ornamental Bantam class:				
Sebrights.................	26	22	22	20
Rose-Comb................	26	22	22	20
Booted...................	26	22	22	20
Brahma...................	30	26	26	24
Cochin....................	30	26	26	24
Japanese..................	26	22	22	20
Polish....................	26	22	22	20

us say, the Hamburg or the Polish. On the contrary, if this fellow
wanted to establish a specialty in fancy eggs for a select trade,
and the trade called for white eggs, he would do well to raise
Leghorns, and not Cornish or Red Pyle Games. If the market

requirements demanded brown shelled eggs, he must forget about
Leghorns and turn to one of the American breeds or English
breeds. New York, for instance, is very partial to white eggs—
they command premium prices over brown eggs; whereas Boston
favors brown eggs, as do other New England cities.

We now come to the fellow who has few if any convictions as

Fig. 9.—White-Laced Red Cornish Fowls.

to breeds or purposes. He will be influenced by his circumstances,
of course, the amount of available capital, the amount of labor he
intends putting into the project, and the locality in which he
hopes to make a start. This prospective poultryman is pretty
much in the same class with the fellow who is possessed of a location, and must make his specialty suit that property.

In extremely cold climates where the winters are long and

severe, it is sometimes best to adopt one of the heavier breeds, such as the Plymouth Rocks, for these are hardy birds. See Fig. 8. If the location is adjacent to water, ducks should be considered. If there is an abundance of pasture land, geese would be profitable. Where there is wide range in a wood lot, turkeys and guineas do well. If the chickens are to be kept in a backyard, in restricted quarters, it is advisable to select a docile breed rather than a nervous, excitable variety. If the flock is to be restrained within a low fence, the meat or general purpose breeds are best. They can be confined within a three or four foot fence, whereas the Mediterranean breeds are high flyers and generally require a seven-foot yard.

The popularity of the breed is another factor worthy of one's consideration. Whimsical and illogical as this may sound, it is nevertheless an important item toward the success of a poultry venture.

CHAPTER III

EGG BREEDS

Definition.—The term EGG BREEDS is used to designate those varieties whose greatest usefulness lies in the production of eggs. Most of these fowls comprise the Mediterranean class. They are mostly small or medium-size birds, of a nervous, active temperament, take flight readily when frightened, excellent foragers, generally poor sitters, and not very dependable mothers. For best results their eggs must be hatched in incubators, and the chicks raised artificially. They mature quickly, and feather at an early age, the chicks often developing wing feathers when but a few days old. The meat of the egg breeds does not rank very high for table purposes, except when the fowls are young. They make fair broilers up to twelve weeks of age. Surplus cockerels are used to this end, though it would hardly pay the poultryman who specializes in broilers to think of adopting these breeds. He would do better to select one of the meat or general purpose breeds.

Broad Generalities.—In discussing a subject of this kind it must be borne in mind that these descriptions are based upon broad generalities—the popularly accepted rules. There are exceptions, of course, since many natural instincts have been outbred in certain strains of fowls by long periods of intense domestication. Chickens are made docile by training, just as wild animals are made to live peaceably in a domestic state. I have seen Leghorns that were a lot more approachable than Plymouth Rocks, though the rule is not the case. I mention this here that the reader will not be misguided into the belief that all egg breeds of fowls are wild, or even semi-wild, uncontrollable birds, because they are not.

Vigor of Mediterranean Varieties.—Until the past decade, most of the Mediterranean varieties were thought to be more or

23

less delicate. This is accounted for by several reasons. Nearly all of the eggs breeds have large combs and pendulous wattles, hence they are sensitive to cold.   It has been found, however, that if the poultry houses are kept dry by adequate ventilation, and the fowls are given suitable roosting quarters, there is little trouble from frost-bite, even in climates where the temperature goes below freezing for months at a time. The fact that the young chicks

Fig. 10.—Single Comb White Leghorn.

feather very quickly, which is a great drain on their systems, was accepted as another reason for making them delicate for a time.

Being imported from countries that are warm and sunny the greater part of the year, it was to be expected that these Mediterranean breeds should be influenced by our northern winters at first. Now, however, they have been bred here so long that they are in the main thoroughly acclimated, and little fear need be entertained as to their hardihood.

The Leghorn is probably the best example of the non-sitting class. See Fig. 10. It is certainly the most popular and the most widely bred of any European fowl. Other egg breeds include the Minorcas, Anconas, Andalusians, Campines and Spanish, all Mediterranean breeds. To these should be added the Hamburg, Houdan, the Redcap and possibly some others. They all lay white-shelled eggs.

Fig. 11.—Single Comb Brown Leghorn.

The most common varieties of Leghorns are White, Brown, Buff, Black and Silver, and most of these color varieties are again sub-divided into Single and Rose-Comb species. The White and Brown Leghorns are the most widely bred, and they were the first varieties known. There seems to be excellent ground for the belief that they were first introduced into America from Italy. There is a story to the effect that in 1834 a sailing vessel from

Leghorn, Italy, brought a small cargo of fowls to this country, which were at once named Leghorns. They were found to be prolific layers, which at once gave rise to their popularity.

Improvement.—Since their first importation the Leghorns have been greatly improved, needless to say, and American breeders are also responsible for creating most of the sub-varieties of the breed. I am of the opinion that the Browns run a trifle smaller than the Whites, and that they lay a smaller egg. See Fig. 11. The Browns are probably hardier, but to offset this advantage, they are harder to breed true to color, especially in large flocks, and they do not make so good a carcass as the Whites when dressed.

The White Leghorn is acknowledged to be the premier in laying, so much so that it has come to be recognized as the standard by which the prolificness of other breeds are judged. It may be said to represent in the poultry world what the Jersey cow does in the dairy—small in size, but great in production. Some of the most successful poultry plants in the country use the White Leghorn exclusively, especially those catering to a fancy egg trade.

The Buff Leghorn is a beautiful variety, and has many admirers. See Fig. 12. It has the same general characteristics as its White cousin, only its plumage is a rich golden buff color. The Black Leghorn is another striking example of this breed, and is a favorite among those who are partial to dark-colored birds. This variety is sometimes mistaken for the Black Minorca, though the latter is larger in size, has a longer body, larger comb, and dark slate or nearly black shanks and toes. The Black Leghorn is a glossy black throughout its plumage; comb, face and wattles are bright red; the ear lobes white, and the shanks yellow or yellowish black.

Silver Leghorns, sometimes called Silver Duckwing Leghorns, are not so widely bred in this country, though they are frequently seen in the show rooms. In point of beauty they are considered very interesting, but they are likewise difficult to breed true to color in large flocks. In other ways they are considered as profitable as the other varieties.

The only distinguishing difference between the single-comb and rose-comb varieties is in the comb. The former has a blade, while the latter has a spike. Rose-comb varieties are preferred by some because there is less danger from frost-bite.

There is no standard weight given for Leghorns, though the average may be said to be 3½ pounds for hens, and 4¼ pounds for cocks. Pullets and cockerels are a trifle under these weights. Some strains run heavier, which is obtained by introducing

Fig. 12.—Rose-Comb Buff Leghorn.

Minorca blood. It is thought that some of the English breeders have crossed a little Wyandotte blood, for their Leghorns are of a much different type.

Correct Circumstances.—The Leghorn, while it is a breed of great merit, should not be confused as the right breed for every man and every place. Put the Leghorn in its proper environment and there is no fowl that will surpass it. On the contrary, if subjected to conditions that are not suitable for it, it will be an endless source of trouble and disappointment. Leghorns are ambitious fowls, tireless foragers. If there is any scratching in sight, whether it be a garbage heap, truck garden, cold-frame, manure pile, or rose-bed, the Leghorn will make strenuous efforts to get into it. Therefore, the person who would keep a garden, for flowers or vegetables, had better see to it that his hen yard is securely inclosed with wire netting if he expects to raise Leghorns.

If the runs are large enough, Leghorns can be confined within

Fig. 13.—Lakenvelders.

a seven-foot fence, as a rule. Once they develop the habit of "yard-breaking," however, they will clear this with ease, for they make a practice of half-flying and half-climbing up the netting.

Nervousness.—As previously mentioned, the Leghorn is a nervous bird. Whenever danger approaches, or anything that they imagine is dangerous or unusual, their first impulse is to fly out of the way. They abominate the feeling of being cornered.

In a large yard, especially if it is square, this condition is not so likely to occur. And if a flock is not made to resort to its wings, in time it seems to forget their real power, in which event they are confined with less trouble. It is a mistake to place rails at the top of a fence, or anything that appears as a perch. These tempt the birds to fly to them.

Small Size.—One argument that is used in favor of the Leghorn is its small size, which will enable one to keep a large number in a small space, such as a backyard or town lot. It is true that Leghorns require smaller house room than some other breeds, which is an important factor on the commercial plant, but at the same time they should have greater outdoor freedom. From my experience, I would prefer one of the American breeds in limited areas. As a rule they are more docile and will stand close confinement with better results.

The eggs from the Leghorn run high in fertility, which has made this breed a favorite in the day-old-chick trade. It is not unusual for breeders to secure 98 per cent fertility. An average of 90 per cent would be considered equal to 65 per cent in the heavier varieties. The pullets often begin laying when 4½ months old, though 5½ months should be reckoned as the average. The cockerels commence to crow at two months, or younger, and are very precocious.

Leghorns, and in fact, other egg breeds, are comparatively small eaters, and the cost of raising them to maturity is about one-half that of the meat breeds. Under proper conditions a flock of Leghorns that has been bred for heavy egg production should average between 130 and 180 eggs a year. Many individuals have scored over 250 eggs in a single year, and even 300 eggs.

General Characteristics.—The Minorcas belong to the Mediterranean class, and are often rated next to the Leghorn in laying qualities. They are bred in two colors, white and black. The Whites have a single comb, whereas the Blacks are bred with a single comb, also rose comb, which is now accepted as a standard variety. The single comb Black is the most widely bred of this family, a truly beautiful and useful fowl. See Fig. 14.

Origin of Name.—Why these fowls are called Minorcas is one of the mysteries so common to the history of poultry. The breed was formerly called the Red-Faced Black Spanish, or Portugal fowl. Some persons got the idea that they originally came from the island of Minorca, hence the name.

Weight.—Minorcas are heavier than Leghorns, and are therefore more to be recommended as table fowls. The standard

Fig. 14.—Single Comb Black Minorca.

weight of the Single Comb Black variety is 9 pounds for cocks, 7½ pounds for hens, 7½ pounds for cockerels and 6½ pounds for pullets. The weights for the Whites are one pound lighter. See Fig. 15.

Popularity.—The Minorcas have long been a popular fowl in all sections of the country, and this popularity has been attained solely on the merits of this breed. They are non-sitters and year-round layers. For table purposes they would be classed as "fair." Their flesh is sweet and juicy, but owing to its being white, and the shanks black or slate-color, it is discounted by American housewives, who prefer a yellow-skinned carcass. While the Leghorns as a class surpass the Minorcas in the number of eggs laid, the latter are considered to lay a larger egg, and to equal the Leghorn in actual bulk of egg material.

Fig. 15.—Single Comb White Minorca.

In recent years, however, so much improvement has been made in the Leghorn in the way of selecting layers of large-size eggs, that I am inclined to think this superiority of the Minorca is more romance than fact. I have seen whole flocks of Leghorns that laid uniformly large eggs, eggs that seemed out of all proportion to the hens' bodies.

Good Breed for Farmers.—Minorcas make a desirable fowl for the farmer; they have an active disposition and are excellent foragers. Perhaps their most striking feature is the comb, which is almost ponderous in size and quite thick. This feature is often raised as an objection because of the susceptibility to frost bite. In climates where the winters are long and severe Minorcas are not to be recommended on this account. This huge comb and proportionately pendulous wattles offer another disadvantage in marketing the fowls for table poultry. Large combs are considered an indication of age, which makes it difficult for dealers

Fig. 16.—White-Faced Black Spanish.

to secure top prices. Then, too, buyers do not want to pay fancy prices for this extra weight, which is, after all, waste. To obviate this drawback the combs of dressed birds are often torn off, but this removal presents an unattractive appearance, and is susceptible of deception.

Black Spanish fowls, sometimes called White-Faced Black Spanish, constitute one of the oldest varieties of domestic poultry, and are probably the oldest pure-bred fowl in the Mediter-

ranean class. Móubray, one of the earliest writers on poultry,
includes this variety in his descriptions of fowls. See Fig. 16.

They were more widely bred in this country a quarter of a century ago than now, probably because of the increased popularity
of the Leghorn. In size they are about equal to the White Minorca, though their mold is somewhat different, being shorter and
perhaps more erect. Their haughty bearing, and the white face
and lobes peculiar to the breed, contrasting with their glossy
black plumage, render them strikingly beautiful birds. This distinguishing white face, rising well over the eyes and extending to
the back of the head, should be pure white and free from wrinkles.
The greater the extent of surface the better. Needless to say,
this is the one difficult problem in breeding Spanish, though from
a fancier's point of view it is this very feature that adds interest.

Black Spanish lay a white-shelled egg, and for productiveness
and other qualifications they can be rated with the Minorca.
The males are said to be wonderful fighters, and to be capable of
holding their own against all comers, save the Games. Recently
they have not been bred in any numbers except by fanciers.
Lack of the more accepted utility qualities is probably the reason
for this falling off in popularity.

The Andalusian, sometimes called the Blue Andalusian, the
prefix "blue" being superfluous, since there is only the one variety
of this breed, has the rather unique distinction of wearing the
national colors. See Fig. 17. The face, eyes, comb and wattles
are red, the ear-lobes are white, and the plumage is a beautiful
light and dark blue.

There is considerable confusion concerning the origin of this
breed. Judging from its name, it might have come from Anda-
lusia, a province in Spain, which is celebrated for its bulls for the
ring, though Mr. Weir, an English authority on poultry, who
visited this section with the view to learning something about this
breed, failed to find any convincing evidence.

Anatomically the Andalusians and Minorcas are noticeably
alike, though the former are somewhat smaller, the standard
weight for cocks being 6 pounds, for hens and cockerels 5 pounds,

3

and for pullets 4 pounds. For farm purposes they are an admirable breed, good layers, non-sitters, active and vigorous. The chicks are hardy and mature early, and the pullets begin laying at five or six months of age. They are rather difficult to breed a uniform color in large flocks, because the plumage is likely to show many shades of blue, from light gray to a slate-black.

The Jersey Blue, once very popular as a farm fowl, is some-

Fig. 17.—Andalusian Fowls.

times confused with the Andalusian, but this is a mistake. The Jersey Blue was the counterpart of the Andalusian in disposition and color, but favored the Brahma in size and shape. Jersey Blues were large fowls, and bore indications of a cross between Asiatic and Spanish breeds.

The Ancona, or Mottled Ancona, is one of the least common varieties of the Mediterranean class. See Fig. 18. In shape and size and general characteristics they are the same as the Leghorn. They are hardy, quick to mature, and are prolific layers of white-shelled eggs. Recognition of their virtues as EGG MACHINES is becoming more widespread every year. The color of the plumage should be a beetle-green or lustrous greenish-black, with about every fifth feather tipped with white. This mottling should be uniform throughout, with no tendency to lacing. It is this uni-

Fig. 18.—Single Comb Mottled Anconas.

formity that makes for careful selection at breeding times. Like the Leghorn, there is no standard weight for the Anconas. They are Italian fowls, and are sometimes referred to as such.

The Hamburgs originated in Holland and derived their name from the city of Hamburg. See Fig. 19. They are one of the oldest standard bred fowls, and were first known as the Dutch Every-Day Layers, or Dutch Everlasting Layers. They are in the front rank of egg producers to this day, but lay rather small eggs, and for this reason they are outstripped by the Leghorns. They are small in size, and by some are considered as ornamentals There is no mistake concerning their beauty, the Hamburg, especially the Silver Spangled variety, is one of the most beautiful, striking domestic fowls.

There are six varieties of this breed, the Silver Spangled, Golden Spangled, Silver Laced, Golden Laced, White, and Black.

Fig. 19.—Golden Spangled Hamburgs.

The first named variety is probably the most popular. Hamburgs are economical fowls to keep, beside being small eaters, they are great foragers. The only serious objection to them is the smallness of their egg. By careful selection this may be remedied and the size of the egg improved. One of the distinctive features of the Hamburg is its rose comb, which should be developed into a straight spike. There are no standard weights given for Hamburgs.

The blood from the Hamburgs has been used in establishing some of our most useful American breeds. Though they breed remarkably true to color and shape, especially for fowls that have such a wide range of color, the problem of securing perfection demands the skill and patience of the most inveterate fancier. In this country. they are bred more for exhibition purposes than anything else.

Fig. 20.—Sicilian Buttercups.

French Breeds.—There are three varieties of poultry listed in the French class—Houdans, Crevecoeurs and La Fleche. Of these, only the Houdans can be classified as an egg breed, and, in fact, they might just as well be called a dual-purpose fowl, for they are highly esteemed for table purposes. See Fig. 21. They are the most popular French breed in this country, and while not raised in such large numbers as many other varieties of chickens, still they are bred to a fair extent throughout the States.

Houdans are hardy, prolific layers of white-shelled eggs, non-sitters and light feeders. They have small bones and the flesh is tender and of a delicious flavor. It is white, however, which is discredited in America as prime table poultry. Some day it is to be hoped we will get over this foolish prejudice.

Crested Variety.—Houdans are a crested variety; their crest or "top knot" is their most conspicuous mark of distinction. Another peculiarity is their having five toes, like the Dorking.

Fig. 21.—White Houdans.

Both sexes have a V-shaped comb which rests against the front of the crest. It is often partly hidden by the crest. Houdans have mottled plumage, black and white, with the black predominating. Recently a variety of White Houdans has been developed, which is quite attractive. The standard weight of the Houdan is 7½ pounds for cocks, 6½ pounds for hens and cockerels, and 5½ pounds for pullets.

Campines.—Though comparatively new in America, the Campine (pronounced kampeen) is a very old breed, and derives its name from the sandy plains of La Campine, in Belgium. See Fig. 22. Thus we are accustomed to credit this country with the ancestral dignity of this breed. Lapse of centuries and the absence of authentic records make it impossible to trace the exact origin of the Campines, though tradition has it that they were first taken into Flanders (Belgium and northern France) by Charlemagne from the shores of the Mediterranean. They cer-

Fig. 22.—Silver Campines.

tainly bear a close resemblance to other Mediterranean varieties, especially the Leghorn, in shape and habits, though not so much in color. The color of the Campine is quite distinctive, and probably creates the greatest interest in the breed.

There are two color varieties of Campines, Silvers and Goldens. Both lay a white-shelled egg, and the fowls are precisely the same in shape, size and general characteristics. It is said that the Goldens are likely to lay a slightly tinted white egg, and that they are more difficult to breed true to color in plumage. The ground color of the Silver is white, in the Golden it is yellow.

Both varieties were admitted to the American Standard in 1893, but through lack of interest in the breeds, due chiefly to the absence of a uniform size, shape, and color, they were dropped in 1898. These defects, if they should be termed such, did not

Fig. 23.—White Aseels.

make the fowls any the less valuable for utility purposes, but the early importations were too crude and indefinite for the American fancier.

Even at this date their color qualifications are open to argument, though in the past ten years they have been greatly improved and standardized, especially by the English breeders.

The breeds are now pretty widely bred in this country, and invariably render an excellent account of themselves.

Plumage.—It is generally accepted that the ideal plumage should represent two distinct colors, silvery white (substituting golden yellow for the Golden variety), overlaid by black barrings that possess a rich purplish green sheen. The barrings should be clean-cut, about three times the width of the ground color, and run transversely across the feather, with such regularity as to form the appearance of rings around the bird's body. With due allowance for a slight departure in the breast, wings, and tail, every feather should conform to this idea of symmetry, excepting the neck hackle, which is clear white, or clear yellow, in both sexes.

The Braekel (sometimes spelled Brackel) is a similar breed, frequently confused with the Campine, and is also indigenous to Belgium. It flourished in the more fertile regions of the southern districts, where it is supposed to have gained greater size. Both have the same progenitors, there can be no mistake on this point, but the Braekel, having the good fortune to live in a plenteous land, waxed big and fat, whereas her northern sister, the Campine, residing in a less fruitful section, was obliged to pass a more frugal existence. In consequence her growth was in proportion to her living, small and lean. This training, however, has made the Campine a very thrifty fowl, and given it a hardihood and vigor for which it is celebrated.

The two names, Campine and Braekel, were intermingled—large and small Campines, and large and small Braekels—until about fifteen years ago, when the English fanciers took up the breeds and set about standardizing them, and at the same time eliminating several objectionable features. They found the Campine too small, and by infusing Braekel blood increased the size. The Braekel, however, had a white saddle hackle, similar to the neck hackle, which was objectionable, because it required a double mating to secure any degree of uniformity in the color of the plumage.

By careful selection this white saddle was removed; in fact, the hackle feathers themselves were eliminated, until to-day we have the males feathered on the saddles precisely the same as the females, or what is termed "hen-feathered males." This development of the breed became known as the Improved English Campine, and is the accepted standard to-day.

The carriage and mold of the Campine is alert and graceful, and may be compared to the Leghorn or Hamburg. In size and

Fig. 24.—Long-Tailed Yokohamas.

weight it is also the counterpart of these other egg breeds. The comb of the male is of fair size, with an erect blade; in the female it falls gracefully to one side. Like other breeds of the Mediterranean class, the Campine is rated as a non-sitting variety. The hens seldom become broody, and if they do, they are easily discouraged.

A large prominent eye is peculiar to this interesting breed. It is a brilliant red, and like the proverbial hawk's eye, it is always searching and never failing in its accuracy. It seems almost to be endowed with a kind of supernatural power, an obscure sense of being aware of the existence of an object before the object has actually appeared. This readiness of eye, coupled with a wonderful sagacity, and strong constitutional vigor, serves to make the

Fig. 25.—Silver Laced Polish.

Campine a peerless forager, and easily adaptable to any conditions.

Campines resemble game birds in many respects. They are strong flyers, fast runners and good fighters, yet no fowl is more domesticated than the Campine, and none more docile and responsive. Though quick to scent an enemy, they are almost equally quick to make friends with the attendant or feeder. The writer's tamest pets in the poultry yard were Silver Campines. They haunted the kitchen door for tid-bits, and at the first opportunity would enter the house and beg for food. An English Bull Terrier and these Campines often ate from the same platter at the same time.

A breed that will show the maximum production at the least expense is the desirable commercial fowl, and on this score the Campine extends her challenge to the world. Not that Campines are necessarily smaller feeders than other breeds of the same size, for they are not; but if given the opportunity they will forage for two-thirds of their food. Furthermore, they will thrive outdoors the greater part of the year, and all year, if the climate is at all temperate. They resemble the guinea in this respect.

Energy of Campines.—In any kind of weather, rain, wind, heat, and cold, when most other breeds are content to remain snugly indoors, the Campine is abroad, wrestling for its daily keep. In fact, from close observations I have concluded that if a Campine seeks shelter, it is an ill omen, similar to that drawn from the guinea's entering the hen house at night, that a terrific storm is in the making, and that it is high time to make things snug and secure.

It is always unwise, and sometimes inhuman, to neglect birds or animals of any kind; yet there are various degrees of negligence. In most cases where chickens are raised in the backyard or on the farm the owner is unable to devote any more time to the flock than is required by the bare necessities—feeding and watering. The point to be emphasized is this: where it is impossible to give frequent attention to the needs and requirements of a flock of chickens, that flock should consist of a breed that is competent to shift for itself. The Campine is an ideal fowl for this purpose. Literally speaking, it will thrive where many other birds would starve.

Redcaps.—There is another breed frequently classed as an egg-variety, though it is seldom found in the poultry yards of this country—the Redcap. This is the modern name for a very old English breed, commonly bred in Yorkshire, Lancashire, Staffordshire and Cumberland. Its exact origin is unknown. Redcaps were called by different names in different localities, such as Copheads, Corals, Rosetops, Redheads, Derbyshire Redcaps and Yorkshire Everlayers.

The colors of the Redcaps are red, brown and black, the red a mahogany tint, and the black a bluish black. Each body feather ends with a black spangle, shaped like a half-moon, in which respect they resemble the Hamburgs. They have a large rose comb, terminating in the rear in a well-developed, straight spike. The cock, especially, is a very handsome bird. The hens lay large-size white-shelled eggs and are generally prolific. The standard weights call for 7½ pounds for cocks, 6 pounds for hens and cockerels, and 5 pounds for pullets.

Of late years, probably due to the improvement and increased popularity of some other breeds, the Redcaps have lost caste, notwithstanding their many useful, practical qualities. In habit they are alert, given to roaming, and somewhat wild unless thoughtfully managed. Their flesh is light and of a good flavor, though not so rich as some of the meat varieties.

CHAPTER IV

MEAT BREEDS

Definition.—The term "meat breeds" is intended to designate those varieties of chickens whose greatest usefulness lies in the production of meat. Do not be misled by this definition. It does not mean that these meat breeds are only useful for the production of meat alone; nor that others of the dual-purpose and other classes are undesirable as meat producers. The term simply means that these breeds excel in this branch of the poultry industry, reasons for which will become apparent in the following paragraphs.

Largest Fowls.—As might be expected, the meat breeds are the largest fowls, the heaviest, broader and deeper in the body, with a full breast, heavy limbs, and relatively short legs and neck. They are mostly of Asiatic origin, and are popularly conceived to be rather poor layers, persistent sitters, weak fliers, docile and easily controlled. They lay large brown eggs, which are not likely to run so high in fertility as the lighter breeds, consequently they are seldom used in the day-old-chick trade. Because of their size their development is slow, or rather a longer time is required for them to reach maturity. Leghorns mature in from five to six months, sometimes earlier; whereas the Asiatic Breeds take from eight to ten months, often longer. The chicks do not feather quickly, and are often almost nude at the age of two months. This feature has its advantages and disadvantages. Chicks that start to feather as soon as they leave the shell, such as the Leghorns, are often weakened by this rapid growth of plumage. On the other hand, chicks that are bare of feathers are sometimes affected by cold weather in the early Spring hatches.

The leading varieties of the Asiatic class are the Brahmas,

46

Cochins and the Langshans. The Brahmas are conceded to be the most popular, and are divided into two varieties, Light Brahmas and Dark Brahmas, of which the former are the most widely bred.

The Light Brahma is the largest chicken. See Fig. 26. The standard weight calls for 12 pounds for cocks, 10 pounds for cockerels, 9½ pounds for hens, and 8 pounds for pullets. These weights are often exceeded; I have seen specimens that weighed from twelve to seventeen pounds, regular giants, they seemed.

Fig. 26.—Light Brahmas.

History.—It would take a large volume to hold the history of the Light Brahma. It was probably the first breed of poultry to be popularized in this country, where it was greatly improved by American fanciers. Though fundamentally an Asiatic, it is really an American output. They were first known as Brahma Pootras, Gray Shanghais, Chittagongs and Cochin Chinas. In fact, the early breeders were disposed to give them high-sounding, fanciful names, for the sake of the benefit these names might have in helping them to sell stock at fancy prices. Records show that in many cases fabulous prices were received. The *hen fever* ran high about this time, which was in the early fifties. There was a *craze* for pure-bred poultry, and the Light Brahma occupied the center of attraction. In later years their popularity abated somewhat with the advent of other breeds, though to this day they still remain high in the esteem of poultry lovers who prefer a heavy fowl.

Shape.—The Brahma is different from the other meat breeds, and must not be confounded with the Cochin or the Langshan. It has a long, deep body, with full, broad and round breast, carried well forward, which is characteristic of prolific birds. It is by far the best layer of the Asiatic breeds. Numerous hens have made enviable records in laying contests, though the average flock production should be placed at about ten dozen per year.

The plumage of the Light Brahma is white and black, with the white predominating. Any other color is a disqualification. The body plumage is white, the tail feathers are black, with the sickles a greenish black. The neck hackle is white with a black stripe running down the center of each feather and terminating in a point. The shanks are well feathered, with the feathering extending down the middle toe. This feathering may be white, or white marked with black.

Feathered shanks and toes is probably the most distinguishing feature of the meat breeds, for all of the Asiatics have them.

The Light Brahma has a small pea comb; its face, wattles and earlobes are a bright red. The shanks and toes are yellow.

Brahmas are excellent mothers; they will hatch and rear large broods of chicks. In fact, their maternal instincts constitute a drawback to the poultryman who is after eggs in large numbers. For capons the Brahma is in a class by itself; it has size, shape, a small comb and all the other qualifications that make for prime table poultry.

Dark Brahmas are not so popular, and never have been. This is due to the great difficulty of breeding them true to a uniform color. The head and neck of a Dark Brahma male are similar to the Light for hackle, but the neck other than the hackle should be black. The back is silvery white, the breast is black, the thighs are black, and the fluff either black, solid, or very slightly mottled with white. The saddle feathers are similar to the neck hackle, and as they approach the tail the stripes become a broader black until they merge into the tail coverts, which are a glossy, greenish-black. The wing coverts are greenish-black, the secondaries and

Fig. 27.—White Cochins.

flight feathers are mostly black, and the shank feathering is black, or black mottled with white.

The weights of the Dark Brahmas are one pound lighter than the Light Brahmas, or about the same weights as the Cochins. The plumage of the Dark Brahma hen is a white ground, closely penciled with a dark steel gray. This produces a beautiful effect, if it is correct; but unless extreme care is taken in the mating, the plumage is likely to be a dingy color, and lack uniformity.

The Cochins probably rank next to the Brahmas as meat breeds, and are bred in four colors: Buff, Partridge, White and

4

Black. See Fig. 27. The Buffs are the most widely bred; they are, indeed, beautiful birds, and have a color that is golden throughout in both sexes. They have the purest buff color of any of the buff breeds of poultry, and have been used quite extensively in improving this color in other breeds. They are bred with loose feathers, so that the general effect is that of a ball of feathers. The shank feathering is more profuse than the Brahmas. And unlike the Brahmas, the Cochins have a small, single comb. In

Fig. 28.—Partridge Cochins.

disposition it might be said without fear of controversy the Cochins are the least restive of all fowls. They have a quiet, sluggish nature, and are the most determined sitters. They stand confinement well, and may be restrained within a three-foot fence.

The Partridge Cochin is a beautiful bird, but like all penciled varieties, it is difficult to breed true to color. See Fig. 28. The plumage arrangement of the Partridge Cochin is not unlike the Dark Brahma, except the colors are red and brown instead of

steel-gray and black. In breeding penciled or parti-colored birds
it is so often necessary to mate them so close in order to secure
the correct characteristics, that the productiveness of the birds
is likely to be slighted. In consequence a solid-color bird is the
more practical one for the farmer or general market poulterer.
They can be raised in large flocks with the least amount of attention paid to color, and all the attention bestowed on their utility
qualities.

Black and White Cochins.—In this respect the Black Cochin
or the White Cochin is the more desirable variety. The Black is
of a rich, glossy, greenish-black throughout its entire plumage,
and the White is pure white throughout. A flock of Black Cochins present a handsome sight, and being dark they do not soil
so readily, as do the Whites. On the other hand, the Whites
dress better for market purposes, for there are no dark pin feathers
to mar the clean appearance of the flesh.

Langshans are the smallest and the most active of the Asiatic
breeds, also the most rangy-looking birds. See Fig. 29. They
are bred in two colors, White and Black, and the latter is probably
the most widely bred in this country.

The Langshan is distinct from the Brahma or Cochin in shape.
The male, especially, has a very majestic carriage, tall and stylish,
not the least gawky, a splendid leader for the flock, attentive to
the hens, and an excellent forager. Langshans are good sitters
and mothers, and having a gentle disposition, they are ideal fowls
for the farm. They are fair layers, particularly during the winter
months, and the chicks are hardy and grow well. Langshan
chicks mature earlier than the other Asiatic breeds.

To the inexperienced eye some confusion exists between the
Black Langshan and the Black Cochin; but this should not be.
The Cochin is a stocky bird, with heavy-looking neck and legs,
whereas the Langshan is very erect, with a high tail and sweeping
curve to the neck. The Langshan fluff is moderate and close,
while the Cochin fluff is extreme and loose. Then, too, the feathering on the shanks of the Langshan is not so profuse, and the
shanks are longer in proportion. The comb of the Langshan is

relatively larger than the Cochin's comb, well up in front, and arch-shaped. The wattles are longer and more pendulous.

The quality of the flesh of the Langshan is all that could be desired in a sense; it is fine-grained, tender and nicely flavored; but it is white, a feature that is not so acceptable to American housewives. The skin of the Cochin is yellow. Another objec-

Fig. 29.—White Langshans.

tionable feature is the bluish-black shank in the Langshan. In the Cochin it is yellow.

The plumage of the Black Langshan is a glossy, metallic black throughout; in the White it is pure white throughout. The standard weight of cocks for both varieties is $9\frac{1}{2}$ pounds, for cockerels 8 pounds, for hens $7\frac{1}{2}$ pounds, and for pullets $6\frac{1}{2}$ pounds.

CHAPTER V

DUAL-PURPOSE BREEDS

Definition.—The terms "dual-purpose" or "general-purpose," for the expressions are used interchangeably, are intended to designate such fowls as may be found useful and profitable in the production of both meat and eggs, and if need be—under conditions that require natural incubation. In other words, instead of possessing qualities of a particular nature, such as intensive egg production, "dual-purpose" birds are adapted to the common and more general conditions of the country—the farmer and backyard poultryman.

The farmer and backyard poultryman want hens that are good layers, of course, but they also want fowls that produce an abundance of meat, so that when the hens' days of usefulness in the egg basket are over, they can terminate their utility on the dinner table.

The "dual-purpose" breeds meet these requirements. Most of them belong to the American class, in addition to which there are breeds of like type, such as the Orpingtons, Dorkings and Faverolles. We will take the American breeds first. There are no finer specimens of poultry in the world than these products of American fanciers.

American Class.—The American Standard of Perfection of 1910 admits six breeds in the American class, as follows: Plymouth Rocks, Wyandottes, Rhode Island Reds, Javas, Dominiques and Buckeyes. The first three named breeds are the most widely bred, and of these three the Plymouth Rocks are undoubtedly the most popular. It has been said, and there seems to be excellent ground for the opinion, that Plymouth Rocks,

53

notably the Barred variety, are the most widely bred fowls in
America. They are business birds from the ground up; intensely
practical and utilitarian, at the same time they possess enough
caste and beauty to satisfy the most fastidious fancier. In any
climate, north, south, east or west, and in any locality or under
any circumstances that will permit other fowls to live, there also
will the Plymouth Rocks thrive and be of profit to their keeper.

Fig. 30.—Barred Plymouth Rock.

There are six varieties of Rocks, differing only in the color of
their plumage: Barred, White, Buff, Silver-Penciled, Partridge
and Columbian. See Figs. 8, 30, 31. Popular fancy delights in
the sentiment that the Plymouth Rock was named after the landing place of the Pilgrim Fathers. This fancy exemplifies the enduring qualities of the breed; but it must not be construed as
establishing their age. The name was first given to a nondescript

breed about 1849, but not until twenty years later was the real Plymouth Rock established.

The Barred variety was the original Plymouth Rock, and to it rightfully belongs the title of the pioneer of American fancy poultry. See Fig. 30. There were two other breeds of fowls produced before the Barred Rock, namely, the Dominique and the

Fig. 31.—Columbian Plymouth Rocks.

Java, but at that time they were not bred to anything like a fixed standard, as were the Rocks.

The weights of the Rocks show a betwixt-and-between fowl, cocks 9½ pounds, cockerels 8 pounds, hens 7½ pounds, and pullets 6 pounds. Neither too small for meat purposes, nor too large for egg production, hence the name "dual-purpose." For the farmer or market poulterer these fowls are favorites, being of medium-size, well proportioned, with a deep, full, round breast. They are hardy, mature in about eight months, and are excellent layers the year round.　The eggs are brown, as with all American breeds; the hens are good sitters and excellent mothers.

The other varieties of Rocks followed in the wake of the Barred, of which the White, Buff and Columbian, in the order named, are the most widely bred.　It is doubtful, however, if all of the other varieties combined, equal the popularity of the Barred Rock; it seems to stand supreme.　Besides being a thoroughly practical

Fig. 32.—Silver Wyandottes.

fowl, it is highly esteemed by fanciers for exhibition purposes. No class is offered to keener competition in the show room.

The Wyandottes stand next to the Rocks in popular favor. There are eight varieties, differing only in the color of their plumage, as follows: White, Silver, Golden, Buff, Black, Partridge, Silver-Penciled, and Columbian.　The Silver Wyandotte is the original and the foundation of all the other varieties.　See Fig. 32. It was admitted to the Standard as the Wyandotte, and later, as the other varieties were brought out, it was called the Silver Laced Wyandotte, which has been shortened to Silver Wyandotte.

The White Wyandotte is by far the most popular variety. See Fig. 7. In fact, it is conceded to be the most popular white fowl of all-round capabilities in the world. The popularity of the Plymouth Rocks was the main stimulus of the origin of the Wyandottes, which came into being about 1875, though they were not

Fig. 33.—Columbian Wyandottes.

admitted to the Standard until about eight years later. They have been a huge success ever since. They weigh about a pound less than the Rocks, are prolific layers, easily cared for, and stand confinement well. For table poultry, especially broilers, they are ideal. They have plump, round bodies, and the flesh is sweet and of excellent flavor. Furthermore, the skin is a rich yellow, so much sought after by the average housewife.

All Wyandottes have rose combs, which is an indication of Brahma blood in their make-up. The Rocks have a single comb, of moderate size. Often, the comb is the only distinguishing feature between some varieties of Wyandottes and Rocks. For example, except in shape, which is not always so apparent to the inexperienced eye, the Buff, White, Silver-Penciled, Partridge and

Fig. 34.—Single Comb Rhode Island Reds.

Columbian varieties of Wyandottes and Rocks are virtually the same, only for the difference in the comb. See Fig. 33.

The Rhode Island Reds probably come third in popularity among the American breeds, and this popularity is growing by leaps and bounds. See Fig. 34. They are of comparatively recent origin, in the sense that they have only been admitted to the Standard since 1895; yet they are the result of over fifty years persistent breeding toward a very definite end. That goal was to produce a utility fowl of red plumage, of the greatest hardihood and the most enduring qualities, and a fowl that would tend to great fecundity and all-round, general qualifications under all conditions and in any climate.

The Reds are the result of out-crossing, rather than out-breeding, which probably accounts for their great vigor and productiveness. Out-crossing is the mating of breeds that are en-

Fig. 35.—Black Javas.

tirely foreign to each other; for instance, a Brahma mated to a Cornish fowl. Out-breeding is mating fowls of the same breed, but not related by blood.

In the make-up of the Reds new males were used each year, but always of some red breed, such as Malay Games, Red Pit Games, and Brown Leghorns. The Red is truly a composite fowl. Many breeders were of the opinion that red plumage stood for stamina, and it is certain that this belief has been confirmed in the Rhode Island Red, at least.

The Reds are divided into two varieties—single comb and **rose comb**. Except for these head points they are identical. In the past few years another division has been created—the Rhode Island Whites, which gives promise of making a splendid showing. There does not seem to be so much demand for another white breed, however, in view of the great popularity of the White **Rock** and White Wyandotte, not to mention the white varieties of numerous other breeds.

Fig. 36.—Buckeyes.

The Reds are ideal birds for farmers, not only because of their size and prolificness, but because their plumage is well adapted to farm conditions. Being dark in color, it does not soil so readily. The weights for the Reds are the same as for the Wyandottes, though the mold or contour of the former is not so round.

The Javas, see Fig. 35, of which there are two varieties, Black and Mottled, the Dominiques and the Buckeyes complete the list of the purely American breeds. None are bred so extensively as the breeds described above, though all of them are well suited to the class of general-purpose fowls.

Buckeyes, named after the state of Ohio, where they were originated, were obtained from Barred Rock crosses on Buff Cochins, with an infusion of Cornish and Pit Game blood to give the flesh quantity, and the offspring vigor. See Fig. 36. In 1905 they were admitted to the ranks of the Standard of Perfection.

The name of the Dominique is closely interwoven with our early history of poultry, yet the real origin is very obscure. Some claim that the breed is a product of the Island of Dominica, but apparently this idea is purely fanciful, since a breed of these characteristics was never found there. A more reasonable theory is that the name just evolved—as an ambiguous term to cover a mixed origin. The similarity of the Barred Rock has no doubt been responsible for the falling off in the popularity of the Dominique, for it is no longer bred by so many poultrymen. In fact, it is seldom seen to-day. The same holds true of the Javas and the Buckeyes.

For poultrymen specializing in eggs, we were glad to doff our hat to the sprightly little Leghorn; it holds first place among all comers in the egg class. But, for all-round, general poultry business, which includes all of the different branches of the industry —eggs, meat, broilers, roasters, capons, feathers and maternal instincts, not to forget caste and beauty—the American breeds are at the top of the heap, and not likely to be supplanted.

There are numerous other varieties closely resembling the American breeds in shape, disposition and size, and are commonly classed as general-purpose fowls.

English Breeds.—First, let us take the English breeds. Without question the most popular of these are the Orpingtons, of which there are ten or more distinct varieties,—single-comb White, single-comb Black, single-comb Buff, single-comb Spangled, and single-comb Diamond Jubilee; also, rose-comb varieties of the same colors. See Fig. 37. In the past few years a couple of other varieties have been originated, notably the Blues, but these are so recent and bred in such limited numbers that we will not devote special attention to them. As a matter of fact, the Whites, Blacks and Buffs are the only varieties bred in large numbers in this country, and the Whites are the most popular of these.

Orpingtons.—This remarkable family of fowls, and I qualify them as such because of the perfection and enormous popularity that they received in such a brief space of time, were originated

Fig. 37.—Single-Comb Black Orpingtons.

by one man, William Cook, of Orpington, England, whence they take their name.

Mr. Cook had a definite object in producing these birds. He found that most of the old varieties of English poultry were inbred too closely for egg production, and that no one breed combined laying and table qualities to any marked degree. Then, too, he noted the success of the Plymouth Rocks in America, which were then coming to the fore, and he was determined to produce a like general utility fowl, one that would answer the prime requisites—eggs, table and show qualities.

The first Orpingtons to be produced and exhibited by Mr. Cook were the Blacks. This was in 1886. In 1889 the Whites were brought out, and in 1894 the Buffs. Later the Jubilee and Spangled were developed.

Composite Birds.—The Orpingtons are an amalgamation or composite bird, which is largely responsible for their productiveness and vigor. We all know of the hardihood and other excellent qualities of the Barred Plymouth Rocks. The black sports, or "off colors" of this American breed were made the basis of the Black Orpington. To this was added Minorca blood, for productiveness and to intensify the color, and finally Langshan blood, which was calculated to give the breed size and a superior flavor to the flesh. Several years were required to eliminate the feathered shanks of the Langshan. To this day stubs are likely to appear on the shanks of some specimens, indicating a throw-back, though as a whole the breed develops with unusual dependability as to type and color.

Origin.—Mr. Cook realized that poultry raisers were more or less partial to buff colored fowls, therefore he set about producing the Buff Orpington, having the same characteristics as the Black as to shape, size and so on, but of buff plumage. The Cochin was the basis of this variety, crossed with Golden-Spangled Hamburg blood, and Dark Dorkings. He took the Hamburg for its laying qualities, and the Dorking for its length of breast and the quality of its meat. The latter had a fifth toe that required years to eliminate, which was only one of the many problems that had to be solved, for here again, in the Cochin, was the feathered shank. And the Orpington must have a clean shank.

In producing the White Orpington Mr. Cook turned to the White Leghorn for color and productiveness, and to the Hamburg. For size and table qualities he used the White Dorking. With this combination it is no wonder that the White Orpingtons should have gained prominence as good layers. See Fig. 38. The chief difficulty in this cross arose from the tendency to throw cream-colored plumage, especially brassiness in the neck and saddle feathers. Even now this is a problem. There are few flocks that are entirely free from this defect, though each year satisfactory progress is being made. The females breed true without difficulty; the trouble lies with the males, because of their hackles.

The standard weights for Orpingtons are 10 pounds for cocks, 8½ pounds for cockerels, 8 pounds for hens, and 7 pounds for

Fig. 38.—Single-Comb White Orpingtons.

pullets. These run about a half-pound heavier than the **weights** called for in the Plymouth Rocks. In shape the Orpington resembles the Wyandotte or Cochin more than it does the **Rock,** since it is a round, short-legged, short-necked, chunky **sort of** fowl. The plumage, too, is more fluffy than the Rock, thereby giving the Orpington a more rotund appearance.

All Orpingtons lay exceedingly well, and they are exceptionally good winter layers. From the writer's experience, which **seems**

to be borne out by the experiences of others, though the opinion
may be denied by some, the eggs are not so large as they might be,
nor so uniform in shape, texture nor color as is to be desired by
poultrymen catering to a fancy egg trade. The eggs have not
the "egg-shape" nor uniformity of Rock eggs. They are rounder
and more elliptical in contour, and are given to a polished surface

Fig. 39.—Silver-Gray Dorkings.

rather than a dull, matt surface, which makes the characteristic
"bloom" of a Rock egg so desirable.

I have found this virtue about the Orpington, however, it
matures earlier than the Rock, and can be made to put on more
weight at the least expense. Furthermore, the Orpington is not
so prone to put on fat, which means that they are better able to
stand forced feeding for egg production.

The habits and demeanor of the Orpingtons are practically the
5

same as the American breeds. They are quiet birds, easily made
pets, are confined within low fences, become broody, sit and hatch
well, and make excellent mothers. The chicks are hardy and grow
rapidly, and make good broilers at an early age.

The meat of the Orpington is delicious. It is soft, juicy and
abundant. But, unfortunately for American markets, which cater
to yellow-skinned poultry, the skin of the Orpington is a pinkish-
white. In the Blacks it is likely to be a bluish-white. **Moreover,**

Fig. 40.—Rose Comb White Dorkings.

the shanks are pink instead of yellow. In the Blacks they are
bluish-black. It is an absurd notion, but these qualities are sometimes interpreted to be indications of cold storage poultry.

Poultry growers are gradually educating the public on the
fallacies of its prejudices, and in time they will succeed, in which
event the Orpington will rank among the best of the meat-pro-
ducing breeds. In Europe, and we are inclined to concede the
honors to its chefs, notably the French, the white-skinned fowls
are acknowledged to be more highly esteemed.

The Dorkings, of which there are three varieties—White, Silver-Gray, and Colored, constitute another favorite English breed, and one of the oldest of domestic fowls. There are no accurate records to show its exact origin, but the supposition is that it was carried to England by the Romans.

Weight.—The Dorking is not so heavy as the Orpington, but it is highly prized for table meat. See Fig. 39. The flesh is white and possesses a very delicate texture and flavor, and there

Fig. 41.—White Faverolles.

is an abundance of meat on the breast, which is broad, deep and full. The weights given for Dorkings are somewhat variable; the Colored Dorkings are heaviest: 9 pounds for cocks, 8 pounds for cockerels, 7 pounds for hens, and 6 pounds for pullets. The standard weights for the Silver-Gray Dorking run about a pound under the above, while the weights for the White Dorking are about a half-pound under the Silver-Grays.

Fifth Toe.—The most distinguishing feature of this breed is the presence of a fifth toe, or supernumerary toe, extending a little behind, above the foot and below the spur, similar to the fifth toe of the Houdan.

Dorkings could not be rated as a popular breed in this country. They are rather indifferent layers, and while nice looking fowls, there are too many other breeds of superior qualities.

Faverolles.—Excepting in France, very little was known of the Faverolles until 1896, when they were taken up by English

Fig. 42.—La Fleche Fowls.

breeders and later, by Americans. They are bred in several colors, white, salmon, ermine and black. The Whites and the Salmons are probably the most popular in this country. See Fig. 41. They are considered good layers, and weighing about the same as the Dorkings, they are valued as table poultry.

The odd feature of the Faverolle is the growth of feathers, resembling a beard and mutton chops, around the throat and ears. This whiskering is one of the fixed characteristics of the entire breed. Another feature is the fifth toe, like the Dorking, also, a booted or feathered shank, such as is found on the Brahma. For those who prefer the unusual, the Faverolle is to be highly recommended. Because so little is known about it, however, its sale for breeding purposes is likely to be restricted.

Crevecoeurs and La Fleche fowls are two other French breeds little known in this country, though they are widely bred in their native homes. See Figs. 42 and 43. They are kept for general

Fig. 43.—Crevecoeurs.

farm purposes, but are best for the table. The former are the better layers. The weights of both breeds are about the same as the Dorkings.

The plumage of the Fleche fowls is a glossy, greenish black. Their chief claim to distinction is the peculiar comb, which is in the form of two well defined spikes, resembling horns. Crevecoeurs have a similar comb, only it looses its distinction by reason of the crest of feathers growing on the top of its head, like the Houdan or Polish.

CHAPTER VI

ORNAMENTAL FOWLS, GAMES AND BANTAMS

Ornamental Varieties.—Some varieties of poultry are purely ornamental in character and purpose. They have no particular virtues as to egg production, neither are they superior for table purposes. The Bantams are in this class. They are raised simply for the interest attached to their oddity or beauty. Other breeds are deemed fancy, by reason of some unusual characteristic, or scarcity, though in reality they may be good layers, or splendid table poultry. Custom has placed them in the ornamental class, because few are adapted to the farm or general commercial use. It is easily understood that the more variegated the fowl's plumage, or the more eccentric its shape and feathering, the more difficult, almost impossible it is to breed them to any degree of uniformity in large flocks. Ornamental breeds almost invariably require special matings, and years of experience to know how to make such matings, hence their appeal to the fancier.

No one will deny that the work of raising fowls for purely ornamental purposes is most interesting, and some fanciers have found a big outlet for their products, thereby making their work profitable, but these cases are the exceptions and not the rule. Those who would enter the poultry industry for pecuniary gain had better start with one of the breeds described in the earlier chapters of this Analysis of Chickens, such as the egg breeds, meat breeds, or dual-purpose breeds.

The Polish varieties are generally regarded as strictly fancy chickens, though they are known to be one of the oldest breeds of pure-bred fowls. Their ancestry has been traced back to the

70

sixteenth century. The eight varieties of Polish are: White Crested Black (see Fig. 44), Bearded Golden, Bearded Silver, Bearded White, Buff Laced, Non-Bearded Golden, Non-Bearded Silver, and Non-Bearded White. (See Figs. 25-45-46.)

It will be seen that there are two distinct sub-breeds of Polish, the plain, or non-bearded varieties, and the bearded ones. All have crests, or "top knots," which is their chief mark of distinction, in addition to which the bearded varieties have a thick,

Fig. 44.—White-Crested Black Polish.

full beard of feathers running under the beak from eye to eye in a graceful curve. The plain varieties are without this beard.

The White-Crested Black is the most extensively bred in this country, and the Bearded Silver variety probably comes next. They are beautiful fowls, all of them, and by some are considered good layers. As with all crested varieties of fowls, their "top knots" are really against them. This head feathering obstructs the vision, causing them to be timid and suspicious, an easy prey to vermin, and much subject to colds if the birds are allowed to run in the wet. No standard weights are given for Polish. They are medium-size birds, about that of Leghorns.

The Sultans, as the name implies, are from Turkey, and might with propriety be classed with the Polish, except the former have additional peculiarities. Sultans, in fact, possess about every peculiarity possible for a fowl of its size. A compact crest surmounts the head, more profuse even than the Polish, and they are full bearded. For a comb they have two small spikes, re-

Fig. 45.—Bearded Golden Polish.

minding one of horns; their legs are feathered and booted, their hocks are vultured, and they possess a fifth, or supernumerary toe. They have an abundant neck hackle and a large tail, which is erect and contains many flowing sickles. The color of the plumage is pure white throughout. They seem to thrive well, but are too small for practical purposes.

The Game is one of the most interesting of the ornamental breeds, and perhaps the most widely bred. It is noted for its vigor and courage, and were formerly raised for fighting. They are still raised for this purpose in countries where cock-fighting is permitted.

The beauty of the Game is unquestioned. It is a tall, slim bird, very erect in carriage, with long legs, and short, close feathering. The carriage of the Game is peculiar to it, and is spoken of as "station." Specimens with the highest "station"

Fig. 46.—Non-Bearded White Polish.

are the most desirable. It is customary to remove the comb and wattles of the cocks, a practice that is termed "dubbing." This adds to their sleekness and general fighting trim.

Games are not without their practical qualities, though they are seldom bred for general farm purposes. They are fair layers, and their flesh is excellent, the meat being fine-grained, tender and juicy. The chicks are said to require considerable care, but this is probably due to weakened constitutions from too much close breeding for strictly ornamental purposes. There is no reason why the Game should not be as easy to rear as the Leghorn.

The varieties of the Games are: Black-Breasted Red, Brown

Fig. 47.—Silver Duckwing Games.

Red, Golden Duckwing, Silver Duckwing (see Fig. 47), Birchen, Red Pyle, White, and Black.

Game Bantams.—For every Game there is a Game Bantam. See Fig. 48. The color of the plumage in the Bantam, its markings, shape and carriage correspond precisely to the Game that bears its name. The Bantam's diminutive size is the only dis-

tinguishing feature between the two. · Bantam cocks average
twenty-two ounces, and the hens twenty ounces.

Oriental fowls comprise the Cornish, Sumatras and Malays,
and the Malay Bantams. Cornish fowls, sometimes spoken of
as Cornish Indian Games, are really an English product, having
been originated in Cornwall, whence their name. They were
produced from Black-Breasted Red Games crossed on Red Aseel

Fig. 48.—Red Pyle Game Bantams.

fowls imported from India. This cross produced what is known
as the Dark Cornish. There are three varieties: Dark, White,
and White-Laced Red. See Figs. 9–49. The Whites were produced from "sports" from the dark variety, crossed with White
Aseel. See Fig. 23. The White-Laced are of Yankee origin,
using both the Dark and White varieties, with some infusion of
White Georgia Game blood.

Cornish.—As layers the Cornish fowls do not rank very high, but they make excellent table poultry. They are frequently crossed with the meat breeds for this purpose, especially for capons. The shape of the Cornish creates the impression of massiveness and great muscular strength, also pugnaciousness. They are stockily built birds, with heavy thighs, legs set far

Fig. 49.—White Cornish Fowls.

apart, and a full round breast and broad shoulders. Then, too, they have the characteristic feathering of the Game—closely set, thin hackle and small tails, which gives them a rather ferocious appearance. The standard weights of the Cornish are 9 pounds for cocks, 8 pounds for cockerels, 7 pounds for hens, and 6 pounds for pullets. It will be seen from these weights that they are very worthy birds for the table.

Malays are little known in this country except for crossing with other breeds to infuse vigor and size. They have about the same weights as the Cornish, and are strong and powerful looking. They are reputed to be extremely savage, and in battle often actually tear their opponents to pieces. The plumage of the Malay is very close, like other Games, only perhaps more

Fig. 50.—Silver Sebright Bantams.

scanty, and the color is red or maroon and black. The head is long, with a projecting crown, which gives the cock a cruel and fierce expression. The wattles and earlobes are small.

Sumatras, or Black Sumatra Games, differ from the Malays, in that they are of a gentle disposition, though once started in a conflict, there is no Game that will show greater staying power than the Sumatra, especially if it is in defense of its mates and young. The quality that really removes the Sumatra from the Pit Game class is its long, flowing tail, with an abundance of sickles and coverts. The plumage is of a rich, greenish black throughout. Sumatras and Malays are little bred in America except for exhibition purposes, and even this is more or less restricted.

Fig. 51.—Rose-Comb White Bantams.

The Malay Bantam, or Black-Breasted Red Malay Bantam, should be the same in color, shape and general characteristics as the full-size Malay. Cocks should not weigh over 30 ounces, nor the hens over 28 ounces.

Other Bantams.—In addition to the Bantams previously described, there are numerous other varieties, almost all of which are made to imitate in miniature the standard size breeds of fowls. There are: Sebright Bantams (see Fig. 50), in two varieties—Golden and Silver; Rose-Comb Black Bantams and Rose-Comb White Bantams (see Fig. 51), which are counterparts of the Hamburgs; Booted White Bantams; Light Brahma Bantams and Dark Brahma Bantams, which are miniatures of the regular Brahmas; Cochin Bantams (see Fig. 52) in four varieties—Buff, Partridge, White and Black; Polish Bantams in three varieties—Bearded White, Buff Laced, and Non-Bearded;

Fig. 52.—White Cochin Bantams.

and the Japanese Bantams (see Fig. 53) in three colors—Black Tailed, White, and Black.

Bantams are raised almost exclusively for pleasure, though they are sometimes used to hatch the eggs of Pheasants and other fowls, since as a rule Bantam hens are good sitters and mothers. Some of them are good layers, and for their size they lay unusually large eggs. These, however, are not marketable as prime eggs, for they are too small.

Any of the Bantams will make delightful pets for children. With their cute and saucy ways they are a constant source of amusement. Where is there a small boy who has not sometime yearned for a pair of these little feathered friends?

Silkies.—We could continue to enumerate other varieties of ornamental poultry, but most of them are so rare, that to devote space to them would serve no practical end. Silkies are fowls of small size, whose chief peculiarity consists of very soft, web-

Fig. 53.—White Japanese Bantams.

less feathers, which are exceedingly loose when in prime condition, and stand out from the body in all directions. They are purely ornamental birds.

Frizzles are another grotesque member of the poultry family. The ends of their feathers curl backwards, giving them a frizzled look, hence the name.

Long Tailed Yokohamas represent still another odd variety. See Fig. 24. They are Japanese birds, having very long tails, sometimes attaining a length of fifteen or eighteen feet, and are quite beautiful in coloring. Occasionally specimens are exhibited in American poultry shows.

Sicilian Buttercups have appeared in the poultry exhibits from time to time, but they never won any particular favor. See Fig. 20. They have an odd comb, resembling a cup. In Sicily, their native home, they were known as "Patera Opulentæ," meaning sacred cup of riches, and were formerly used in religious sacrifices.

6

CHAPTER VII

A BUSINESS ENTERPRISE

Principles.—Fine feathers usually make fine birds, for the same reason that up-to-date business methods make satisfied customers, than which there is no greater asset to the poultryman, be he conducting a large or small industry. *Fine feathers* indicate quality—careful breeding of known reliability, proper feeding and good care generally. Up-to-date business methods bear the hall-marks of ambition to please, of progressiveness, of painstaking, workmanlike ability, of superiority and dependability. Business to-day demands certain conventionalities, and those who do not appreciate the fact, and who remain in the *rut* of a past generation, thinking that they can do things as their grandfathers did, are sooner or later destined to become relegated to obscurity.

Conducting a poultry farm is no different from any other enterprise in this respect; if the poultryman wants to make a success of his business, and derive other than laborer's wages from his investment, he must conduct his operations on what have come to be recognized as the standards of modern business. He must produce commodities which are in popular demand, not *has beens;* he must exercise good salesmanship, by using every means at his command to get his commodity before the public; following which he must keep his products up to their representation, and never tolerate a *sag* in quality; and above everything else he must give good value, and wherever possible—just a little bit more than the other fellow. Perhaps not in a reduction in price, for under-cutting is sometimes accompanied by retaliative measures; but in the excellence of the product, or the manner in which it is packed and distributed, in the service—the prompt-

Fig. 54.—Large colony houses, used first as breeding pens, later to grow the young stock.

83

ness and courteous treatment accorded a customer; these are values, more or less intangible, no doubt, but nevertheless potent. Furthermore, the buying public is quick to appreciate them.

Food Products.—Primarily, the poultryman should always bear in mind that he is producing food products, not coal or iron castings, consequently his wares have an esthetic appeal. There is much agitation these days over sanitary conditions in factories where food stuffs are made, and it is right that there should be; the slogan is *cleanliness* to a most exacting degree. We do not

(Courtesy Monmouth Poultry Farm)

Fig. 55.—A centralized plant.

look for purity in articles of food that are made in dirt-infested, antiquated factory buildings, where the workers must toil amid sordid conditions,—filth, improper ventilation and unhealthful environments generally. Pure food commissions are empowered to regulate these affairs. If it is important that milk should be produced and sold under certain restrictions, it is also important that poultry and eggs should be produced likewise. At least the commercial aspects are the same.

The buildings on a poultry farm, their yards and all the accessories should be built with the idea of maintaining them in a strictly sanitary condition. Dirt, disorder and dereliction must not be tolerated. By this it is not necessary to erect unduly expensive buildings, nor to install elaborate equipment; for the margin of profit in the business will not warrant the expenditure of unnecessary capital. Besides, unnecessary capital brings no return on the investment.

(*Courtesy Million Egg Farm*)

Fig. 56.—There is no better spot for poultry than a grove of trees, providing the sun is able to shine on the house.

Show Place.—Every poultryman should aim to make his farm a show place, where visitors can be allowed at regulated times, and where prospective customers can be shown the articles that are for sale, and note for themselves that what has been said about the articles is true. None of us like to buy "a pig in a poke," if we can help it. And do not think that a show place should consist of *show* in the sense of elegance, a mere spending of money.

Simply constructed buildings, in an orderly arrangement, neatly painted or white-washed, and having an air of practical utility, make the most impressive showing, provided they are clean and stocked with vigorous, healthy-looking poultry. By all means avoid an accumulation of junk or rubbish lying about the premises; it is bad for the fowls, and a constant eyesore. Have a place for everything, and everything kept in its place. The appointments necessary for the convenience of the fowls and their caretaker are really very simple, and most of them can be bought for a small outlay, or they can be made at home.

Home made devices should not necessarily mean makeshifts—odds and ends, broken china, discarded kitchen utensils, old buckets, rusty pans and other receptacles that have long since passed their age of usefulness, and which only serve to clutter up the yards and houses. Visitors and customers observe these things, and their opinions are formed accordingly. Nothing is more enbarrassing than to have to make excuses for the appearances of everything; and besides excuses are futile. All the excuses in the world fail to make a reason that will justify shiftlessness; it is inexcusable.

Visitors.—Notwithstanding the poultryman's time is very much occupied by routine work, and that visitors are sometimes rather troublesome to entertain, also that their presence is disturbing to the birds unless precautions are taken, experience has proved that one of the best selling methods is to get the public's interest in your work. Every family living in the community, or who might be visiting the community, is a prospective customer if their interest is aroused, and they are assured of a courteous reception when they seek information. If you have something to sell, you must let it be known, following which you must be perfectly willing to exhibit your goods.

Sign.—The first step toward publicity is to erect a neat sign, setting forth the name of the farm or that of the owner, whichever is used to trade under, together with any other advice, clearly and concisely worded, such as the names of the breeds raised, and whether hatching eggs, breeding stock, day-old-chicks, market eggs or table poultry are for sale. If you wish to allow visitors, or have a particular place for customers to call, mention the facts or directions. If necessary state that visitors will be welcomed on certain days or between certain hours. Place the sign in a conspicuous position, preferably at the entrance

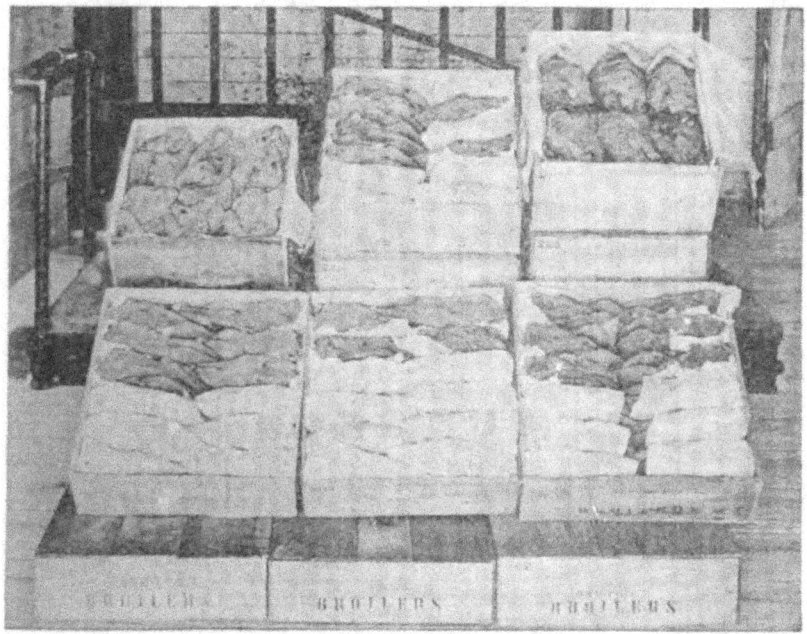

(Courtesy U. S. Dep't Agriculture)

Fig. 57.—Box-packed poultry, well-graded and ready for shipment.

to the grounds, and endeavor to make it as attractive as possible, yet in harmony with its surrounding.

If there is a wind-mill tower on the premises, and your business warrants the display, have your name neatly lettered on the rudder, or on the water tank, or on the roof of the barn, so that it can be seen for long distances, especially if it can be seen from a railroad or trolley line over which many people travel.

If there is likelihood of the fowls being disturbed by the intrusion of strangers, a question that is largely determined by the arrangement of the yards and buildings, there is no harm in putting up a sign to the effect: Visitors are invited to inspect

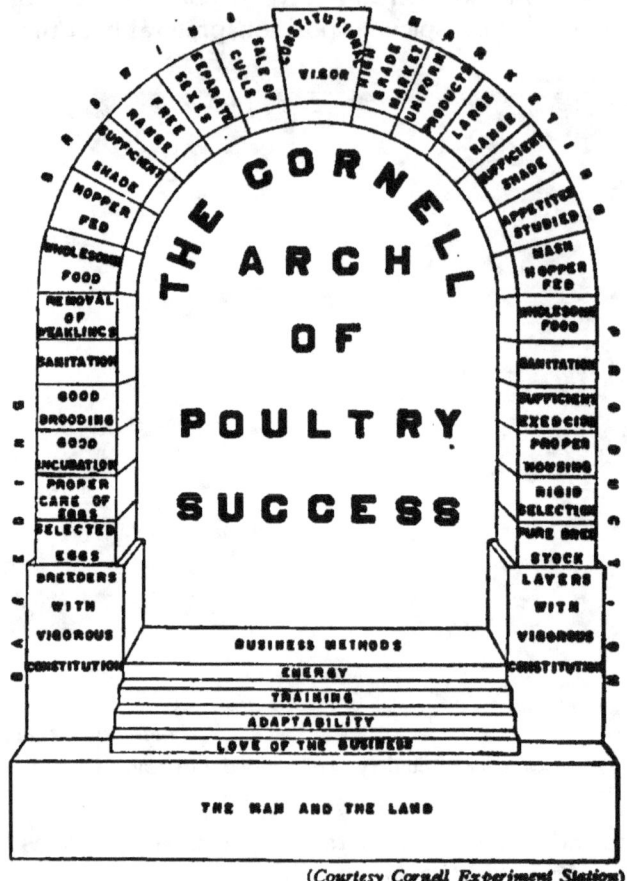

(*Courtesy Cornell Experiment Station*)

Fig. 58.—Elements which make for success in poultry raising.

the farm, but they are requested not to enter the buildings, or to frighten the fowls in any way.

Printed Matter.—Another factor that plays an important part in the farm's publicity, and one that inspires confidence in its establishment, is the use of a neatly printed letter-head and bill-head in all business transactions. On farms of any size there is quite a little correspondence, inquiries are received, asking for prices on stock, eggs and so on, and if these are answered in a poor handwriting on scraps of paper, or if bills are rendered in some obscure, *back-woods* style, they are sure to create a very unfavorable impression. It is unfortunate, perhaps, but we are frequently judged by these apparently trivial details.

Mail.—It should be a hard and fast rule, that all mail be answered promptly, preferably the same day it is received, and it should be answered courteously and fully, paying the same attention to a small order as to a large one. The person who writes for a setting of eggs to-day, or this season, may be in the market for a thousand eggs next month or next year. If bills are paid, or money is received on deposit, it must be acknowledged, with thanks, immediately. Indifference to these matters has a far-reaching effect. Business men are accustomed to extending and receiving certain formalities, and they expect them; it is a part of our great commercialism.

Printing is so cheap these days, there is seldom any excuse for a farm being without printed stationery, which should include a letter-head, bill-head, envelopes, shipping tags and labels, if such are used in place of tags, as, for example, on egg cases. There is usually a printer in every town of any size, who will get up some ideas at a small cost; or, stationery may be obtained through mail order houses, such as the publishers of agricultural journals and weekly papers. Aim to have the printing as attractive as possible, on fairly good quality paper, and to include the name of the farm, its location, its products, any of its most salient features, and wherever possible a trade-mark. Do not use anti-quated wood cuts, meaningless ornaments, poor half-tone reproductions of the owner, or his house, or one of the hackneyed, conventional electrotypes of a trio of fowls; people are not interested in such things, because they convey absolutely no mark of originality or distinction.

Trade-Mark.—Try to think up an original idea for a trade-mark or brand, and incorporate with it something that is significant, an idea that means something besides printers' ink; either the name, the farm's specialty, or its policy will do nicely.

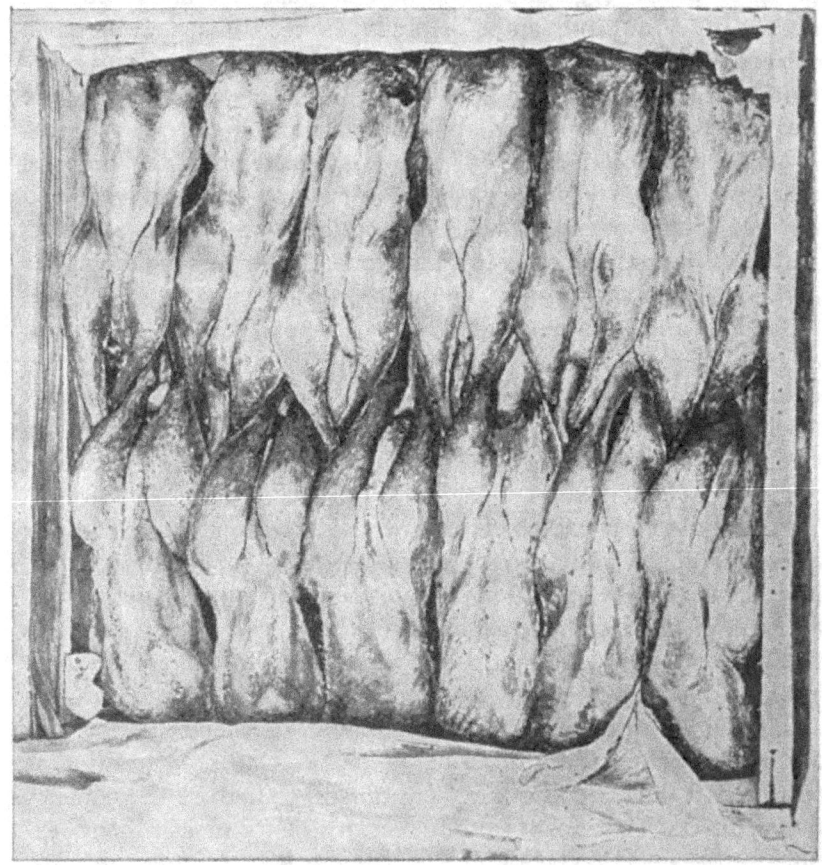

(Courtesy U. S. Dep't Agriculture)

Fig. 59.—Broiling chickens, packed breast up, twelve to the box.

Thereafter, this trade-mark should be embodied in all of the stationery, shipping tags, labels and any advertising literature or price lists that may be gotten out from time to time. If necessary have the design copyrighted or registered, or later, some un-scrupulous fellow may try to take advantage of an established publicity.

Reputation.—At first a trade-mark has very little significance, but as soon as a farm gains a reputation for straightforward deal-ings and a uniform quality in its products, its trade will look for some means of identification, and will insist upon having the goods of known reliability, even if it has to pay more for them. A well-known duck breeder, who makes a specialty of market ducks of prime quality, inserts a neatly printed tin tag in the web of the foot of each duck. Patrons of high-priced hotels and cafés where these ducks are served, have come to recognize that this tag, which is not unlike the tin tags inserted in plug tobacco, stands for quality, and the proprietors of these places, realizing the importance of the name, see to it that the tag is left in the foot when the duck is cooked and served.

The egg case is another advertising medium that should not be overlooked. If gift crates are used, those of light material that are not to be returned to the shipper, the poultryman should have a stencil made, giving the farm's name and address, and apply it to each side of the crate before it is packed. Another method is to paste an attractive label or sticker on the ends of the crates, similar to those seen on orange crates. They are not expensive to have printed, and while they require a little trouble gluing them on, the benefits to be derived therefrom will more than compensate for the time expended. It is a job that can be done on rainy days or at odd moments, and need not intrude itself upon routine work.

If returnable crates are employed, and these are desirable for certain classes of trade, they should be painted a serviceable color, and neatly lettered with the farm's name and address. In addition to this, it may be desirable to include the farm's specialty, such as: DAY-OLD STERILE WHITE EGGS, or SELECTED FARM EGGS.

The proprietors of many high class stores like to offer their eggs for sale in the original carriers, and will take particular pains to give them a conspicuous display, especially when they can afford to recommend goods from a reliable poultry farm. It is therefore a good plan to have the farm's name lettered on the inside of the lid, so that when it is thrown back the lettering will

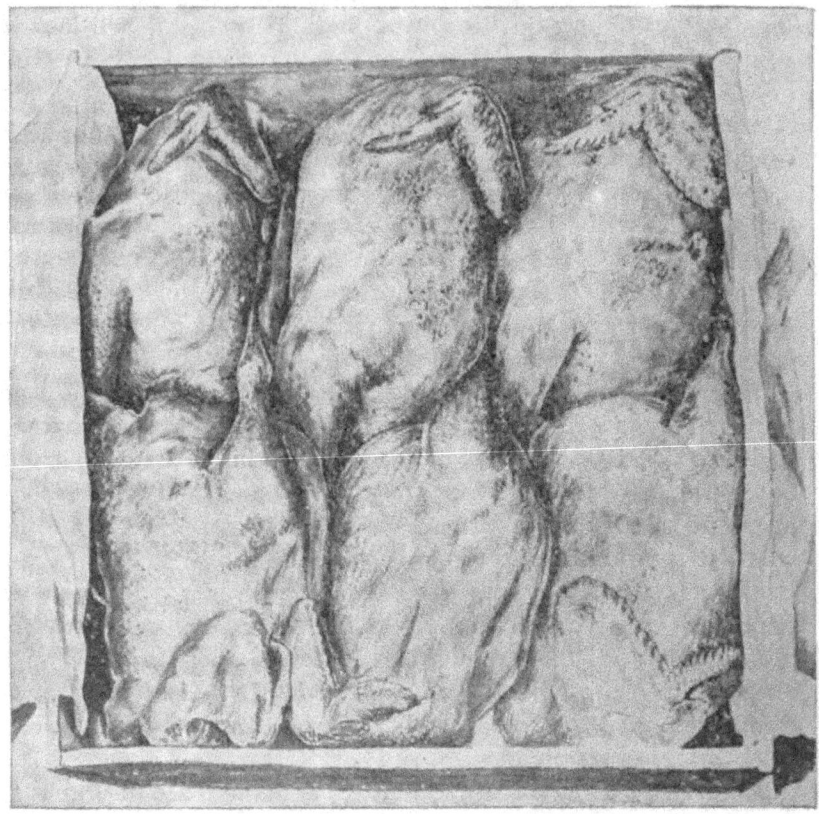

(Courtesy U. S. Dep't Agriculture)

Fig. 60.—Roasting chickens, side-packed, six to the box.

attract attention. The interior of the crate must be kept scrupulously clean, of course, and to aid in this it is well to shellac or varnish the inside of the crate when it is new. A hard oil finish on the outside of the crate is also effective.

Packages.—Many poultrymen are rather careless about the appearance of their packages, and pay no attention to them, any more than to see that they are securely packed. This is an error in judgment. Not only will the crates come under the observation of the consignee and his customers, if shipped to a wholesale

(Courtesy U. S. Dep't Agriculture)

Fig. 61.—Dressed poultry packed in cartons.

firm, but they will also be noted by many persons at the express depots and while in transit. This fact is especially true of crates of live fowls; there seems to be a sort of fascination about them for most people, who will go out of their way sometimes to inspect the tags, and note the name of the sender and to whom they are being shipped.

Publicity.—Summing up the entire situation, it behooves the poultryman to take advantage of every means at his command to gain publicity for his plant and its products; he should leave no stone unturned to keep in the *limelight*, and in so doing he should be careful to conduct his affairs in a dignified, reputable, business-like manner—paying the same strict attention to details that are found among progressive merchants generally.

CHAPTER VIII

KEEPING RECORDS

Know Where You Stand.—To have an accurate understanding of one's position or progress in any line of work it is essential to keep accounts. Conducting a poultry farm is no different from any other enterprise in this respect. To raise chickens intelligently and profitably, one must consider, precisely, such factors as labor, feed, the number of eggs laid by different flocks, cost of equipment, housing and so on. Haphazard, hit-or-miss, guess-work methods belong to the time when chickens were raised as a by-product of the farm, merely to supply the home table with a few tempting viands. Such methods are inexcusable to-day; they are shiftless.

Imagine the confusion that would exist if the general run of business houses attempted to conduct their affairs without some system of book-keeping. It is hardly likely they would survive a week. Yet it is safe to say the general run of poultry raisers are very lax in this respect, many of them keeping no records whatever, not even a memorandum of their feed bills. They have no way of telling whether their hens are an asset or a liability, or what it costs to produce a dozen eggs or a pound of meat. For all they know, it may be cheaper to buy poultry products at the store. .

Leaks.—Farms specializing in poultry products, progressive poultrymen, must have a definite knowledge of the performance of their flocks, and what it costs to maintain that performance. Experienced breeders, those who helped to make poultry raising a billion dollar industry, and thus take front rank in the country's industrial activities, laid the foundation of their success on the *leaks* and shortcomings that were detected by some system of records. Every day valuable information is brought to light by the simple recording of a series of experiments; witness our agricultural experiment stations.

Eliminate the drones, the nonproducers, and cull the undesirables and defectives, is the twentieth century farmer's slogan. Get rid of the *boarders*, either as a class or as individuals; install breeds or stock of known reliability, those that are good

(*Courtesy U. S. Dep't Agriculture*)

Fig. 62.—Wagon load of live poultry unloading at a Western packing house.

tenants. Except for resolutions of this kind the two-hundred-egg-hen would still be a myth; our flocks would never have progressed beyond the average of five dozen eggs per year. In fact, raising chickens could never have been made profitable, not even as a *side line*, except for consistent efforts of careful selection for many generations, which established a standard.

Simple Records.—Unless detailed information is desired, it is not necessary to employ a highly involved system of book-keeping, but a simple record of costs and sales, the revenue derived from eggs, meat, and other sources, such as feathers and manure, and the cost of production, the feed, labor, repairs, improvements and general overhead expenses. A good record should be complete, concise and convenient; above all else it

Fig. 63.—Suburbanite's poultry plant.

should be accurate.. The best way to insure accuracy is to keep the account up to date by a few minutes' work each day.

To those who are unaccustomed to keeping accounts, it may seem rather difficult on first thought to *keep tabs* on a flock of hens. In reality it is quite easy. Some subjects are rather obscure at first, but if they are initiated, one by one, a simple general scheme will evolve, one that will be easy to follow thereafter, and prove unquestionably helpful.

Back lot poultrymen, those who raise but a few hens a year,

7

mostly for home consumption, need not concern themselves with but two columns of figures—a debit and a credit column. The debit column, though it may be the least desirable, will be considered first; it is a *necessary evil.*

Feed bills are the chief occupants of this column. Be sure to enter every item. The cost of the grain is seldom the only item. What about sacks, freight on the feed, hauling and so on? These are chargeable. Include all the labor required to care for the birds, for cleaning, feeding, watering and general supervision, whether it is performed by the owner of the flock, or assisted by

(*Courtesy Wisconsin Experiment Station*)

Fig. 64.—Fowls on range are stronger, more thrifty and less liable to contract disease.

his wife and children. Perhaps only a few minutes are required several times a day; estimate their total in hours and multiply by a fair wage rate.

Original Investment.—It is not fair to charge the cost of the poultry building and all equipment, also the value of the flock, to the expense account. They represent capital. Figure the original investment, and on this it is fair to charge an interest rate of six per cent per annum, which should be added to the debit column. Against the value of the buildings, fencing and equipment should be charged another six per cent for depreciation.

Repairs and any other incidental expenses are also chargeable.

On the credit side of the ledger must be entered items covering the sale of any eggs or stock, the products consumed on the home table, and an allowance made for the value of the manure as a fertilizer, or perhaps feathers. As much should be charged for the eggs used in the home kitchen as would have to be paid for eggs of a similar grade in the retail store.

Information.—Those engaged in the poultry business as a

(*Courtesy Purdue Experiment Station*)

Fig. 65.—"A good hatch." Note the tray of empty egg shells on top of the incubator, and the leg bands on the chicks.

specialty and who raise fowls on an extensive scale, will find it to their advantage to gather as much additional information about their flocks as possible. They should ascertain the laying capacity of each flock, and what it costs to feed each flock. These facts enable you to estimate the cost of producing a dozen eggs at different seasons of the year. If itemized records are kept of the feed consumed by different flocks and by broods of chicks, it is possible to figure out the cost of raising certain breeds to maturity, and of keeping different varieties for a year.

Of what practical benefit is it to know these things? Simply this, the question constantly arises, which is the most profitable breed of chickens? Or, beginners will ask, which is the more profitable specialty—meat or eggs?

It is impossible to furnish a reliable answer to these queries with a general statement. The fact of the matter is, there is no one best breed, and no one branch of poultry culture can be said to be the most profitable. There is no greater proof of this than the reports from the Egg Laying Contests for the past six years. Some breeds are more desirable for certain purposes. Even so, there is as much difference between different strains—different blood lines, of the same variety, as there is between totally different breeds.

The only way to be sure of the best breed for a certain purpose, or the best strain of a certain breed, is to experiment with it and keep an accurate record of its behavior. Never take too much for granted in any line of work, especially in the poultry industry. Not that there is any attempt at willful misrepresentation, but it does not necessarily follow that every one will succeed with certain conditions because some have done so. Do your own thinking; conduct your own investigations; establish your own rates and records. In no other way will it be possible to really know your business.

(*Courtesy Kansas Experiment Station*)

Fig. 66. — Aluminum leg bands. Small one is for chicks, and for inserting in the web of the wing.

Breeding Records.—To return to our subject, if one wishes to breed from none but the heaviest producers, or from specimens of a particular type, records must be kept of the hatching and brooding. A single great performance really conveys nothing to the poultryman so far as breeding progress is concerned, unless it can be definitely located in the preceding generation and later, in the succeeding generation. It is the heavy layer with the faculty for producing heavy layers, or the fowl of superior type with the capacity to beget offspring of the same superior type, that are sought. Such matings are proved by records established for several generations.

In the breeding of other kinds of pedigreed livestock, such as horses, cows and dogs, permanent records are maintained by the officers of a society or association of breeders. In other words, the stock is registered; and an examination of the records of a certain society will disclose the ancestry of any animal of note.

Because the poultryman must keep his own pedigree, it becomes none the less important.

The trap nest and the numbered leg band are the only positive means of determining the exact laying ability of a hen, which hens lay the best shaped eggs, which the largest sized, which the strongest in point of fertility, which are the best winter layers, which pullets begin early and lay the greatest number of eggs in succession, the number of times they become broody, and many other facts of vital im-

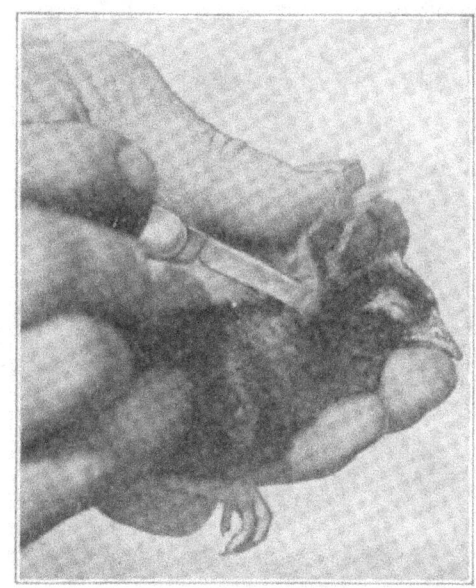

(Courtesy Kansas Experiment Station)

Fig. 67.—Making incision for wing band.

portance to the poultryman. But whether or not this information is considered of sufficient value to warrant the additional time, trouble and expense of operating the trap nest system is the debatable question that must be determined by every poultryman for himself.

The use of the trap nest is described in another chapter.

Fertility.—A fair idea of the fertility and hatchability of the eggs, and the stamina and growth of the chicks may be ascertained by flocks, where the poultryman does not wish to trap nest his birds. Let us assume the breeding pens consist of units of from twenty to a hundred fowls, and that they have been mated with the view to some definite purpose. As the eggs are collected from these pens let the collector mark the number of the pen on the receptacle in which the eggs are gathered; later, when the eggs are selected for incubation this number is marked on the egg shells and they are placed in the incubator or under hens. A card is made out for each hatching, and on it is marked the date the eggs are set. When the eggs are tested for fertility a report is made on the card of the number of clear eggs removed, and to what pen numbers they belong. The same idea is carried out at hatching time—the unhatched eggs are counted and credited to their respective pens.

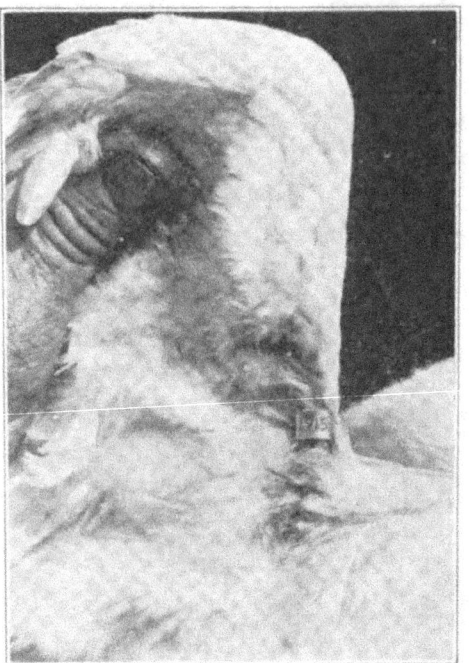

(Courtesy Kansas Experiment Station)

Fig. 68.—Wing band in place on a mature bird.

Marking Chicks.—This data will keep a pretty good line on the fertility of the different pens, and the hatchability of their eggs. If the operator wishes to go further, and follow the progress of the chicks in the brooder, he can mark them when they are removed from the incubator, by leg bands, such as are used for pigeons. See Fig. 65. In keeping track of the eggs in the incubator they are given ordinary treatment up to the eighteenth day, or when the eggs are turned for last time. Then, by means of wire baskets, pedigree trays, or mosquito-netting sacks, the eggs are segregated according to the numbers on their shells, and when the hatch is completed these numbers are designated on the leg bands placed on the chicks.

This method permits the breeder to follow the progress of the chicks from different pens, and to note the results of his selection for given matings. It also keeps a check on the mortality and on early development. When the chick is five or six weeks old,

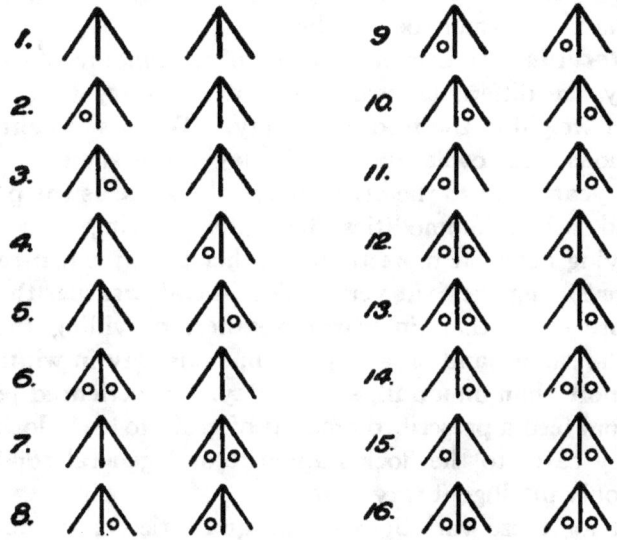

Fig. 69.—Sixteen different methods of marking the toes of chicks.

and has outgrown the size of the first leg band, it must be removed and a larger one substituted, or the original band may be secured to the web of the wing, where it will remain throughout life.

Wing Bands.—Attaching the band to the web of the wing is accomplished without serious discomfort to the bird. See Fig. 67. Select a spot free from blood vessels of any size, pierce it with the point of a sharp knife and adjust the band.

Toe Marking.—The advantages of having the fowls marked are numerous; it is always possible to tell the bird's age, and if the records are properly kept there will be no likelihood of inbreeding. Another method of marking the chicks is to punch a small hole in the web of the foot. Special punches may be purchased for this purpose. The chicks should be marked the day they are hatched, as the web is then soft, does not bleed so much as later, consequently there is little risk of the other chicks pecking the toes, as they would do when older. By different combinations there are sixteen markings possible, and a chart should be kept illustrating the marks. See Fig. 69. In my experience leg-banding is better than toe-punching.

Feed Records.—In keeping a record of the amount of feed consumed by the different flocks it is not necessary to go to the trouble of weighing the feed every day. To do so might prove very tedious. A fowl's appetite is never the same; it varies from one year's end to the other, much the same as the prices of grain and other commodities fluctuate. During a period of heavy laying hens eat more than at other times; when they are molting or sitting they eat very little; in cold weather they consume more grain than in warm weather, providing their egg yield is the same; and on a bright sun-shiny day in winter they will eat more than on a dull, stormy day. Experienced poultry-men seldom feed a prescribed amount of grain to each flock every day; they cater to the flock's appetite and general conditions. It is the only intelligent way to feed.

Except for these variations in the quantities of the feed and in the prices of the feed, it would be a comparatively simple task to figure the cost of the feed for a given pen.

On most farms it is customary to feed the layers a certain amount of scratch grains in the morning, just enough to keep them at work in the litter, and all they will clean up in the late afternoon. At the same time a dry mash is kept before the flock all day, together with oyster shells, grit, charcoal and beef scrap, unless the last two of these articles are included in the mash. Naturally, the quantities vary considerably. Furthermore, the numerous kinds of feed are purchased at varying intervals, in

Fig. 70.—Laying house well situated on sloping ground.

105

different quantities, and at such times as they may be bought advantageously.

Obtain a Rate.—The only simple way to surmount these irregularities is to reduce them to a unit basis or rate—the cost of a pound or quart for a given period, as, for example, the cost of a pound of dry mash for the month of October is $0.0195, or the cost of a quart of scratch grains for November is $0.0198. If large quantities are handled the unit may be raised to a hundred pounds or a hundred quarts.

(*Courtesy Missouri Experiment Station*)

Fig. 71.—Outdoor feed hopper for growing stock.

A scheme of this kind is in use on a farm of my acquaintance, and it works out very nicely. Every consignment of feed received is apportioned to dry mash, scratch grains, chick feed or whichever way the meals and grains are to be used, in two totals, weight and price. At the end of each month the total costs are added, and divided by the sum of the total weights, and the quotients are the rates per pound for each classification for that particular month.

The rates are then applied to the quantities consumed by the various flocks, whose records are kept daily in each house or pen, then totaled for the month.

The manner of keeping the records of the feed consumed is rather unique. A card or slate is fastened near the door of each pen, and as the attendant goes about distributing the feed, an entry is made of the quantity of each kind of feed, dry mash, scratch grains, shells and so on. Large hoppers are employed for the storage of dry mash, some of them having a capacity of two and three hundred pounds, hence there may be only five or six entries for mash in a month. Similar devices are installed for shells and grit, which are replenished on an average of once a month.

Buckets of known uniform capacity are used for distributing the feed and other supplies, which obviates the necessity for weighing and feeding. Let us say, a certain-sized bucket contains forty pounds of scratch grains: If the feeder throws half of it to one pen and a quarter to another, and then four buckets

Fig. 72.—Substantial set of poultry buildings.

to a larger house, he enters twenty, ten and a hundred and sixty pounds respectively on each slate.

The same bucket will probably hold about thirty pounds of dry mash, and about eighty pounds of grit or shells. At the end of the month all slates are brought to the office, totaled, entered in a book and wiped clean for the next month.

Tell-Tales.—If the record of a particular house indicates a falling off in feed along with a decrease in eggs, or if the egg record does not compare favorably with the cost of the feed, it proves substantially that something is wrong, or that a certain flock is not up to standard.   An inquiry is then held and the reasons ascertained.   The system also acts as a check on the feed; since the total amount of feed placed before the fowls must compare with the amount of feed purchased.   This is a feature not to be ignored; we have known feed to be wasted, sold short weight and stolen.   The question of keeping records on a poultry farm is in harmony with this age of time clocks and cash registers.

CHAPTER IX

PRINCIPLES OF POULTRY HOUSE CONSTRUCTION

The question often arises, which is the best system of poultry management?

Many persons have been led to believe that such systems exist, even to the extent of being patented, and for the use of which it is necessary to pay certain sums of money. The truth is, that while hundreds of books and pamphlets have been published on this subject, purporting to be secret methods and systems, many of which contain extravagant claims, there is no such thing as a hard and fast system of poultry culture.

On the contrary, the only principle which might be said to constitute a system, and upon which all of the so-called systems and secret methods agree, is—that the management or care shall be systematically done. Beyond this every poultryman must evolve his own individual methods and practices, those which are best suited to a given purpose in a given locality. It is important that one should have a definite purpose, for in that way only is it possible to determine the most suitable location, the best type of housing, the most profitable breeds, and so on.

Location.—While fowls can be kept almost anywhere and everywhere, they do best in congenial locations. Soil conditions and the arrangement and construction of buildings have much to do with their health and profit. We must also consider the means and inclinations of their owner.

Briefly, the ideal location may be summarized as follows: Choose a soil which is light enough to provide good natural drainage, yet heavy enough to grow grass, and a site having a south or southeast exposure, protected from prevailing high winds.

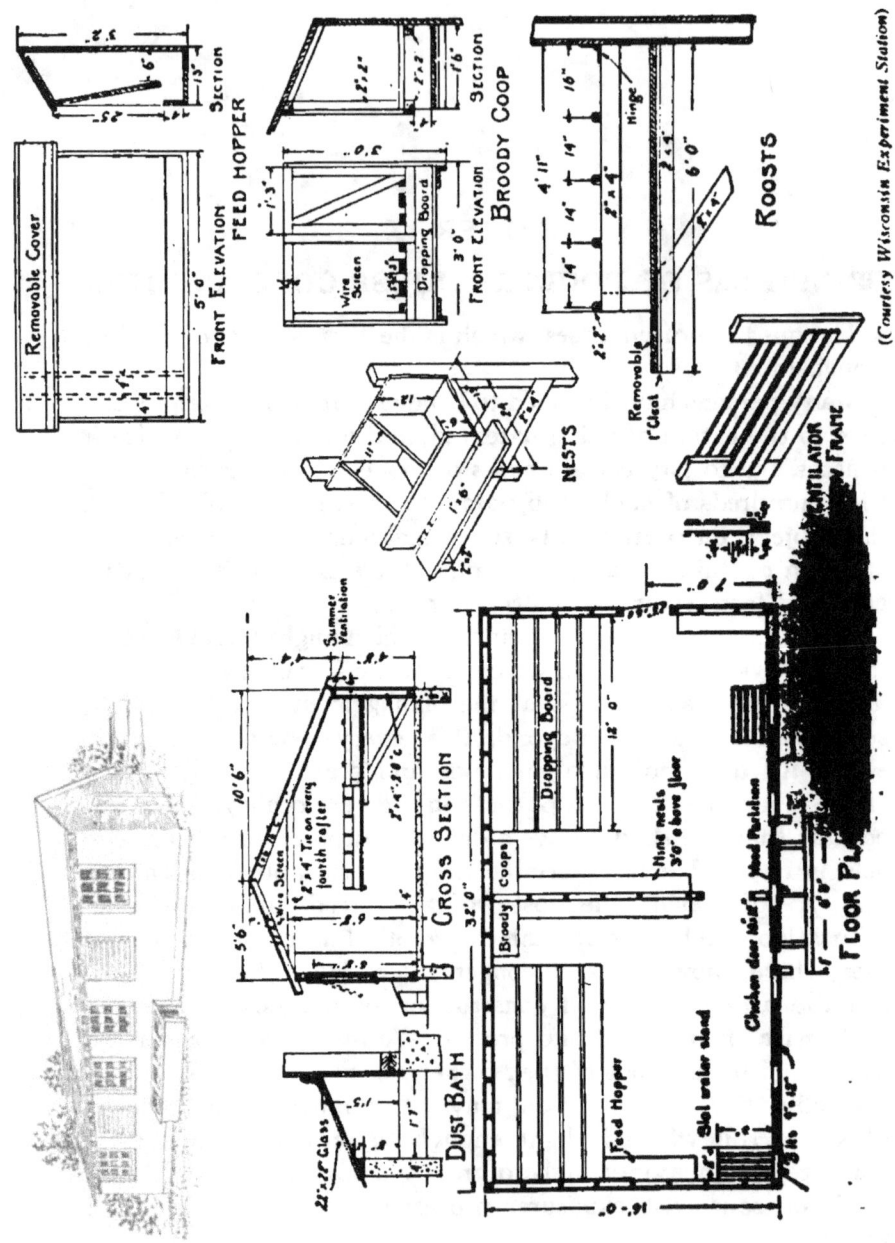

Fig. 73.—General arrangement and details for medium-size hen house.

(Courtesy Wisconsin Experiment Station)

110

A dry porous soil, such as sandy or gravelly loam, is preferable to a heavy clay soil, because the former is easily kept sanitary. A purely sandy soil, however, is not desirable, because it will not support the plant and insect life which poultry should have.   If it is impracticable to select a naturally dry soil, it should be made sanitary by underdrainage.

Build the houses in the lee of a wind-break if possible, and on an elevation having a natural drainage away from the buildings. When a direct southern exposure is not obtainable, aim to have

(Courtesy Monmouth Poultry Farm)

Fig. 74.—A site like this means well-drained, sanitary yards.

the buildings face the southeast rather than southwest, for fowls seem to enjoy morning to afternoon sun, and other things being equal the quarters should be warmer.

In the manner of housing fowls there are two general ideas— the colony plan, which consists in placing small houses for small flocks far enough apart so that they will have an abundance of range, and with little chance of intermingling, and the more intensive plan of keeping the birds in long continuous laying houses. See Fig. 76.   This latter arrangement of housing may consist of a series of separate pens under one roof, connected by an alley-way at the rear, or by doors or gates between the pens, or it may be one long house capable of accommodating units of from 500 to 1000 birds.

The advantages and disadvantages of the two ideas are numerous. Birds on free range require less scrupulous attention to cleanliness, no expense for fencing, and they will pick up the greater part of their green and animal food. Moreover, should sickness break out there is less likelihood of its becoming an epidemic. On the other hand, the colony plan involves considerably

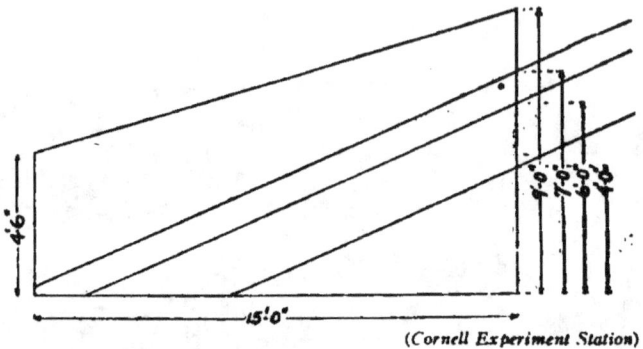

(*Cornell Experiment Station*)

Fig. 75.—Diagram illustrating the angle of the sun's rays during December. Sunlight should be permitted to reach the rear of the building, where it is most needed.

more labor in the performance of routine work, such as feeding and watering, cleaning and gathering eggs, than the continuous house plan, which is especially true in stormy weather.

The cost of building houses on the colony plan is much higher per bird, not only because a number of smaller houses require so many additional end walls, but being smaller the allowance of floor area per bird in the colony house should be almost double that required in the long continuous house.

The relative merits as to productiveness are debatable, although there is a tendency to accept the idea that small flocks produce the greater egg yield. To offset this, however, it costs more in labor to produce a dozen eggs by the colony plan than by a more intensive arrangement.

The ideal type of poultry house is not necessarily the most expensive building. It should be serviceable above all things, fairly roomy, well ventilated and yet free from direct drafts, capable of being flooded with sunlight, and dry and sanitary at all times. It should be built wherever possible with the view to simplicity, economy and convenience. To spend large sums on it is almost as grave an error as to slight it, for money expended for unneces-

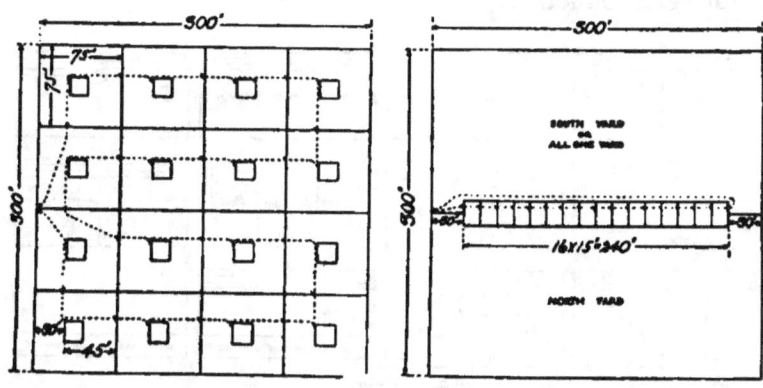

(*Cornell Experiment Station*)

Fig. 76.—Housing plans. Diagram on left indicates colony house system; dotted line shows the distance travelled to reach all the buildings. In the right-hand diagram the same number of pens are brought together in a continuous house; note the amount of walking saved.

sary purposes on a poultry farm is dead capital and brings no return on the investment.

It is impossible, of course, to meet all conditions or suit all tastes in one or two types of houses, but if one gains a familiarity with the fundamental principles of poultry house construction, it is then a simple matter to incorporate those principles into a type suitable to any tastes, conveniences, soils and climatic conditions.

Warmth.—A warm house, or at least warm sleeping quarters, is one of the prime requisites for winter egg production; yet arti-

8

ficial heat is not to be recommended.   In fact, it has been tried
very thoroughly, and with unsatisfactory results.   Birds so kept
quickly lose their vitality, and sickness and other troubles develop.   It is better to build the house substantially, and thus insure it against drafts and dampness, for these are the poultryman's
greatest foes.

Egg production is really the result of a secondary circulation,
hence if the fowls require all their surplus energy and vitality
to combat improper conditions and to keep warm, there is none
left for egg production.

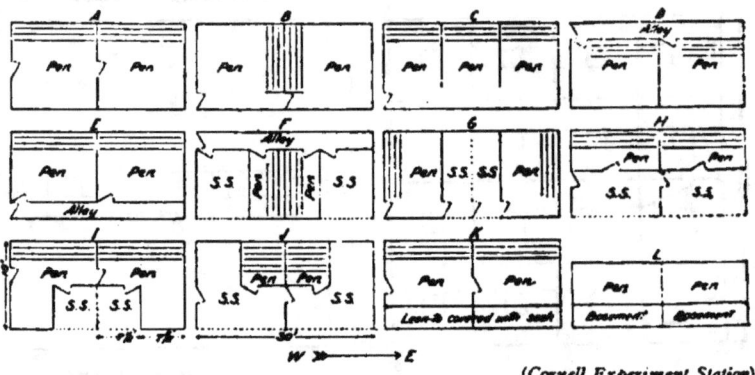

(Cornell Experiment Station)

Fig. 77.—Types of pens, illustrating arrangement of roosts, partitions, alley-
ways and scratching sheds.

The size of a poultry house is largely influenced by the breed
and number of birds kept in each flock.   If the fowls are kept in
small flocks, more floor space per bird is needed, and the reason
is quite apparent.   In the small flock, say twenty birds in a house
10 by 12 feet, each hen is confined to a very narrow area, although
she has an allotment of six square feet; whereas in a large flock
of 500 layers, housed in a building 16 by 100 feet, each bird would
have but a trifle over three square feet of floor space, yet it would
have the freedom of roaming and scratching over the entire area,
and would not be oppressed with the feeling of constraint.

Ordinarily, the heavier breeds require about one-half again as much floor space as the Leghorn and other Mediterranean classes; but the smaller breeds, being more active and more nervous, are more apt to become unduly excited and panicky when crowded in a small pen.   It is difficult for the attendant to work in a small house or pen without getting the flock into a condition of unrest and excitability for fear of being cornered, and such is not conducive to egg production.

The situation may be summarized about as follows: Allow 6 to 7 square feet per bird for all classes, if the houses are small.

Fig. 78.—Poultry house under construction by students at Purdue University.

Provide 5 to 6 square feet per bird of the Plymouth Rock or Wyandotte type, in houses where the flocks are made up of large units.   Allow 3 to 4 square feet per bird for Leghorns and similar classes, when they are kept in large flocks.   This is the practice on many of the largest commercial farms.

To obtain the greatest amount of floor space at the least cost, the building should be designed as wide as possible, yet not exceed a point where long timbers are required, for these are certain to be expensive.   Sixteen feet is a good width for the shed roof type of house; the timbers required are stock lengths, therefore sold

at regular rates; and the front wall need not be so high but that
the sun will reach the rear wall, where it is most needed on account
of the roosting compartments.

There are three general types of roofs for poultry buildings—
the single pitch or shed roof, the gable roof, or double pitch,
having equal or unequal slopes; and the half-monitor style of
house, which consists of a shed roof in the rear of the house,
covering about two-thirds of the building, then from the front
eaves a wall is built for perhaps a distance of three or four feet,
consisting mainly of windows, and from which there extends

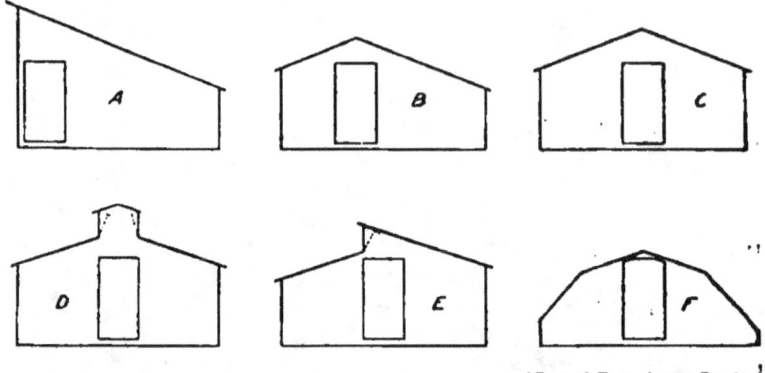

(Cornell Experiment Station)

Fig. 79.—Types of roofs for poultry houses.   A, shed roof; B, two-thirds span;
C, gable; D, monitor; E, half monitor; F, hip roof.

another plane or shed roof, covering the front third of the building. See Fig. 79. Houses of this type can be erected wider than
sixteen feet and not require exceptionally long timbers, and it is
also practicable to have the sun's rays reach the rear of the interior without a high front wall.

The single-pitch roof is the easiest to build, and is probably
the most generally used on that account. It furnishes the highest
front wall, and sheds all the rain water to the rear.

All poultry buildings should be built as low as possible, not
only to save material, but to conserve warmth in cold weather;
yet they should be built with standing room in all sections where

the routine work is performed.   If the rear wall is built of sufficient height for the attendant to perform his work without bump-
ing his head, and the front wall is to be kept as low as practicable,
the roof will necessarily have to be of comparatively low pitch.
Since shingles do not wear well on roofs of low pitch, houses of
the shed roof type are usually roofed with a good grade of ready-
to-lay patent roofing.   These roofings, providing they are of
known reliability, have rendered satisfaction, and are to be
recommended.   They require few repairs, but an occasional

(Courtesy Cornell Experiment Station)

Fig. 80.—Interior of continuous poultry house in course of construction.
Note the framing for pen partitions, ceiling against the rear wall in way of the
roosting compartments, and the dirt floor.   The sills are laid on concrete walls.

painting, and are very economical in the amount of labor involved in laying them.

When to Build.—Whenever possible it is best to build during
the spring or early summer, for the building then has time to dry
out during the hot days.   This is especially true of houses which
are intended to have dirt floors, or those having cement floors
and foundation walls.   Much of the sickness attributed to dampness will be avoided in this way, also considerable trouble and
annoyance caused by wet, mucky litter.

Kinds of Floors.—The floor may be of earth, wood or cement; location and soil conditions are the determining factors. See Fig. 81. It is highly important that the floor be dry, otherwise it will be impossible to keep the litter dry and sweet enough for the fowls to work in. Straw and similar materials absorb moisture very quickly, whereupon they give off foul odors and are very apt to contaminate the scratch grains thrown into them.

Earth Floor.—There is no better floor for poultry than an earth one, providing it is practicable to keep it dry, and it is also the most economical. A light sandy loam is best. A dirt floor should be about a foot above the outside ground level, hence the

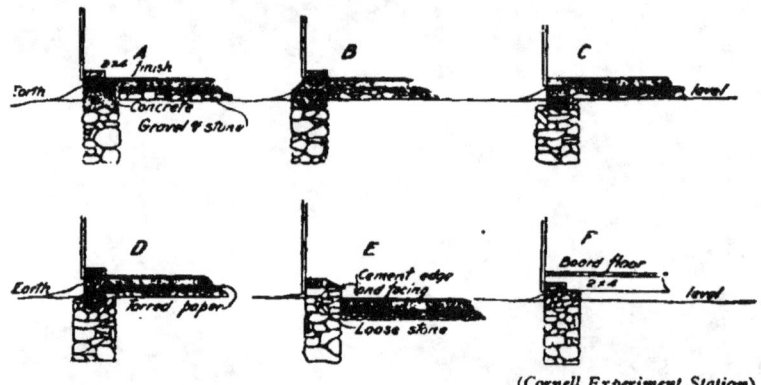

(*Cornell Experiment Station*)

Fig. 81.—Methods of constructing foundations and floors.

best method of construction for such a house is to erect a concrete wall for a foundation. This should be built deep enough to prevent heaving by frost action, and extend about a foot above ground, after which it is filled in with soil, preferably sand, before the balance of the structure is erected.

A wooden sill is laid on the top of this wall, upon which the frame studding is built, with the weather boards or siding carried three or four inches down on the outside of the concrete. An arrangement of this kind promotes great durability, for there is no part of the frame structure in contact with dampness and therefore likely to deteriorate.

Board floors are usually short-lived unless a free circulation of air is allowed under them, in which case it is well to build the house on piers two feet from the ground, or on a wall having adequate openings for ventilation The piers should be built of concrete, stone or brick for permanence. If posts are used, they should be charred or treated with a wood-preserving compound to prolong their life.

Another objection to the wood floor which is built close to the ground, it offers a refuge to rats and mice and, perhaps, other animals. These pests are likely to occur on any farm, and if means are not provided to combat them, they will rapidly become a serious nuisance.

Cement floors are the only absolutely vermin-proof ones; they are easily cleaned and durable, but apt to be cold and hard on the fowls' feet, unless covered with a thick layer of sand and litter. In constructing a cement floor the ground should be excavated for a depth of eight or ten inches, and filled in with broken stones or cinders to make a good foundation, which also acts as a sort of French drain and keeps the floor dry. The concrete slab should be about two or three inches thick, poured over the broken stones or cinders after they are well tamped to a solid bed, and mixed in the proportion of 1 part Portland cement, 2½ parts crushed stone or cinders, and 5 parts clean sharp sand. It is advisable to pitch the floor to a drain, which will greatly facilitate house cleaning.

Floor Joists.—If the floor is of wood, built upon piers, the sills should be of fairly heavy timbers, running the long way of the house, which support the floor joists; these latter of 2 by 8, or 2 by 10 material, and spaced about 20 inches on centers. If a single floor is to be installed, it should consist of a good grade of matched flooring; otherwise, if a double floor is contemplated, the rough flooring may be of 1 by 12 inch sheathing boards, laid diagonally across the joists, and over-laid lengthwise of the house with 1 by 3 inch matched flooring. The finished floor should be *blind nailed*, so that no nail heads project to hamper the use of a shovel or scraper in cleaning. Where necessary a layer of

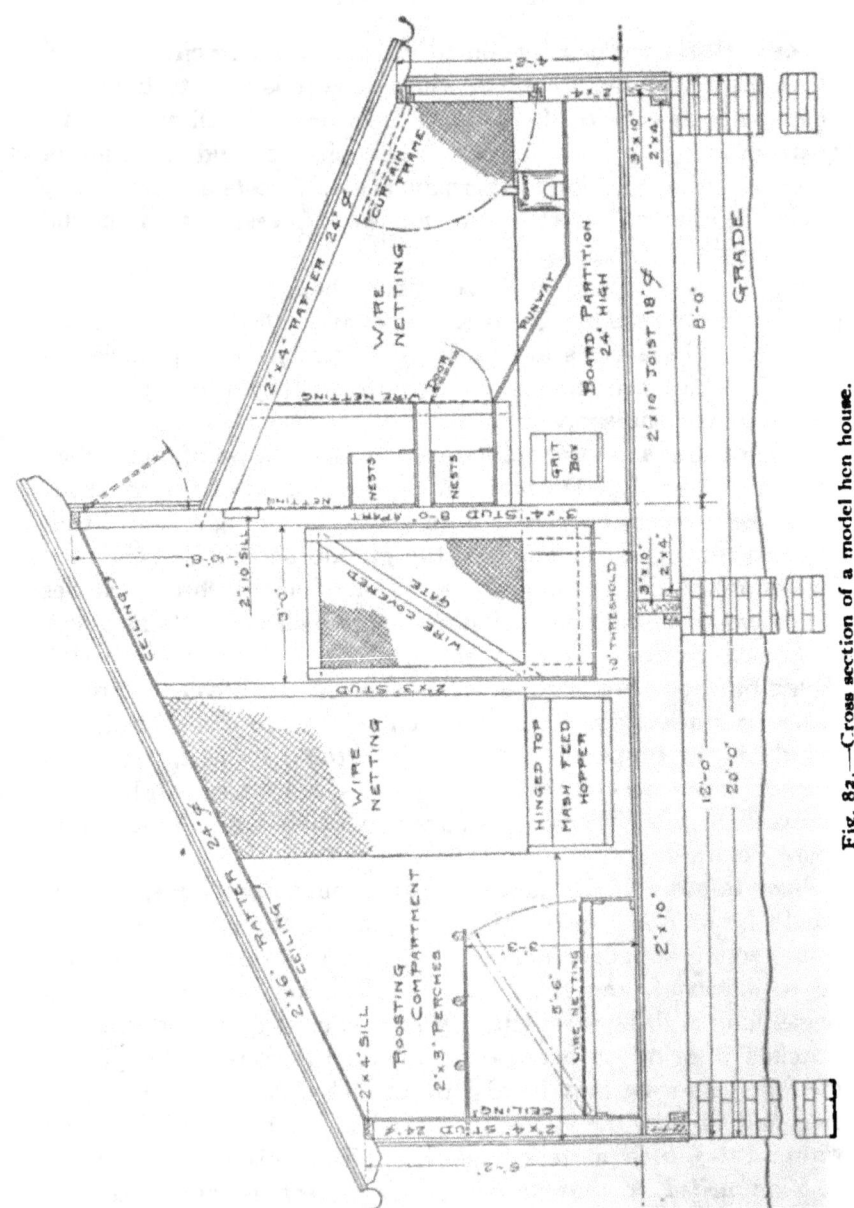

Fig. 82.—Cross section of a model hen house.

120

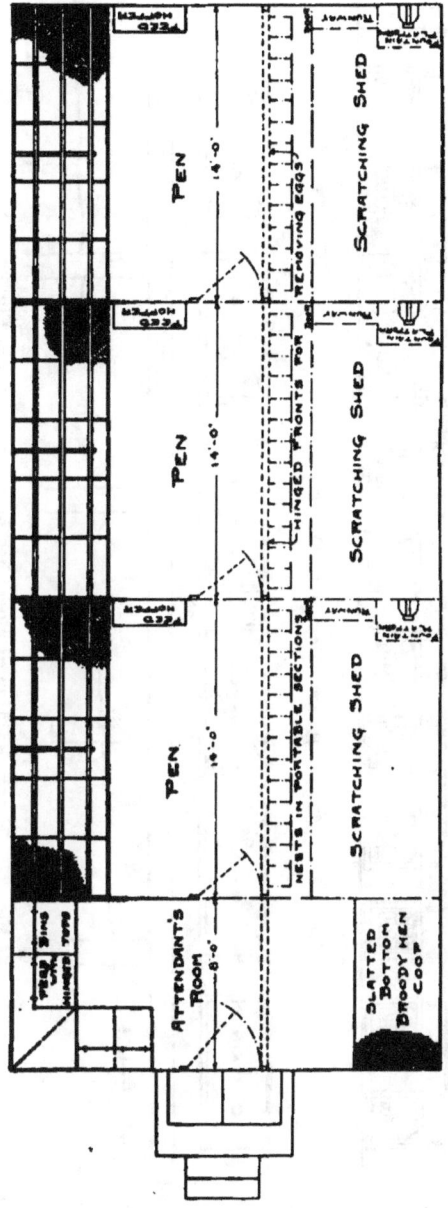

Fig. 83.—Plan of model hen house, as shown in Fig. 82.

121

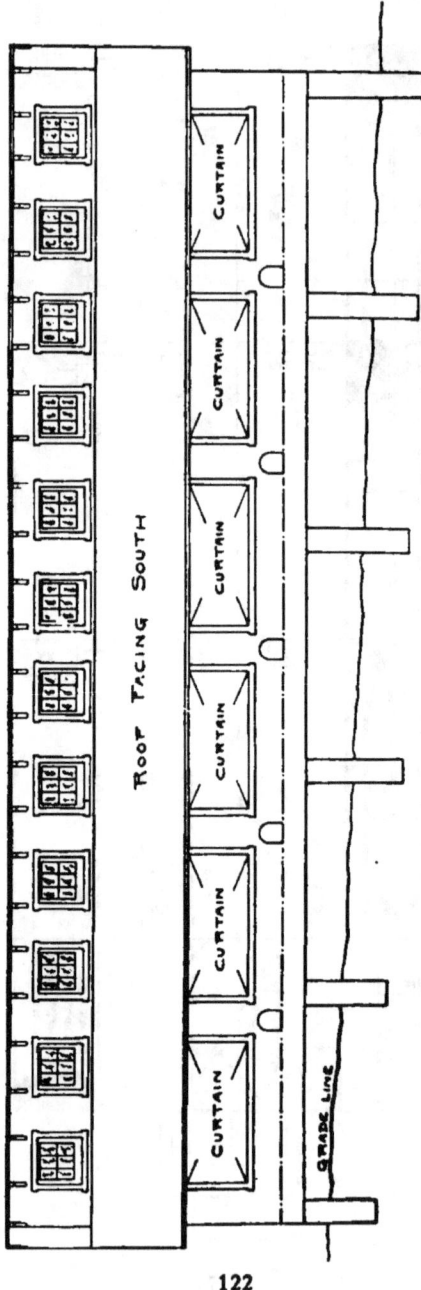

Fig. 84.—Front elevation of model hen house, as shown in Fig. 82.

122

felt building paper may be inserted between the two floors, which will add greatly to the warmth of the building.

The framework is constructed mainly from 2 by 4 inch lumber, as are the rafters of buildings less than fourteen feet in width. The walls may consist of one thickness of matched boards covered with 1-ply smooth-surfaced patent roofing, similar to the roof covering, novelty siding, weather boards, or rough sheathing shingled. Or, as is the custom in extremely cold climates, the walls may be made double, with a dead air space between, or

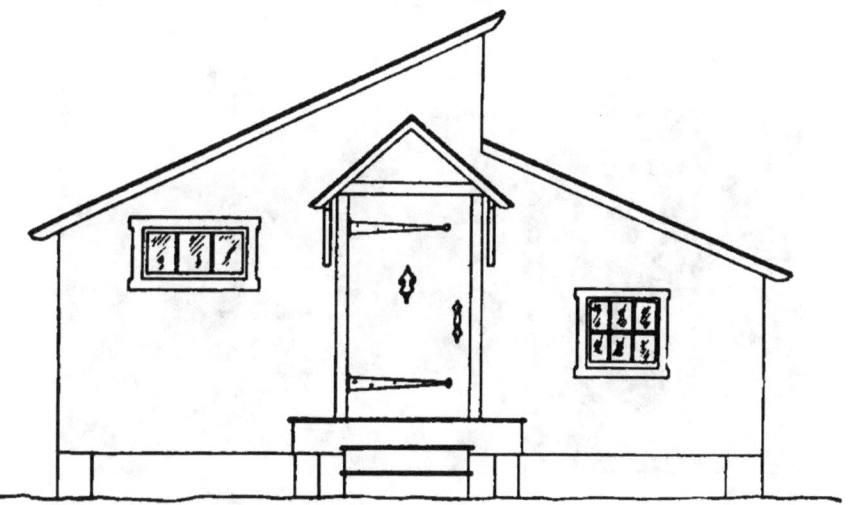

Fig. 85.—End view of model hen house, as shown in Fig. 82.

filled in with straw, hay or shavings for insulating purposes. These precautions, however, are not necessary unless the temperature goes below zero for long periods at a time.

Another practice is to ceil the inside edge of the studding in way of the roosting compartments, and also the under side of the rafters. This makes a very neat interior, and helps to carry off any foul odors from the perches. Some poultrymen advocate the use of felt or tar paper on the inside of laying houses, lining the walls with it; but in the writer's experience this is very unsatisfactory.

It affords a breeding ground for lice and mites, and it cannot be cleaned. Then, too, it is inflammable.

Openings.—The best arrangement is to have all the openings in the house on one side—the front wall—so that by keeping the other three sides tightly closed, drafts are prevented.

Too much glass in a poultry house makes it cold in winter and hot in summer. In recent years curtain frames have taken the

(*Courtesy Cornell Experiment Station*)

Fig. 86.—Framing for a continuous poultry house. Note the concrete walls; building is intended to have a dirt floor.

place of glass windows to a great extent. In fact, many farms use curtain frames exclusively, although this practice makes for a dark interior, when on very stormy days it becomes necessary to close the curtains. The semi-open front house should consist of one-third board partition for its front wall, commencing at the floor line, one-third curtain frame openings, and one-third windows and ventilators. See Fig. 84.

Curtain Frames.—A medium-weight, unbleached muslin is

the proper material for the curtains, not heavy duck; the idea being that they should be porous and permit a circulation of fresh air without direct drafts. The frames may be hinged at the top and made to swing inward or outward, or they may be made portable and held in place by wooden buttons, operated from the outside. Inasmuch as they are in use for only about four months in the year, and then only at night or at such times when there is a severe storm from the south, it is best to make the frames portable, and to store them elsewhere when not needed.

The position of the glass windows in the upper section of the

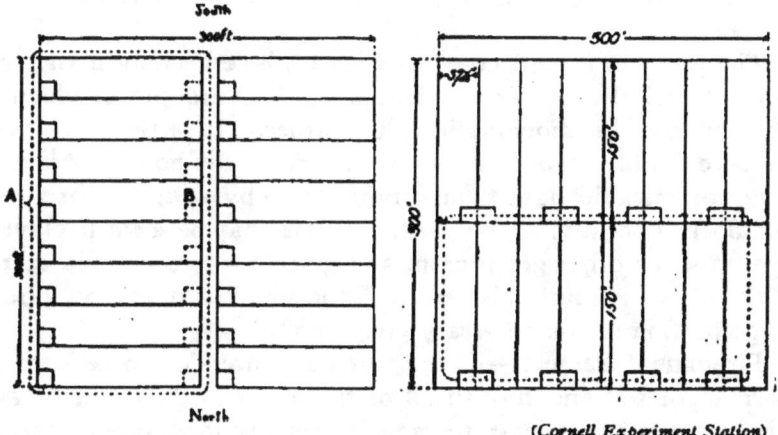

(*Cornell Experiment Station*)

Fig. 87.—Methods of arranging colony houses and yards to save unnecessary steps.

front wall should allow the maximum amount of sunlight to enter the building. They may be made portable, similar to the curtains, and stored elsewhere when not in use, or they can be opened by sliding to one side, or by swinging outward, which also serves the purpose of shedding driving rains from the south.

Doors.—A small opening should be cut in the front wall as a means of egress for the birds, and fitted with a door. The main entrance door should be located at the end of the building, and be of generous proportions. There is no economy, only inconvenience, in a small entrance door. It should be large enough to permit the passage of a wheelbarrow or push-cart for cleaning, replenishing the litter and other routine work. In houses fifty feet in length and longer it will be found advisable to have a door at each end of the building. With this convenience a great many unnecessary steps may be avoided.

Yard Space.—Chickens do not require unlimited range to give results, providing the deficiencies of a small enclosure are met by supplying them with an abundance of green food, animal food, and so on from other sources. Yet it is a bad plan to attempt to keep a large flock in a very small yard, unless considerable labor is expended in spading or plowing up the soil at frequent intervals.

The earth in a yard crowded with fowls, especially if the soil is heavy, will become contaminated by their droppings, which is particularly objectionable in wet weather. It is therefore best to have double yards, one on either side of the house, or if this is not practicable, have them arranged side by side; so that when the flock is occupying one yard, the other may be sown to clover, rye, rape, or other green crop, and given time to make a start. This cultivation not only sweetens the soil, but it will provide a large portion of the necessary green food.

Permanent Pasture.—If the yards are intended to be kept in permanent sod and furnish all of the green food, it will be advisable to allow at least 100 square feet per fowl, otherwise the birds will destroy the entire growth. If yards are intended for exercise only, and the greens are supplied from other sources, about 25 square feet per bird is sufficient.

In any event, it is well to remember that the nearer square a yard is made the less it costs to fence a given area and the flock is more easily confined. Obviously, the small yard requires a higher fence than the large one, although the question of height is largely determined by the breed one keeps. The meat breeds, such as the Brahmas, may be confined within a 3- to 4-foot fence, general-purpose fowls within a 4- to 5-foot fence, and the egg breeds within a 6- to 7-foot fence.

Erecting wire netting is sometimes attended by difficulties, which are generally due to lack of experience with this commodity, and not to the netting itself. It is contrary material to work, yet if a few principles are followed, the task may be made comparatively simple, and one that can be done single-handed.

In purchasing poultry netting, even the best grades, it sometimes happens that one selvage is slightly longer than the other, which will be responsible for trouble in hanging it, unless precautionary measures are taken. To ascertain if such is the case,

Fig. 88.—Wire-covered yards for cross-breeding experiments at the Kansas Agricultural Station.

unroll the netting on a level stretch of ground, and if instead of lying in a straight line, the netting describes a slight curve, it is because one edge or selvage is longer than the other, perhaps but a few inches, which will not interfere with its efficiency, if the defect is borne in mind.

Netting having uneven edges should always be hung with the shortest selvage—the selvage on the inside of the curve—at the top. Otherwise, if the longest selvage—the one on the outside of the curve—is placed at the top, the upper section of the netting, after it is hung, will sag between the posts in spite of every effort to remedy it.

There is one exception to this rule, which will explain the principle of it: If a fence is to be erected on ground which is rolling, and the contour of the ground is such that the highest ground is in the center of the curve, grading away uniformly on both sides, it is quite likely that this curve will conform to the curve in the netting, in which case the wire may be hung with the longest edge uppermost.

Top Rails.—A poultry yard should never be constructed with a rail at the top, unless the top is to be covered with netting also, or unless the fence is to be built very high for a particular breed. To build rails only invites the birds to fly and alight on them. Fowls are not so apt to attempt to fly over plain wire, though now and again the more venturesome members of a flock will try to elude the mysterious barrier by climbing up and over it.

If a rail is necessary at the top of a fence, as over gateways, it is well to erect a piece of netting over the top of the rail, which will baffle and discourage those who attempt an aërial escape.

Base boards 12 inches wide, securely nailed to the fence posts, make the best bottom for a poultry fence, and assist greatly in stretching the netting, but they are also expensive if large areas are to be inclosed. They are not absolutely essential; in fact, equally good results can be obtained without them, for which a method is herewith described.

The fence posts should be well planted in the ground, from 10 to 12 feet apart, and braced at the corners, or in way of gates, to take the strain of stretching the wire. Then, commencing at a corner, unroll the bale of netting on the ground for its entire length, ascertain if it is straight, and decide which is the best edge for the top.

Start to hang the netting by the top selvage at the proper height from the ground, driving one staple—no more—in each post, until the entire length is hung, all the while stretching the selvage away from the starting point. Be careful not to walk on the netting unnecessarily, or to handle it in such a way that it develops bulges or sagged places, and do not attempt to stretch the middle of the netting. To do so will only result in a distorted sagged section which can never be straightened without

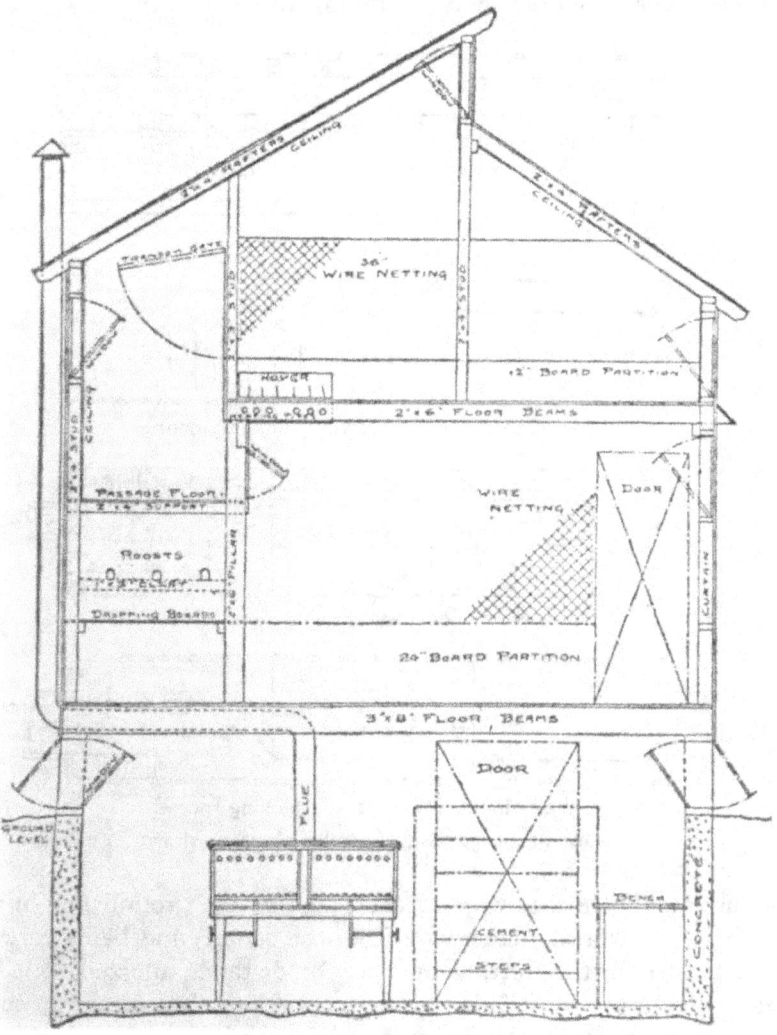

Fig. 89.—Section of intensive broiler plant.

great difficulty, if at all.   It is also a bad plan to try and carry the netting around a corner without cutting it and making a new place of beginning, especially if the posts are round.

Stapling.—When the netting is entirely hung by a single staple at each post, commence at the middle of the length of wire and

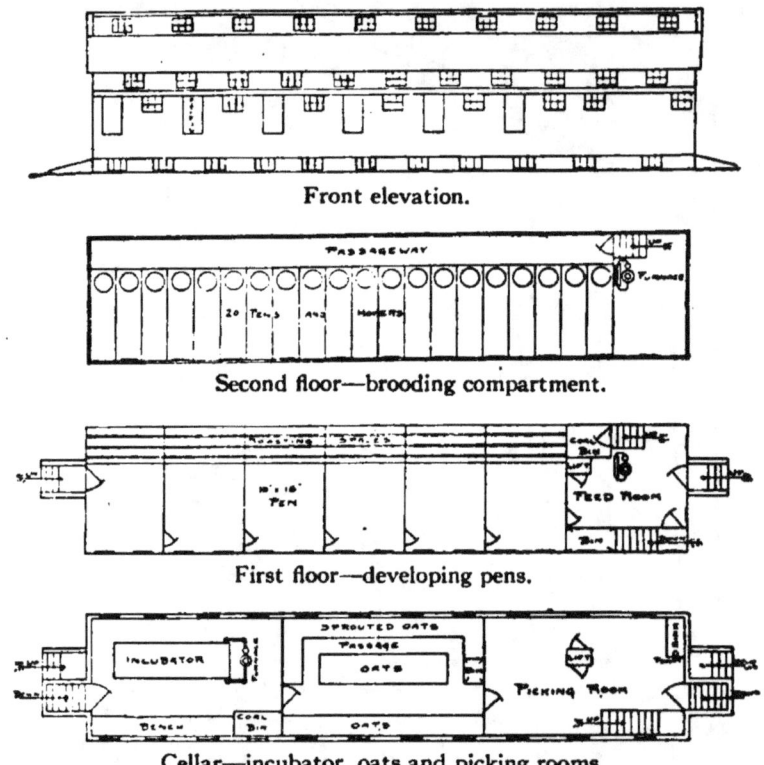

Front elevation.

Second floor—brooding compartment.

First floor—developing pens.

Cellar—incubator, oats and picking rooms.

Fig. 90.—General arrangement of broiler plant, as shown in Fig. 89.

secure the bottom selvage at each post at the ground line or an inch below, working each way from the center, and being careful not to stretch the netting beyond the distance allotted to it by each post interval.   This method apportions the correct amount of netting over the entire area, whereupon it is then a simple matter to return and staple the wire to each post individually, stretching it slightly here and there as it is required.

To hold the bottom of the netting in place and make it hug the ground closely, drive stakes at intervals of three or four feet— two stakes between the posts are generally sufficient—and secure the netting to them by a staple. Discarded wooden fence palings are splendid for this purpose, especially if they are tapered. Drive the largest end in the ground first, for a distance of about eighteen inches, whereupon they are less likely to heave upward by frost action.

The life of these stakes will be prolonged if they are first dipped or soaked in a wood preservative, or else charred.

Fig. 91.—Poultry house at Wisconsin College of Agriculture.

By the use of stakes in this manner, especially in a light soil, it is possible to sink the netting six inches into the ground without difficulty, which in some respects is better than the wooden base boards, the bottoms of which soon rot away, or under which the fowls are able to dig their way to freedom. Moreover, with a little practice one will soon develop the skill and judgment of being able to drive the stakes with just enough tension on the netting to stretch it perfectly flat and tight.

Chick Runs.—For inclosing yards intended for chicks, a course of inch mesh wire netting should be used at the bottom, twelve or eighteen inches high is sufficient, to which the coarser mesh netting is fastened by pieces of pliable wire bound around the two selvage edges.

CHAPTER X

INTERIOR ARRANGEMENT OF BUILDINGS

Convenience.—In the design and construction of poultry buildings not only the health and comfort of the fowls must be carefully considered, but the convenience of the caretaker should receive equally thoughtful attention. If not, and the arrangement of a building or plant is inconvenient, or its facilities are meager and of such a character that the performance of one's work is made unnecessarily tedious and laborious, it is quite likely to have a demoralizing effect upon the attendant's interest and ability. In consequence, some of the routine chores are apt to be overlooked or slighted, and in due time carelessness takes the place of thoroughness.

The interior fittings of a poultry house—the arrangement of nests, perches, feed hoppers, watering devices and so forth—are no less important than the construction of the building itself.

Cleanliness.—The paramount issue is cleanliness, or rather, let us say, facilities which will obtain cleanliness at the least possible effort. It is a subject that admits of much argument and varying principles. Some methods achieve their end at too great a cost for labor; others simplify labor at the expense of sanitation. Various degrees of cleanliness are maintained either by an intricate or simple operation, and the same ease or effort may be expended upon the feeding, watering and egg collecting. If the methods are so antiquated or so involved as to require an unreasonable amount of labor, the efficiency of such a system is defective and should be corrected at once.

Roosting Compartments.—The warmest part of the building, that which is freest from drafts, should be selected for the roosting compartments, which is usually against the rear wall. Each

132

Fig. 92.—Interior of continuous laying house divided into pens. Note the abundance of litter, swinging doors, and platform for water crock, shell box and mash hopper.

133

fowl should be allowed about ten inches linear perch room, and all the perches must be of the same height, or the birds will fight and struggle for the highest ones. Running the perches the long way of the house, that is, parallel with the rear wall, is generally the most economical, convenient arrangement, and their height from the floor is determined by the breed of poultry kept, and whether dropping-board platforms and nests are intended to be installed under the roosts.

When dropping-boards are used, which are advisable, the roosts

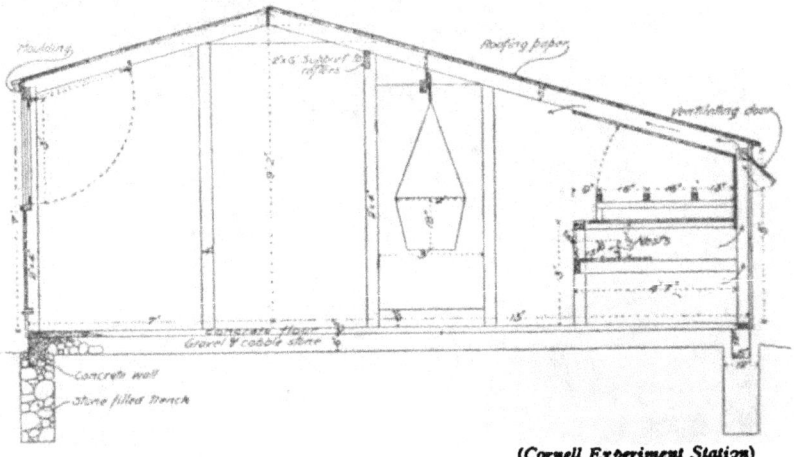

(*Cornell Experiment Station*)

Fig. 93.—Cross-section of laying house 20 feet deep, built on continuous plan, any length desired.

should be located high enough to make their removal unnecessary when cleaning the boards, and yet not so high but that the fowls can fly to them without difficulty. When fowls jump to a hard floor from a considerable height, especially to a concrete floor which is scantily covered with litter, they are in danger of bruising their feet, causing a very painful condition which later develops into bumblefoot. On this point the dirt floor is desirable; it is so resilient that cases of bumblefoot or corns are virtually unknown.

Height of Boards.—As a general rule the dropping-boards for the Asiatic classes should not be over eighteen inches from the floor; for the general-purpose or American breeds, twenty-four inches; and for the Leghorn and other light weight Mediterranean classes, forty inches. The perches should be located

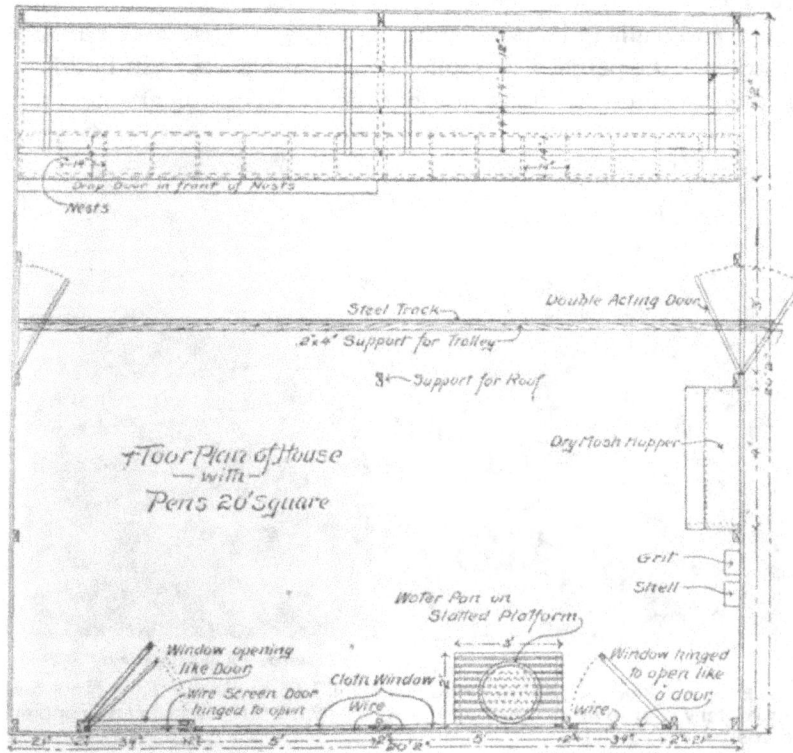

Fig. 94.—Floor plan of 20-foot laying house, as shown in Fig. 93. View shows the width of a single pen, 20 feet wide.

about twelve inches above the boards, and so constructed as to permit them to be readily removed and cleaned.

Perches.—A 2- by 3-inch timber stood on edge, with the upper edge slightly rounded, makes an excellent perch. If these are supported at the ends by U-shaped wooden sockets, they may be made to span ten feet without need for an intervening support. Spans greater than this should be supported in the middle by a third cleat, a piece of 1- by 3-inch stuff stood on edge, otherwise the perches will sway heavily and disturb the sense of security in the fowls.

The end walls in houses sixteen feet long and less may be made to support the perches, dropping-boards and nests. In buildings of the continuous laying-house type it is customary to erect

(*Courtesy Purdue Experiment Station*)

Fig. 95.—Floors of poultry buildings should be unobstructed. Note how this battery of trap nests is arranged, together with the perches and dropping-boards.

transverse partitions, at intervals of twelve or sixteen feet, extending five or six feet from the rear wall, which are designed to prevent currents of air from forming dangerous drafts in the roosting compartments, and which serve the additional purpose of a foundation for the perches and so forth.

Construction of Boards.—The dropping-boards should be made fairly heavy and rigid, for it must be remembered that they will be made to carry considerable weight, not alone the weight of the fowls, but perhaps the nests as well. They should consist of matched boards—8-inch roofers or 6-inch tongue-and-grooved fence boards are just the thing, and secured to their framework at right angles to the rear wall, never lengthwise of the house. The reason for this is apparent: there is a tendency for the boards to warp and curl, which would seriously interfere with scraping

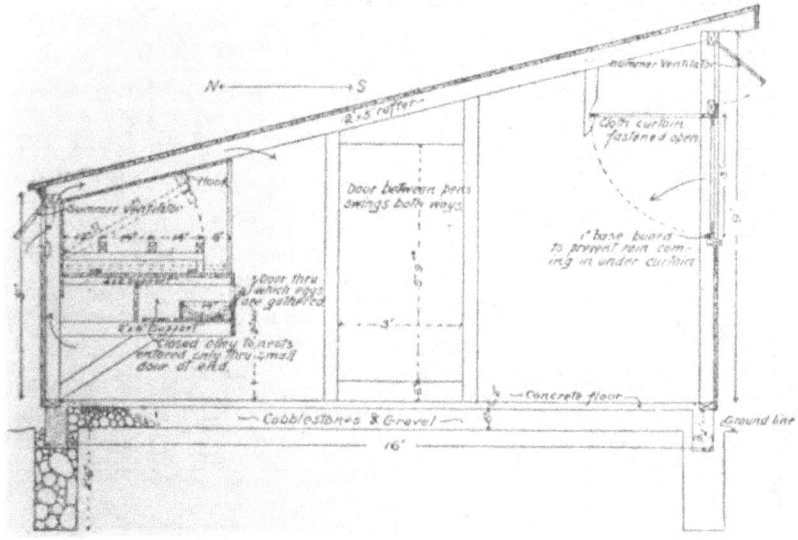

(*Cornell Experiment Station*)

Fig. 96.—Cross-section of laying house 16 feet deep, built on continuous plan, any length desired.

the boards at cleaning time if the hoe or similar implement had to oppose these cracks and irregular places.

Roosting Curtains.—When the roosting compartments are arranged as described above, it is a simple matter to fit the fronts of them with curtains, which may be lowered on extremely cold nights and hooked up to the ceiling in the daytime, a practice, however, which is not necessary except in very cold climates. Heavy unbleached muslin or a light duck tacked to a wooden batten or pole makes a good arrangement.

Still another point concerning the dropping-boards: they should be arranged with the view to scraping them into a push-cart or wheelbarrow, preferably the former. No part of the poultry-man's equipment is more generally useful and a greater labor saver than a well-built push-cart with fairly high wheels. It need not be an expensive one with springs.

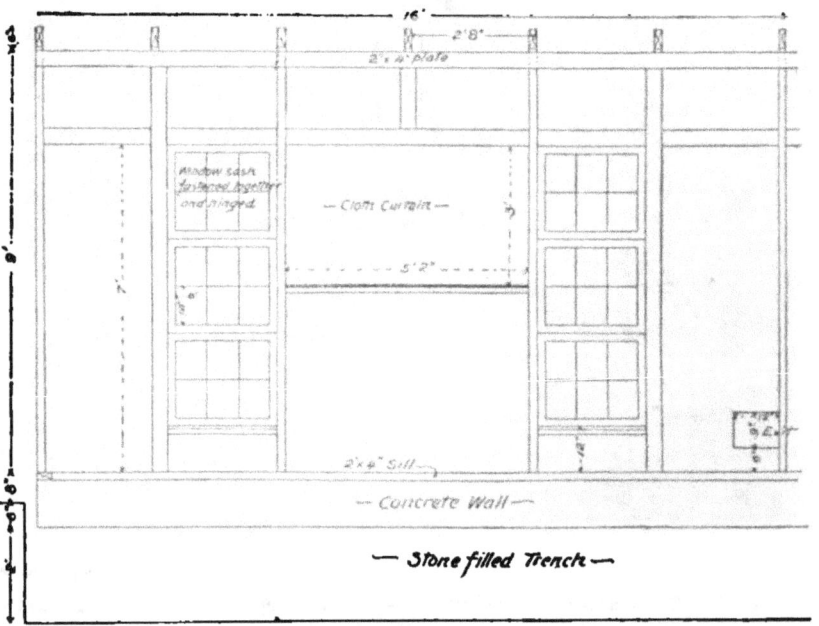

(*Cornell Experiment Station*)

Fig. 97.—Front elevation of laying house shown in Fig. 96. View shows the width of a single pen, 16 feet wide.

Nests.—Next in importance to the roosting compartments is the arrangement of nests, of which there should be one to every four or five hens. Under natural conditions the hen seeks a secluded spot in which to lay, hence it behooves the poultry house designer to imitate the natural environment as much as practicable. The nests should be situated in a more or less dark place, easily accessible, and in such a way that they do not obstruct passageways or intrude upon the floor area. An unobstructed floor area is a vital consideration, and it should not be robbed in any way unless absolutely necessary.

A good location for the nests is underneath the dropping-boards, suspended from them in light portable sections or batteries, which may be removed conveniently for cleaning. This practice, however, is not possible in houses intended for the heavier breeds, where the dropping-boards are built close to the floor, but it is admirably well suited to Leghorns.

The nests should be arranged so that the hens enter from the side toward the rear wall, and have a shelf upon which they may alight before entering a particular compartment. Each nest should be from 12 to 14 inches square and about 12 inches high, with solid partitions between them to prevent the hens from fighting. The side from which the fowls enter should have a front piece—a batten about three inches high—to keep the nesting material in its proper place; the side of the nests exposed to the interior of the house and from which the attendant is supposed to gather the eggs, consists of a hinged lid, arranged to operate with the greatest facility.

Portable Nests.—The advantages of nests built in portable sections are numerous. It is necessary to clean them frequently in order to secure clean eggs at all times, and to replenish the nesting material. Moreover, they should be sprayed or scrubbed with a disinfectant at regular intervals to avoid a pestilence of lice and mites. The easiest way to perform this work is to remove the nests outdoors, dump out their old contents, wash or spray them, place them in the sun to dry, refill with fresh straw, excelsior or other material and return them to their proper place.

Built-in nests, those which are not portable, are very difficult and tedious to clean, and should not be tolerated.

Cleanliness.—It is also important that the nests be protected against the fowls roosting in or on top of them, thereby soiling them unnecessarily. Soiled eggs are very bothersome, and even when thoroughly cleaned they are apt to be detected as washed eggs, and will be discounted accordingly. Furthermore, the

Fig. 98.—Interior of laying house, typical of commercial egg plants.

140

nests should be built so that it is convenient for the caretaker to remove any setting hens. In the brooding season, especially on the commercial egg farm, it is necessary to go over all the nests every night and remove all hens showing any broody inclination. Therefore, unless the nests are suitably arranged, this detail in the day's routine will consume a great deal of time and is likely to prove a severe tax on one's patience.

Broody Hen Coop.—On farms where incubation and brooding are entirely artificial, the hen is not made responsible for the rearing of next year's pullets, hence her maternal instincts should

(Courtesy Purdue Experiment Station)

Fig. 99.—The elevated walk in front of this house, under which the birds pass to gain entrance to the yards, enables the attendant to observe the different pens without entering the building.

never be allowed to develop. When a hen has laid a series or clutch of eggs, a rest period follows before another clutch develops, and at certain seasons of the year this dormant period is usually accompanied by broodiness. This condition must be broken up at once or there will be a heavy falling off in the egg yield. When a hen is allowed to remain on the nest she eats and drinks very little, draws heavily upon her internal storehouse for sustenance, and in a few days she becomes thin and emaciated.

For economical reasons alone it is imperative that this self-imposed starvation be intercepted at the earliest possible moment. Moreover, a hen that is removed from the nest on the first or second day of her broody spell is much more easily discouraged than if she is allowed to remain longer, and thus have her inclination become confirmed.

There are many ways of breaking up broodiness, but common sense dictates that the proper method shall not subject the hen to cruelty or privation. To do so only further retards egg de-

(*Courtesy Missouri Experiment Station*)

Fig. 100.—Poultry house with slatted openings for ventilation instead of curtains.

velopment, and thus defeats the whole idea. A sitting hen should be induced to eat and drink freely and to exercise, for only in this way can her productive organs be stimulated.

Slatted Bottom.—If only a few hens are to be considered, a good plan is to construct a packing box or coop with a slatted bottom, and to raise it about six inches from the floor. The hens' feet slip between the slats, the birds are unable to squat in a comfortable position, and the sensation of currents of air under them instead of eggs is disconcerting and harmlessly annoying, so that in a few days they are disgusted with the whole maternal idea and are only too glad to rejoin their comrades in the laying pens. It is understood, of course, that food and water must be kept before the hens while they occupy their place of incarceration. In long laying houses, where a great number of sitting hens may have to be discouraged at one time, it is advisable to set aside a section of the roosting compartments at one end of the building for this special purpose, and to build it with a removable slatted floor, and with a wire netting front, so that the section may be converted into a brood coop as desired. At other times it may be used for an ordinary roosting space.

To return to the subject of nests, if it is impracticable to erect the nests under the dropping-boards, as in the case of houses intended for the heavy breeds, they may be built in transverse tiers at the ends of the building, and in several sections in the interior of the house. But, as far as possible they should be built upon posts, with inclined runways leading to the nests, to obviate obstructing the floor space. In the monitor type of laying house they may be conveniently located along the line of columns supporting the roof, and where the front roof plane joins the main structure.

Some poultrymen are opposed to locating the nests under the dropping-boards because of their proximity to the danger of lice and mites. This argument does not seem to be well taken, however, inasmuch as if the dropping-boards are infested with vermin to that extent, the conditions are as bad for sleeping as they are for laying and should be remedied at once.

A dust bath is as essential to the well-being of poultry as is the soap and water variety to the human. Consequently, unless the poultry houses have dirt floors, which will answer the purpose nicely, provision must be made for a space devoted to dusting, preferably where there is sunlight, for the fowls seem to relish the combination. They like to dust themselves and then recline luxuriously and bask in the sun. Any sort of a board partition about twelve inches high will answer the purpose, and about twenty-five square feet should be allowed for every hundred birds.

CHAPTER XI

POULTRY HOUSE APPLIANCES

Interior Equipment.—In laying out the interior arrangement of a poultry house attention is first given to the location of the roosting compartments, perches and nests. See Fig. 101. After these are installed the poultryman must turn his inventive ability to ideas for equipping the house with feeding and watering devices.

These appliances should be arranged in the most convenient places left vacant after doors, windows, curtain frames, roosts, and so on have been located, and in such a way that they are not only easily accessible to the fowls, but equally handy for the caretaker to attend.

Water fountains, shell boxes and mash hoppers require daily attention, consequently they should be simple in design, easy to clean and especially easy to replenish, otherwise a great deal of time will be needlessly consumed. They should be elevated from the floor as much as possible, not alone to avoid obstructing the floor area, but to prevent litter being scratched into them. This is especially true of the water fountains, hence it is sometimes best to locate these receptacles on a platform built about eighteen inches above the floor, and to which the fowls will jump when they want a drink.

Poultry supply houses display such an array of equipment and of such a wide variety and completeness, the layman is ofttimes amazed at the thought and consideration expended on the up-to-date hen. From a gapeworm extractor to a mammoth incubator or a coal-burning brooder stove, there is an endless assortment of appliances. Many of these devices can be bought more cheaply than one could make them at home; in fact, it would be

144

impossible to make some of them without special machinery and costly tools.   There are some men, however, for whom the pleasure of making can never be equalled by the satisfaction of buying a ready-made article, hence there is a splendid opportunity for them to exercise their ingenuity and skill as an amateur carpenter and mechanic.

Water Supply.—If left to her own inclinations a hen would rather drink from a cow's hoof-print in preference to a fountain

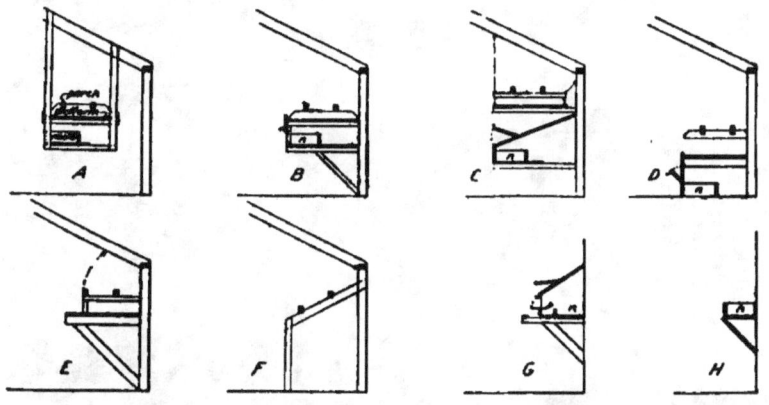

Fig. 101.—Different arrangements for roosts and nests. *A*, Roosts and nests suspended from rafters, clear of rear wall; *B*, roosts and nests built on shelf attached to rear wall; *C*, a complicated arrangement; *D*, nests are too low for convenience; *E*, no provision for nests; *F*, no dropping-boards— manure falls to floor; *G*, wall arrangement for nests; *H*, open nest, which permits fowls to roost upon it.

of clean, fresh water; yet that is no reason why she should be allowed to do so, any more than she should be permitted to eat putrid animal matter, which this perverse creature is sure to do if given the opportunity.   Nature, it seems, has not seen fit to modify the fowl's instincts to conform to a civilization more intensive than was her original state; consequently, inasmuch as we have taken upon ourselves the responsibility of surrounding the hen with more or less artificial conditions, and induced her to be several times more productive than normal, it behooves us

10

to go even further and be responsible for every detail that makes for her safety and well-being.

An unlimited supply of pure drinking water must be kept before the flocks at all times, and on large commercial egg farms this is a factor worthy of serious consideration. If there is water pressure on the premises, either from an overhead tank or from a municipal water supply, the problem is greatly simplified, for it

Fig. 102.—Interior of well-designed laying house. Note how curtains are stowed against under side of rafters, and the partitions between the roosting compartments.

is then only necessary to run pipe connections to the various buildings and fit them with non-freezing hydrants. The water may be allowed to trickle continuously into a trough equipped with an overflow leading to the outside of the house, or fountains of generous capacity may be filled at the hydrant and distributed throughout the building.

Installing a water system does not necessarily require the services of a skilled plumber. On the contrary, any one who is handy with tools will find pipe-fitting a comparatively simple task. The most laborious part of the work is digging the trenches, for these should be excavated deep enough for the pipes to lie below frost line—thirty inches or thereabouts. A pipe vise, pipe cutter, set of dies for cutting threads, and a couple of Stillson wrenches are the essential tools.

Pipe.—It is better to use galvanized pipe throughout the system, although black iron pipe will answer the purpose. In laying

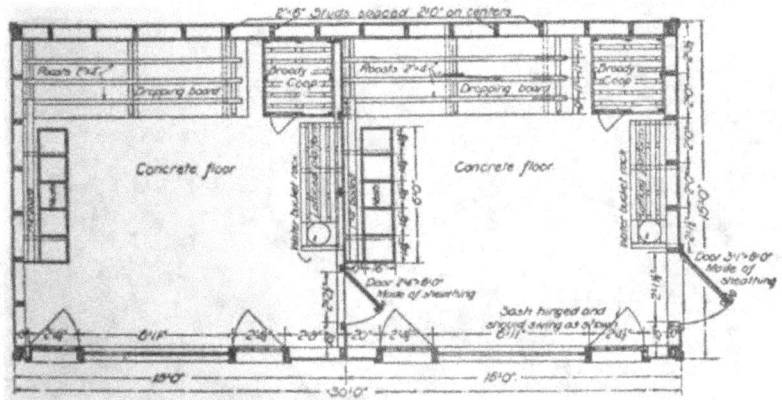

(*Kansas Experiment Station*)

Fig. 103.—Floor plan of a well-arranged poultry house.

out the piping underground, it is well to make provision for extending the system at some future time. That is—put in tee connections here and there, at points where you may want to run off pipe lines to new buildings. These tees can be shut off by plugs made for the purpose, and they will also act as clean-outs should any trouble occur. If these tee connections are not installed, and later one wishes to tap into the main feed pipe, it will be very difficult to do so without taking up a great deal of the system. The reason for this is obvious: The pipe and fittings are all threaded right-handed, and unless unions are installed, it is then necessary to commence at the end of the system, work backward, and take down the entire system.

Make a rough diagram of the proposed scheme first, measure the distances accurately, and, bearing in mind that pipe is purchased in sixteen- and twenty-foot lengths, ascertain the number and kinds of fittings required. One-inch pipe is plenty large

Fig. 104.—Perches arranged in sections which hinge upward clear of the dropping-boards for cleaning.

enough for the mains, reducing to ¾-inch pipe for the branch lines and connections.

In addition to the enormous saving in labor of watering the flocks, a water system having connections in all the principal houses is a very vital factor toward fire protection. A plant thus equipped is usually able to obtain fire insurance at a lower rate, hence this saving should be credited to the original outlay for the water system.

If there is no municipal water supply available, no reservoir and no overhead tank on the premises, and an investment for same is not warranted, the following scheme may be substituted, and it is entirely practicable and satisfactory. The writer worked out the idea and has used it for several years.

The most expensive part of a water system is the tower and overhead tank, both of which may be eliminated. Install a sound oak barrel of about 50-gallon capacity, preferably a charred

(Courtesy Wisconsin Experiment Station)

Fig. 105.—A raised shelf for the fountain insures clean drinking water.

whiskey container (the charred lining serves to clear the water, absorb any impurities and keep it sweet), in each house and on the range where desired, and connect these with underground piping to an ordinary force-pump at the well. Erect a pump-jack, back-geared about seven to one, over the pump, and operate this with a small gasoline engine—one horse power is adequate.

The barrels are left with their heads in, and an inlet pipe fitted with a valve or spigot is inserted in the top of each, also an overflow pipe leading to the outside of the building. Suppose, for example, there are ten of these barrels on a plant; at feeding and egg-collecting time in the evening the attendant first starts the engine, having the spigots or valves to all barrels open. One by one the barrels fill up and overflow, and as they do so the attendant shuts them off as he pursues his other work, until they are all filled, which will require about a half hour's pumping. The cost of the gasoline and lubricating oil is negligible, amounting to but a couple of cents; whereas the saving in labor is enormous. Moreover, among farm hands there seems to be a deep-rooted antipathy against carrying water for livestock, and they will either openly rebel against the task, or purposely neglect to furnish the necessary quantity. At least, such has been the writer's experience and observations.

Trough System.—In the above system the water is placed before the fowls in 15-foot galvanized sheet iron troughs, made from roof gutters, and supported by metal brackets designed for the purpose. The troughs are located against the front walls of the laying houses, about twelve inches above the floors, and covered over by a hinged board to prevent the birds fouling the water. Each trough is fitted with an overflow and a drain for cleaning, both of which lead to the outside of the building. The barrel is bored at the lowest convenient level, fitted with a wooden tap, such as is used for wine, oils or molasses, which is allowed to trickle in a very thin stream at all times, and which keeps the water in circulation and therefore fresh.

The water will not freeze seriously in the barrels of a system of this kind, because having their heads in, the air is virtually excluded from the water; the taps are not apt to freeze because they are wood, and the water in the trough, while ice may form around the metal in the bottom, is always open to the fowls because they are constantly drinking from it. At night the water is drained off entirely, and a plug is substituted for the wooden spigot, consequently the latter cannot freeze.

A barrel of water in a house of 500 fowls is a day's supply, except in very warm weather, when to keep the water cool and fresh it is advisable to fill the barrels twice a day—at noon and in the evening—and to allow twice the ordinary flow from the spigots. The troughs should be brushed out daily with a disinfectant, which is not a tedious task if a cheap scrubbing brush and a small supply of disinfecting fluid is kept handy in each building.

The manner of placing feed before the fowls constitutes another phase of the daily work in which a great deal of time and labor may be conserved by providing the right sort of appliances. It is impossible, of course, to establish rules that will meet the demands of all farms, for feeding principles vary. The most generally accepted practice, however, is to give a light feeding

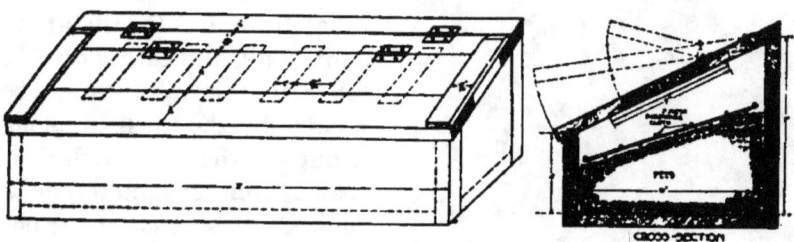

(*California Experiment Station*)

Fig. 106.—Plans for a dry mash hopper.

of scratch grains in the morning, all the grain the fowls will clean up at night, and to keep a dry mash, beef scrap, charcoal, grit and oyster shells before the birds constantly. Or, the beef scrap and charcoal may be placed in the mash.

Automatic feeders may be installed to distribute the scratch grains. These devices insure clean food, save labor and induce exercise, and it is surprising how quickly a pen of fowls will learn the *trick* of operating them. They are constructed mainly of galvanized sheet iron, and those having the largest capacity are usually the most desirable. Many of the largest commercial farms are equipped with them, not alone in the laying houses for mature stock, but in the rearing of young stock on the range and for little chicks in the brooder.

Mash Hoppers.—Assuming that a dry mash is fed, and this is conceded to be the most advanced method of feeding poultry, there is no economy in placing it before the fowls in small quantities, which necessitate filling the hoppers every day. Self-feeding, reliable, sanitary, non-wasteful hoppers of sufficient size to store large quantities, say, two or three hundred pounds, or a week's supply, should be provided.

Fig. 107.—Indoor mash hopper for use against a wall.

There are no limitations to the style and construction of dry mash hoppers, any more than they should be as simple as possible, both for the birds to eat from, and for the attendant to fill, and care should be taken that the mash cannot be wasted by the fowls dragging it out on the floor with their beaks. Neither should they be allowed to scratch in it, nor foul it in any way, as they are prone to do if the hoppers are not fitted with lids or covers. Hoppers having slatted sides with openings just large enough for the birds to enter their heads, and fitted with a lip on the inside to prevent waste, are among the best types. They may be built as long and as high as desired, cheaply constructed, and made for both indoor and outdoor use.

Shell Boxes.—In addition to dry mash hoppers, each pen should be equipped with a box for oyster shells and grit. These need not be large affairs, but convenient and accessible, sanitary and non-wasteful. The supply houses carry a wide assortment of them, built mostly of galvanized iron.

CHAPTER XII

DEVICES FOR DOORS AND GATES

The experienced poultryman never fails to close his hen houses
at night. He will tell you this practice is a part of the daily
routine, quite as important as the feeding and watering. This
closing up does not necessarily mean shutting the windows, curtains and doors with the view to keeping the house warm or protected against storms. In fact, it has nothing to do with the
temperature of the interior, but with its security—against intrusion.

Protection against burglars! Yes, that is correct. But not so
much two-legged gentry, as four-footed raiders. The measure is
inspired by a simple natural law—that poultry is the prey of dogs,
cats, rats, weasels, foxes, minks and other varmints, and against
which fowls must be protected by artificial means.

For the sake of ventilation it is customary to build the greater
part of the fronts of poultry houses in the form of windows, curtains and openings of one type or another. To close these would
make the interior stuffy and humid, unhealthful, and in warm
weather suffocatingly hot. The most approved type of poultry
house is that which is known as the *open-front house.* In good
practice these openings are carefully covered with small mesh
netting, preferably cellar window screening, though inch mesh
poultry netting will do nicely, which not only prevents the escape
of the occupants of the house, but safeguards them against in-
truders.

Thus with netting over all openings the windows and curtains
may be left in any position night or day with perfect safety. But
—what about provision for the fowls to enter and leave the house?

Fowls are Abroad Early.—At the first streak of light from the
East, chickens stir themselves from their perches. At the earliest

153

opportunity they start afield, and how they revel in these morning hours, especially during the spring and summer months, before sun-up, while the dew hangs heavily on the foliage, and everything seems fresh and full of life! During warm weather particularly, this wandering abroad in the early morning is the most beneficial exercise a fowl can take. It not only provides exercise, but
if the flock is given sufficient range, the fowls will forage for a
considerable part of their keep, in the shape of greens and insect

(*Courtesy Million Egg Farm*)

Fig. 108.—"In for the night."

life. All things considered this early morning outing should be
encouraged wherever possible.

Do Not Confine Chicks.—It is also true of young chickens.
The growth of a brood is often stunted by keeping it confined to
the sleeping quarters until the sun is high. Every minute in
which a brood is kept off the grass, when conditions are such that
it should be abroad, is detrimental to the flock's development and
to the keeper's pocketbook. Like grown fowls, chicks are astir at
daybreak, and from the moment it is light enough for them to
even grope about, they struggle to gain their liberty.

Unfortunately, most of us humans have not the same point of view as poultry. We do not struggle and trample one another to get outdoors at four o'clock in the morning. Just about that hour we would be willing to struggle against getting up, and the chances are we would put up a pretty lively struggle against it.

Automatic Device.—Yet the fowls should be at liberty. If you leave a door open all night, though it may be but a hen door, you take the chance of having the house entered and robbed—not robbed of a single fowl, perhaps, but of a dozen, or maybe the entire flock will be killed, just for the sake of killing.

The remedy is to install or equip the house with an automatic device that will permit the fowls to liberate themselves at dawn, in other words, to construct a self-opening door. I have called it the *early-rising door.*

Chickens are creatures of habit. If they are accustomed to drinking at a particular spot, there they will look for water, even to the point of going through the motions of taking a drink. If they are in the habit of leaving a house or coop at a certain door, there they will congregate for admittance. It is due to this characteristic that it is so easy for the poultryman to contrive something; for he has but to utilize the hen's weight to release a catch —and presto! the door is open.

A swinging door or a sliding door can be operated, whichever is the most convenient; in most cases the latter is the easiest to build. In either case the mechanism should be completely located on the inside, so that the door cannot be accidentally released by the *raider.* Above everything else, keep the device as simple as possible, for the fewer the parts the less likely they are to get out of order. The wear and tear is heavy, more especially in the accumulation of dirt, consequently it must be cleaned easily, and without danger of throwing the appliance out of adjustment.

In the drawing shown in Fig. 109 there is shown a simple device for releasing a catch, which is applied to both a swinging and a vertically sliding door. It involves the simplest kind of a principle. The platform, or trigger board "A," is hinged to the threshold or wall of the house, and supported in a horizontal position by means of a light spring "B." The trigger "C," made from a piece of metal, preferably brass, since that material will not rust, engages a trip or cleat "D," when the platform is set. This cleat is also made of metal and is screwed to the door proper. It is not necessary for this part to be of brass, particularly if the trigger "C"

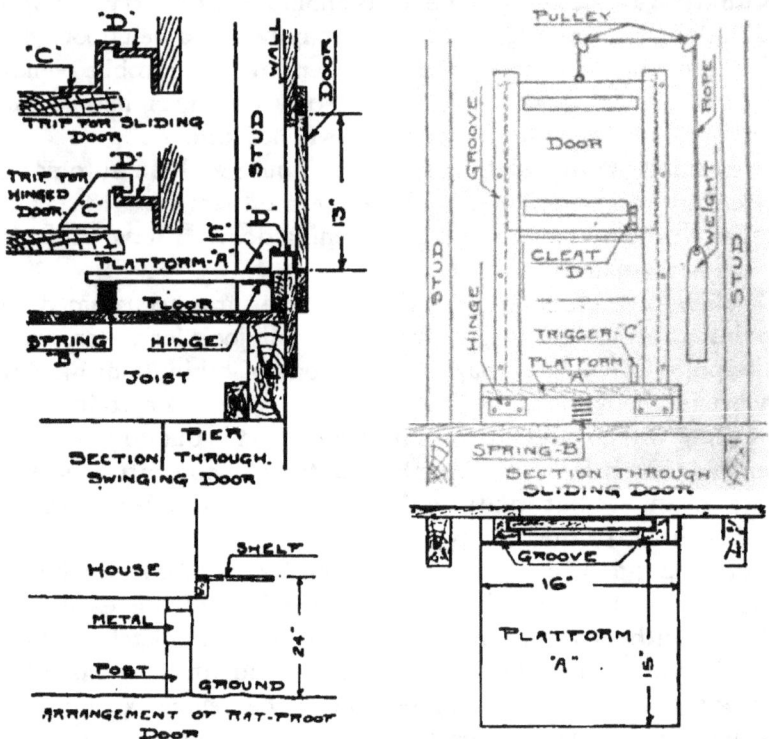

Fig. 109.—Details of a self-rising door.

is composition, since brass against steel makes a good bearing surface.

If a swinging door is used, it should be hinged on the outside of the building, using spring hinges similar to those commonly found on screen doors; but, of course, the hinges must be reversed so that the action of the door is to open outward instead of inward. Or, plain hinges may be employed, and a weight or simple spring can be fastened to the door in such a way that it will be forced open as soon as the trigger disengages the cleat "D." If a sliding door is used, the force to raise it upward may consist of a spring or weight, as shown in diagram. The spring is the cheapest device, but the weight is probably the most generally satisfactory. Two awning pulleys fitted with screw-eyes, screwed into the wall of the house, a short length of cotton rope—clothes-

(*Courtesy C. L. Opperman*)

Fig. 110.—Laying house on a Maryland poultry farm.

line will do, and a cast iron window-cord weight are the articles required.

When the house is closed for the night the *early-rising door* is set, as previously described, which only takes a couple of seconds. In the *small hours* of the morning, when the hens leave the perches and start to roam in quest of freedom outdoors, they are sure to go to their customary place of exit, and to walk onto the trigger platform "A." If necessary, at the beginning they can be trained to walk on the platform by placing a little grain on it, though I have never found this compulsory. Instantly the weight of the hen is transmitted to the platform, it sags an inch or two at the inward end, which releases the cleat "D," and the door flies open. This may surprise the fowls at first, but they soon become accustomed to it. In fact, they will actually learn to operate it, just as they learn to operate an automatic feeder.

Thus, Mr. Poultryman, you can finish your nap in peace; let Biddy do this early-rising work.

It would be impracticable to make the adjustment of the

(Courtesy C. L. Opperman)

Fig. 111.—Serviceable, inexpensive colony house, constructed of rough lumber covered with patent roofing paper.

door so fine that the weight of a month-old chick could depress the platform; but this is not necessary. Chicks move about in crowds, especially when they are trying to get out of a building. Very good, the crowd passes across the trigger platform—and the trap is sprung! Liberty!

In connection with the *early-rising door* it may be desirable to make the exit as near proof against raiders as possible, even in the daytime. Rats do most of their marauding at night, under cover of darkness, but they are not afraid of a daylight attack if conditions seem at all favorable. If rats abound in large numbers and food is scarce, they will become exceedingly bold. I have seen them search for food in broad daylight under the very eyes of an attendant, and fairly defy him to make any defence against their mischief.

Rats appear to be good jumpers, but that is because they half climb and half jump to gain a certain position. They cannot jump straight into the air for any distance—two feet is easily their limit. They are good climbers, of course, yet they cannot pass a vertical surface of metal, because their claws will not grip the slippery metal. Thus, if a house is elevated a couple of feet from the ground, which is desirable anyhow if the house be frame, to prevent decay due to moisture, and its foundations are covered with a strip of tin, as shown in Fig. 109, the rats will not be able to climb into the building. And if the exits are kept clear of any inclined runways, save for a projected shelf, as indicated, which is kept a couple of feet from the ground, the rats will not be able to jump into the building.

Excepting the very heaviest breeds, it is no trouble for fowls to jump two feet, and thus gain entrance to the house. Mediterraneans find it easy to jump three and four feet. Similarly, Mediterranean chicks, whose wing feathers develop at an early age, have no difficulty in flying two or three feet.

Avoid Tedious Methods.—Ultimately the poultryman is bound to discover that the arrangement, accessibility and operation of doors and gates have a great deal to do with the labor expended on a flock. They have an important bearing on the ease and comfort with which the numerous chores are performed, and this factor has a greater influence on the efficiency of a plant than many are inclined to suppose. It is especially vital where hired labor is concerned. Whatever the ethics or circumstances may be, it is a fact that an intelligent worker resents having to perform a task by some tedious method, and in the course of time this worker is likely to slight the task by reason thereof. Finally, it may necessitate discharging the worker for negligence.

Most doors and gates are used constantly, and they are generally used under more or less trying circumstances—either the fowls are being fed and watered, in which case the feeder has his hands very much occupied with buckets, or the eggs are being collected, or the buildings are being cleaned, both of which engage one's hands with equipment. Obviously, all doors and gates should operate freely, which means that they should be properly fitted, hung on strong hinges, and wherever possible they should be equipped with a self-closing device.

Door Check.—There are numerous patented articles for closing doors, some of which close the door securely without slamming, but most of these are too expensive for general use on a chicken plant. Screen door springs are useful contrivances for doors or gates that open one way. Substitutes for them may be made out of a light piece of rope rove through a pulley and secured to a counterpoise weight of sufficient bulk to move the door. We have all seen this principle. It has been used for centuries—long before a steel spring was deemed possible.

For doors and gates between pens and yards it has been found convenient to have them open both ways. In fact, they involve considerable inconvenience if they do not open both ways. This feature requires double-acting hinges. If the door is to be self-closing, which is virtually an indispensable feature, special spring double-acting hinges are necessary, unless the door is equipped with a double-acting self-closing device apart from the hinges. Reliable double-acting spring hinges are rather expensive, and most of them are of such material that if exposed to the weather for any length of time they soon deteriorate and become worthless.

Diagram in Fig. 112 shows two ideas for a home made double-acting self-closing device, based on the simple principle of the rope, pulley and weight employed for the one-way door. In detail 1 a wooden cross-piece "A" is nailed at right-angles to the top of the door, near the end that is hinged. A hole is bored in each end of this cross-piece, from which a light rope is run to a point, "B," thus forming a bridle. The single rope, "C," runs from the point, "B," to a pulley, "D," thence to a counterpoise weight.

Whichever way the door is opened the cross-piece pulls on the bridle, one end of which draws tight as the other end loosens, which in turn lifts the weight. As soon as the door is released a corresponding strain is placed on the bridle by the weight, which closes the door.

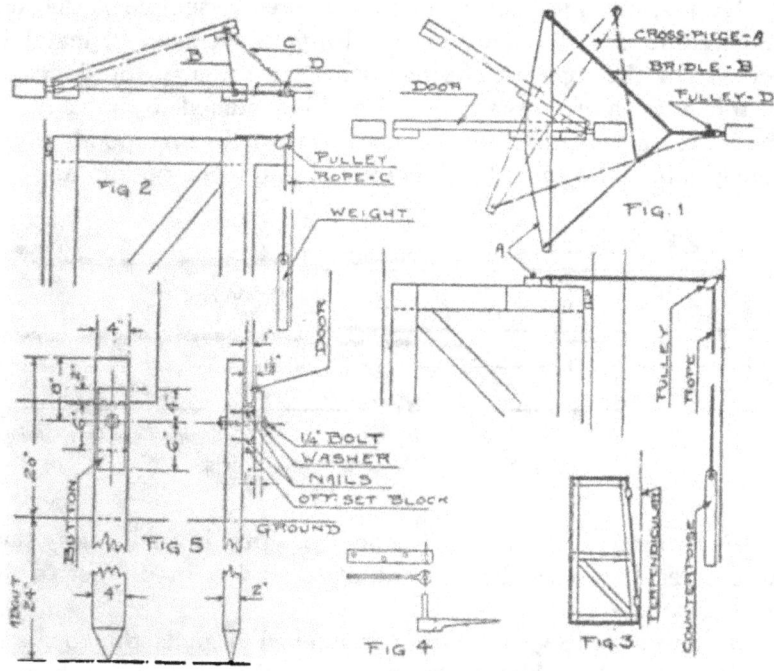

Fig. 112.—Details of a self-closing door.

It is more complicated to describe this device than to build it. The essential idea is to have the weight evenly distributed on both sides of the cross-piece, which really amounts to having the cross-piece and bridle of the same length on both sides of the door. Thus, when the weight is distributed, and the door remains at what we might call a *dead center*, it will be in a closed position.

The size of the counterpoise will necessarily depend upon the

11

weight of the door or gate. The counterpoise should be amply heavy, so that the action of closing will be quick. Sash weights answer the purpose nicely, or narrow canvas bags filled with sand will do for indoor use. Naturally, the doors or gates between partitions should be made as light as possible, which makes for ease of operation. White pine frames covered with wire netting are best.

Detail 2, Fig. 112, illustrates a more direct principle of the rope and weight idea, though it is not always practicable to install it, because of the rope interfering with head room in the doorway. In this arrangement the pulley, "D," is fastened to the center of the jamb, or to a point on the partition which corresponds to the door-jamb. The rope, "C," is rove through the pulley, and one

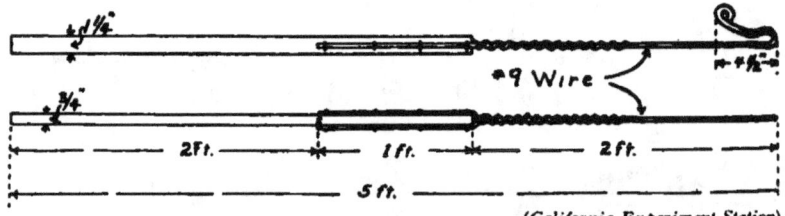

(*California Experiment Station*)

Fig. 113.—Hook for catching fowls.

end is secured to the top of the door or gate, in such a way that it receives the least chafe. The other end is made fast to the counterpoise.

Either way the gate or door is opened it pulls the rope and raises the weight, which reacts as soon as the pressure is removed, the tendency being to draw the point, "B," toward the pulley. There is no leverage obtained by this arrangement, like that shown in detail 1, consequently the door offers more resistance when being opened, which might be considered an objection under certain circumstances.

Hinges Out of Perpendicular.—Doors can be made self-closing by placing the hinges slightly out of the perpendicular, as shown in detail 3, Fig. 112, which alters the center of gravity of the door the more it is opened. It is understood, of course, that the hinges themselves are in perfect line, otherwise they would not operate; it is the line of the door-jamb that is thrown out of plumb.

Regular double-acting hinges are not necessary for partition doors and gates. They are apt to be too expensive. A simple gate hinge, or something fashioned after it, such as indicated in detail 4, Fig. 112, will work just as well.

Keeping a Door Open.— Sometimes it is just as important to keep a door open as it is to keep it closed. For example, in warm weather the doors to poultry buildings should be kept open as much as possible. The interiors need all the fresh air available, which is especially desirable in colony houses for growing stock. Then, too, very often the entrance door is the only means of egress for the fowls, there being no special hen door; in which case it is important that the main door be held open at all times the flock is at liberty.

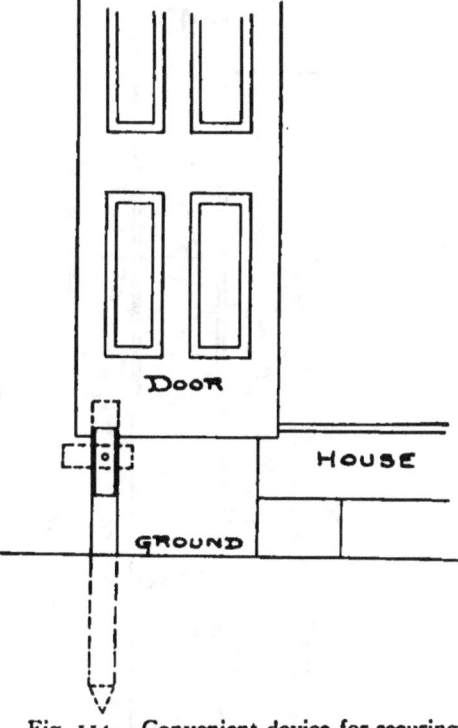

Fig. 114.—Convenient device for securing the open door.

Sudden storms are likely to occur, accompanied by high winds, which will slam and play havoc with doors if they are not firmly secured. If the doors blow shut, the fowls will be unable to seek the shelter of the houses, and such exposure to very young stock is almost certain to result in fatalities. On the other hand, if the door sways violently with the wind, though it does not actually break from its hinges, it will frighten the birds away from the entrance.

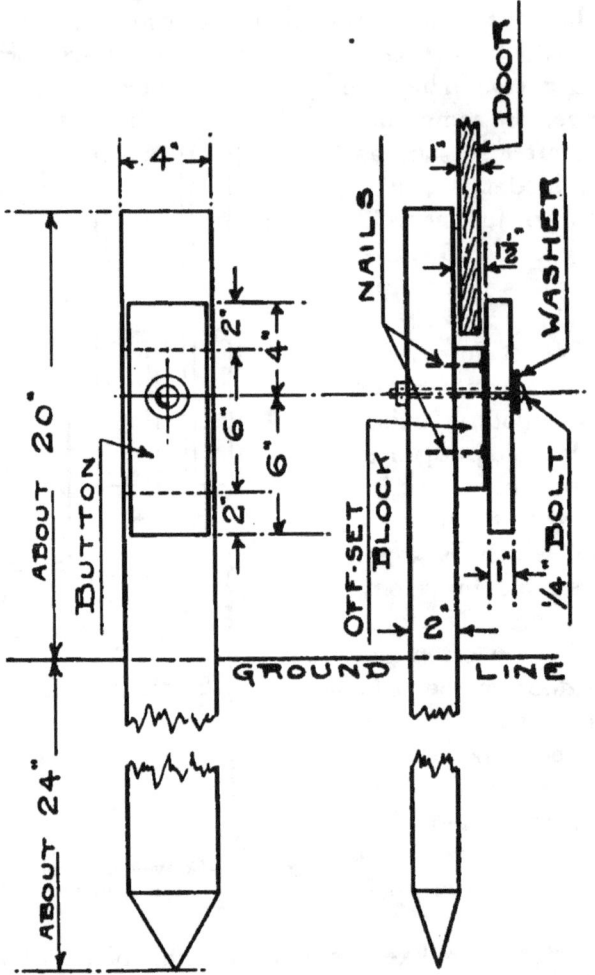

Fig. 115.—Details of stake for securing the open door, as shown in Fig. 114.

Sticks or pieces of lumber, logs, bricks, stones and trash are frequently propped against doors to keep them open. These makeshifts are not only unsightly, but they are insecure and troublesome. Instead of representing an economy, they are actually wasteful, both in the probable damage to the door and livestock, and in the loss of one's time in adjusting them. I have seen doors fastened by an involved system of cord, wire, hooks, leather and other bits of junk that were more intricate to open than a bank vault door, which means lost time—waste.

If a door or gate hinges so that it opens against the building, and there is nothing to prevent it from swinging flat against the wall, it may be secured by a common gate hook. It is not always convenient nor practicable to do this, however, on account of windows, curtain frames and other projections. A similar strong fastening may be made by driving a stake into the ground and attaching the door to it by a gate hook. But this operation requires stooping over, and perhaps the use of two hands, therefore an improvement is suggested in Figs. 114–115. This fastening is operated by a simple movement of the foot, and if the device is securely made there is no chance of the door working loose.

The stake is cut from any timber that may be available, though two-by-four material is about the right size, pointed at one end to facilitate driving it into the soil. See Fig. 115. The rest of the device consists of a button that is movable, and an off-set block, which is nailed to the stake. The off-set block is cut from material slightly thicker than the door, to allow a certain amount of play. A carriage bolt and washer complete the affair.

The hole in the button should be bored slightly larger than the bolt on which it turns; the button, being longer at one end, therefore heavier, will always assume an upright position, which engages the bottom of the door, as shown. By turning the button with the toe of one's shoe to a horizontal position the door may be released or secured. The stake may be removed complete and driven elsewhere when desired.

CHAPTER XIII

TRAP NESTS

While the value of trap nests for breeding purposes is still a subject for argument, there is no controverting the fact that they constitute the only positive means of determining the exact laying ability of a hen. Moreover, their use makes it possible to ascertain which hens lay the best shaped eggs, which the largest sized, which the strongest in point of fertility, which are the best winter layers, which pullets begin early and lay the greatest number of eggs in succession, the number of times they become broody, and many other facts of vital interest to the poultryman. But whether or not this information is considered of sufficient value to warrant the additional time, trouble and expense of operating a trap nest system is the debatable question that must be determined by every poultryman for himself.

Some authorities swear by them as the only means of building up a heavy-laying strain; others condemn their use as too expensive a method of selection for the average breeder. The New York State Experiment Station estimated the cost of maintaining a trap nest system to be fifty cents per hen per year, while the Maine Agricultural Experiment Station states their system can be operated on the basis of one active person to 500 nests for 2500 hens—which, at a wage of $40 per month, amounts to about twenty cents per hen per year. Thus, there is a wide difference of opinion on the subject. There seems to be truth in the arguments for and against the trap nest as a commercial proposition; but for purposes of investigation, and where a breeder of show stock wishes to record pedigrees accurately, they are unquestionably indispensable.

My intention in this chapter is not to discuss the pros and cons for the employment of the device, so much as to describe the construction and operation of a few simple types of trap nests, home-made appliances which I have found to be mechanically satisfactory. There are reliable makes of trap nests on the market, at such nominal prices that in many cases it does not pay the poultryman to make them himself.

Primarily, a good trap nest must be certain in action. It must not only imprison the hen that enters, but it must refuse admittance to all others until a record of the first performer has been secured. It should also offer an inviting, quiet, comfortable retreat for the hen, care being taken that the locking device is not so violent in its action as to engender fright, and that when closed, the interior, especially in summer time, is not suffocatingly warm. Furthermore, the nest should be accessible, convenient for the attendant to open and remove the hen, and with no intricate parts to dislodge and get out of order, or require unnecessary time in resetting the trigger or door.

(*Courtesy Missouri Experiment Station*)

Fig. 116.—Home-made trap nest, similar to the Storrs' nest.

The nest should be of ample size, usually larger than the ordinary nest, so that when the hen has laid her egg and commences moving about in an effort to rejoin the flock, she is not obliged to trample or stand on the egg and thus run the risk of breaking it. The nest should offer easy means of cleaning its interior and replenishing it with nesting material.

Storrs' Nest.—The trap nest employed at the Storrs egg-laying contest, plans of which are shown in Fig. 117, was developed at the Connecticut Agricultural College, and has given entire satisfaction for several years.

It consists mainly of two movable parts, the trigger and the door, both of which are comparatively easy to make and may be fitted to any box of suitable size. The door is 8½ inches high,

11½ inches wide at the top and 11 inches wide at the bottom, and being narrower at the bottom, as it swings shut it has no tendency

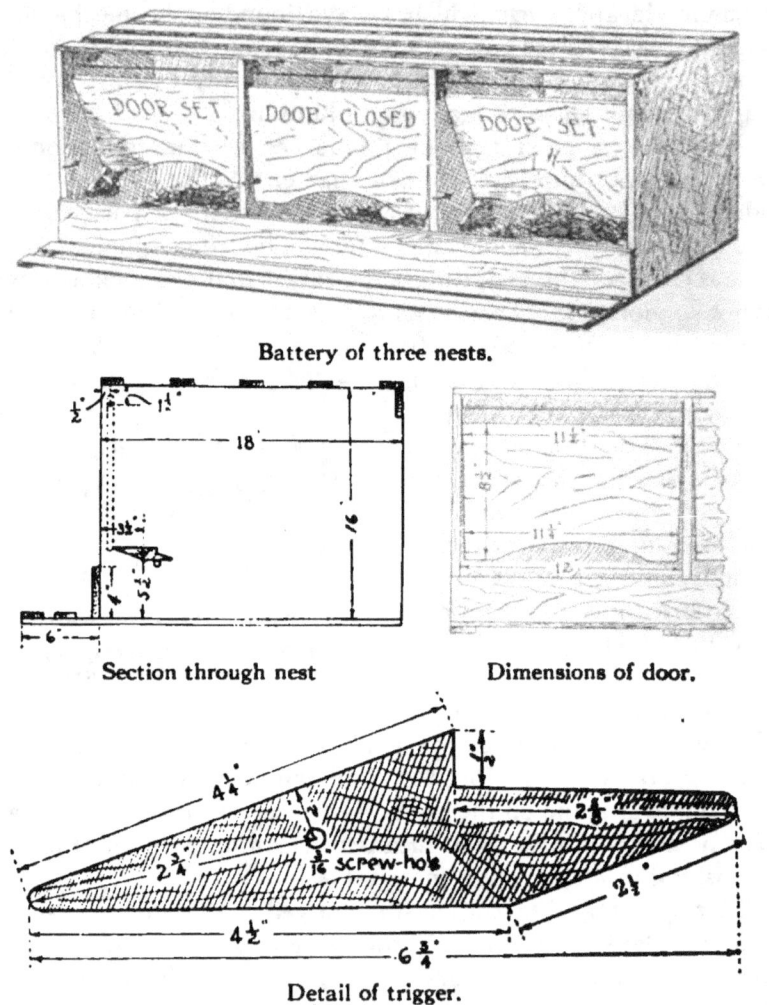

Battery of three nests.

Section through nest

Dimensions of door.

Detail of trigger.

Fig. 117.—General arrangement of the Storrs' trap nest.

to rub or bind against the sides of the nest box.   It is bound with a thin piece of metal, galvanized sheet iron strips cut from stove pipes or other similar material will answer the purpose nicely, which project above the top for about an inch, or far enough to permit of punching a small hole in each end of the strip, through which a piece of fairly rigid wire is slipped, forming a hinge upon which the door swings.

The trigger is a trifle more difficult to make than the door, for it is small and must be accurately cut as shown in the diagram,

Fig. 118.—Corner of a well-designed laying house. Trap nests are arranged under dropping-boards. Note broody hen coop at extreme end.

or it will not balance properly. This is an easy matter with a jig saw or band saw. It should be made from ¾″ or 1″ close-grained stock, the harder the wood the better. The hole bored for the screw that is to fasten the trigger to the side of the box should be slightly larger than the screw itself, to secure free action.

To set the nest, the front end of the trigger is slightly depressed and the door pushed inward until it engages the notch in the

trigger. When the hen enters, she pushes against the door with
her back; this disengages the trigger, allowing the door to close
after her, and by the same operation locks the door from within,
—for the trigger has simultaneously assumed a horizontal posi-

Treadle set with door open to admit hen.

Treadle sprung—door closed.
Fig. 119.—Details of the Maine trap nest.

tion. Small wooden buttons or bent screw-eyes are screwed to
the partitions, turned in such a manner as to prevent the doors
swinging outward when the nest is sprung. To open the nest,
the attendant has but to turn the little button, open the door
outward, and remove the hen.

Batteries of Nests.—The floor and sides of the nest should be solid, the top is slatted for ventilation, and if the back of the nest is to be placed against a wall of the poultry building it may be left entirely open, which will facilitate cleaning when the boxes are removed. It is a good plan to construct the nests in portable sections or batteries of about four each, and if built of light material,—half-inch white pine or poplar will do nicely,—they may be conveniently handled for cleaning by one man. A four-inch board is fastened to the lower portion of the front to stiffen the partitions and retain the nesting material, and if it is found that small hens can walk over this board without disturbing the door, the distance between the nesting material and the lower edge of the door should be reduced accordingly, either by increasing the quantity of nesting material, increasing the depth of the front board or extending the length of the door. A platform or shelf of some kind should be built in front of the nests for the hens to walk upon in entering and leaving the traps.

Maine Nest.—The Maine Agricultural Experiment Station has developed a two-compartment nest which seems to possess several important advantages over the single type previously described; yet because of them it is considerably more troublesome and expensive to build. Diagrams of this nest are shown in Fig. 119, and represent longitudinal views with one side removed to illustrate the construction and method of operation.

The nest is a box-like structure, without front or cover, 28 inches long, 13 inches wide and 16 inches deep, inside measurements,—divided in the middle with a partition extending 6 inches from the bottom. The rear compartment is the nest proper, and contains the nesting material, while the front section is devoted to a treadle for closing the front entrance, and need not have any other bottom. This treadle is hinged to the door, and by a nice balance it can be so delicately adjusted that a weight of less than half a pound on the treadle will spring the trap.

When the nest is set, ready for occupancy, the door extends outward in a horizontal position as shown in upper diagram. A hen about to lay steps on the door and walks in toward the rear of the nest; as she passes the point where the door is hinged to the treadle, her weight causes it to fall, which pulls up the door behind her, as shown in lower view. It is then impossible for the hen to leave the nest or for other hens to enter, until the atten-

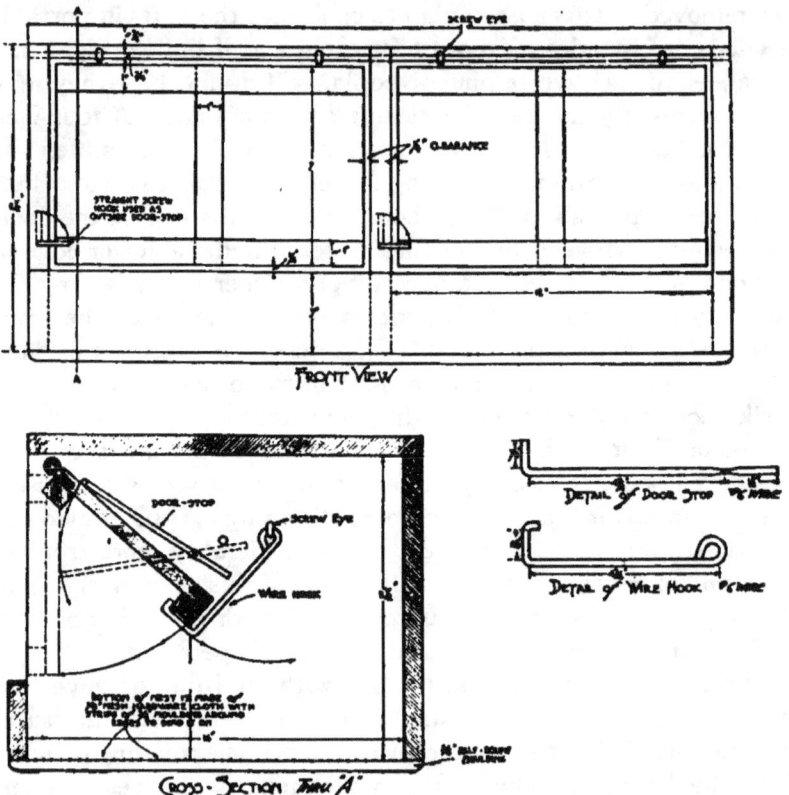

FRONT VIEW

CROSS-SECTION THRU "A"

(California Experiment Station)

Fig. 120.—Plans for a single-compartment trap nest.

dant resets the trap by depressing the door to its original horizontal position. The operation is extremely simple, consists of but one movement, with no triggers or other locking devices to adjust.

When a hen has laid an egg and desires to leave the nest, she steps into the front section, toward the light, entering the screen door, and remains there until released. There is little danger of her trampling on the egg she has just laid, or soiling or dis-

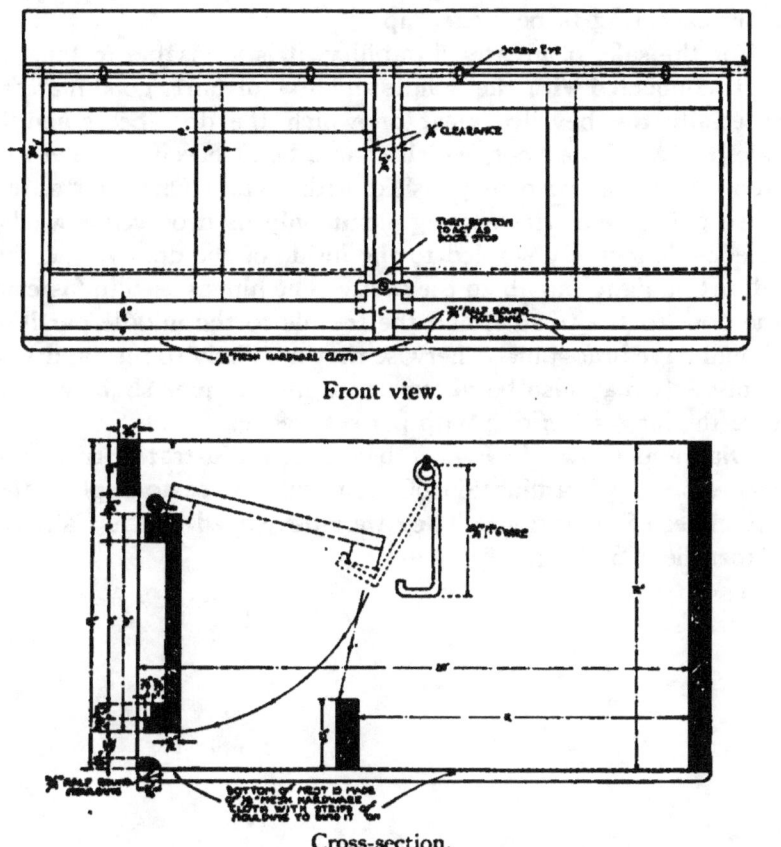

Front view.

Cross-section.

(California Experiment Station)

Fig. 121.—Plans for a two-compartment trap nest.

turbing the nesting material. Dry manure collecting on the treadle is easily scraped off with a putty knife or similar implement, and allowed to fall through the open bottom.

At the Maine Station these nests are arranged in tiers and

operated like drawers.   To remove a hen, the nest is pulled part way out, and as it has no cover, the bird is easily caught. Usually, after having been taken off the nests a few times, hens become accustomed to this handling and remain perfectly quiet, apparently expecting to be picked up.

For the sake of greater durability, it is advisable to build all parts connected with the treadle and door of fairly good material, especially the bevelled cleat on which the door bears and the side frames of the door, which should be of beech or other hard wood that will become polished with wear. One-quarter inch mesh galvanized wire netting, commonly used on cellar window screens, is securely stapled to the inside of the door frame, thus admitting light and air to the nest.   The hinges used in fastening the door to the treadle, and the treadle to the middle partition, should have brass pins, otherwise they are likely to rust and work stiffly.   It may also be necessary to file the pins slightly, to insure the hinges working with perfect freedom.

Diagrams shown in Figs. 120 and 121 illustrate other types of nests.   An examination of them will suffice to explain their principles of operation.   They were developed at the California Experiment Station.

CHAPTER XIV

FEEDING

The feeding operations on a poultry farm, large or small, constitute the largest part of routine work. It is work that is done every day, and several times a day. No other work must be done so punctually and with such care as feeding. Feeding—promptly, properly and economically, makes for success or failure. It is one of the fundamentals of the enterprise.

Need for System.—Before discussing the virtues of different feeds and how to compound them in rations suitable for given purposes, we will review some of the methods of handling feed and placing it before the flocks. The facilities for storage and distribution are highly important, and result either in needless waste of time and labor, or in a corresponding saving of these factors. It is to be deplored that so many farms fail to give this subject the attention which it deserves, and no doubt a great many failures might be attributed to this oversight.

Simplicity of Operation.—The best feeding system is one that reduces and simplifies the number of handlings and operations, and at the same time renders the greatest ease and convenience to the feeder. The first consideration is the location of the feed storage. Obviously, this location should be a central one. The feed house should be as nearly equidistant to the laying houses, brooder houses and colony houses as it is possible to make it; not at one end of the group of buildings, which necessitates long walks in getting the feed to the farthest houses.

If a special feed house is to be built it is a simple matter to locate this structure where it will have the fullest accessibility. On large plants it may be found advisable to have two or more feed houses, for in no other way is it possible to have the feed

175

Fig. 122.—Laying house with feed house in center.

176

close to the flocks. Some poultrymen build a feed room in connection with each long laying house, usually in the center of the
building, and sometimes with overhead tracks to convey the
feed to the birds. See Fig. 122. An important saving in labor
is accomplished by such equipment, and where large quantities
of heavy feed are used, as on duck plants of any size, these measures are essential.

The overhead track system, from which cars are swung and
made to run by pushing them, similar to litter carriers in dairy
barns, is thoroughly practical in every way. Another method is
to employ tramcars operated on rails, which may be of wood or
steel. For indoor use the overhead rail is cheapest, because the
building can be made to support the track. See Fig. 123. Outdoors the tramway is generally the most economical.

It is doubtful if the average size poultry plant would be justi-
fied in expending the capital required to install any sort of a
carrier. Certainly plants under 3000 birds would not be warranted in so doing.

Converting Old Buildings.—Then, again, many poultry farms
are started on premises which have certain buildings erected, as,
for example, a barn or wagon house, which with a few improvements may be converted into satisfactory feed houses. In this
event the poultry houses should be built on sites adjacent to the
barn, and on all sides of it, though not too close, which will interfere with a satisfactory arrangement of yards. Another objection
to having the buildings too close arises from the fire hazard. It
is a good plan to have the buildings at least a hundred feet apart,
preferably two hundred feet. In this way, should a fire occur
in some part of the plant, it can be fought and confined to the one
building.

Avoid Crowding the Buildings.—You will find some farms with
the buildings all bunched up together; incubator cellar with
brooder house and feed room overhead, and leading off from the
feed room are the laying houses, breeding pens and yards for
young stock. Such an arrangement is convenient by reason of
its compactness, but when carried to extremes it is likely to result

12

in grave risks, not alone from fire, but from the evils of too intensive methods—insanitation and disease.

The size, character and appointments of the feed house will be determined largely by the capacity of the plant and its stock. For instance, duck farms need different equipment from chicken farms, and the requirements of the egg farm are different from the broiler plant. Following are some of the points to be con-

(*Courtesy C. L. Opperman*)

Fig. 123.—Feed carrier suspended from overhead track which runs continuously through all the houses.

sidered: Have the building large enough for the storage of at least a month's supply of feed, preferably larger, because during certain seasons of the year it is possible to buy grain at reduced prices, and important savings can be made thereby. Make the building as near rat and mouse proof as possible. When these pests gain access to a feed house there is no end to their thievery. They will consume or spoil hundreds of dollars' worth of feed in a season, and in many cases this loss may go on for months before it is noticed.

Construction.—Masonry or metal is the best construction to obstruct vermin. Hollow tile or concrete makes splendid walls and floors for granaries. If this is too expensive, and frame construction is used, then some provision must be made to keep the rodents from gnawing through the partitions and floors. If the building is raised from the ground and supported on piers, it is feasible to baffle the rats and mice by means of tin so arranged that they cannot climb up the piers because of the slippery surface of the tin. Where this is impracticable small mesh wire netting should be laid over the floor joists and wall studs, through which the rodents cannot pass.

Storage bins are not necessary unless feed is purchased in bulk, or unless the original sacks must be returned as soon as the feed is received. The use of bins for storing unmixed feed involves two unnecessary handlings—the emptying of the feed into the bin, and its removal when required for mixing. Then, too, feed keeps better in sacks, and it may be piled up almost as compactly as in bins. When feed is stored in bulk provision must be made to ventilate it, otherwise it will heat and spoil. If stored in sacks, and piled up so that air can circulate freely around the bags, there is little or no danger from heating.

In the majority of times a charge of about ten cents is made for the bags, which charge is redeemable when the bags are returned in good condition. This is another reason why the feed room should be free from vermin: rats and mice gnaw through the sacks and play havoc with them.

Since most poultry feeds consist of a mixture of grains or meals, it is advisable to have a few bins (barrels will answer the purpose on small farms) wherein these mixtures are stored for every-day use. It is more convenient to remove a bucketful of feed from a barrel or bin than from a sack. And the most convenient bin is one from which the feed may be scooped out with a bucket, preferably the actual feeding bucket. To fill the feed buckets with a shovel or other implement involves needless operations.

Place for Mixing.—If the poultry feeds are mixed on the farm, it will be necessary to provide a suitable place for this work. It is no small task to mix the dry mash mixture for several thousand birds, or even one thousand hens. The mixing will probably be done in lots of from five hundred to a thousand pounds, maybe in ton batches.

How to Mix.—The best place for this work is the feed room floor, consequently it should be tight, constructed of tongue and grooved boards, and smooth, so as to offer no obstructions to the use of shovels or scoops. Supposing that the feed is piled up around the walls of the room, the most convenient way to proceed with the mixing is to empty the required number of bags in a pile. Begin with the ingredient having the greatest bulk, such as bran or alfalfa; dump it, spread it out a bit, and then add the other ingredients, reserving the most concentrated feed, such as beef scrap, until last. When the ingredients are all heaped in a pile, turn the mass over with a broadcasting motion of the shovel or scoop, repeating the operation two or three times, or until the mash is uniformly mixed.

(*Courtesy Million Egg Farm*)

Fig. 124.—Combination automatic feeder, grit and shell box.

If two men do the mixing, one man should rake the mass back

and forward, while the other fellow turns it over. A large, half-bushel grain scoop is the most efficient implement for the shoveling, because it handles such a large quantity in a single operation.

Storage in Hoppers.—As soon as the mash is mixed, it should be stored in a bin ready for use, though a good plan is to have self-feeding mash hoppers in all of the poultry houses of such capacity that they will accommodate the bulk of the mash mixed at each operation. Unnecessary handling is eliminated in this way, for the mash is carried right from the mixing floor to the hoppers, and only a small surplus is stored in the feed-house bins. Savings of this sort are considerable in the course of a year's work.

Unless automatic feeders are installed throughout the plant, it is not practicable to handle the scratch grain mixtures the same as the mash. See Fig. 124. Scratch grains are usually distributed by hand twice a day, morning and evening. Sometimes the grains are fed separately. If mixed, which seems to be the most accepted practice, these scratch grains usually consist of but a few grains, such as corn, wheat and oats (except the chick feeds), which are much easier to mix than the mash, which consists of meals. Dumping the grains into a barrel or bin and stirring them with a potato fork will perform all the mixing necessary.

Need for Power.—On plants of considerable size it is poor economy to mix and prepare the feed by hand. Power should be installed. If electric power is available, this will probably prove the cheapest, otherwise a gasoline engine is necessary. Get a motor or engine plenty large enough to handle all of the machines without danger of over-loading it, and so arrange it that it can be hooked up to each machine with the least amount of time and worry. An overhead system of belts and pulleys is the safest and most convenient arrangement. Provide a tight and loose pulley for each machine, so that the drive can be thrown on and off at will. An engine or motor of three or four horsepower will take care of the average poultry plant equipment.

The machines necessary to lighten the burden of hand labor are as follows: BONE CUTTER, for grinding fresh bones, commonly called green bones, which is fed in place of the cured meat scrap and similar prepared animal feeds.

HAY CUTTER, for cutting clover, rye, alfalfa and other succulent feeds, also to be used for cutting dry hay and straw for litter in the brooder houses.

(*Courtesy Purdue Experiment Station*)

Fig. 125.—Outdoor feed hopper built on skids or runners. Cover is raised, showing partitions for different kinds of feed. Note the small lids hooked against body of hopper, which can be lowered to keep rats or mice from stealing the feed.

VEGETABLE CUTTER: This is really a shredder, and is used to cut and tear mangels, turnips and other vegetables into short lengths. Fowls eat vegetables with greater relish if they are cut into bits, and there is less waste than feeding them whole.

MASH MIXER: This is indispensable if wet mashes are fed in any quantity. Mixers are made in a number of sizes, heavily built and geared, and so devised that they can be made to dump their contents.   They consist mainly of a cylinder-shaped body, in the center of which a shaft is operated.   This shaft is fitted with propeller-like blades or flukes that revolve and agitate the mash.

A COOKER OR STEAMING KETTLE is another valuable adjunct to the feed-room equipment, especially where it is desirable to feed hot mashes during the winter.   The caldron may be used to heat water for scalding feeding utensils, and for scalding ducks for picking.   It will be found useful in many ways where heat is required.

Poultry supply houses keep a full line of these accessories, or they may be purchased direct from the manufacturers.   It will undoubtedly prove worth while for the poultry keeper to investigate them.

Feeding poultry has more angles and arguments to it than is generally supposed.   The broad understanding of the office of food is that it supports life.   In the case of animals this objective becomes more definite.   We feed them that we may convert certain vegetable, animal and mineral substances, which are not palatable to the human taste in their raw state, into finished food products which are edible by man.

In this respect stock feeding may be compared to a manufacturing process.   And this process may be divided into three principal stages: First, growth of the fowl; second, reproduction of the species; and third, the storing of surplus energy in the form of flesh.

These factors are inseparable.   We must supply materials with which to build bone, muscle, tissues and feathers in growing stock.   See Table VII.   We must furnish the fuel for muscular energy, digestion, repairs to tissues and the renewal of organic secretions.   In the young fowl these elements promote growth. When this growth is complete, or nearly so, reproduction commences.   If there is a surplus of material in excess of the requirements of reproduction and maintenance, which there should be in the healthy, normal fowl, this surplus is stored up in the body as a reserve in the event of a failure of the regular supply. Usually this surplus consists of fat.

TABLE VII

DIGESTIBLE NUTRIENTS REQUIRED PER DAY FOR EACH 100 POUNDS OF FOWLS FOR BODILY MAINTENANCE

	TOTAL DRY MATTER	ASH	PROTEIN	CARBOHYDRATES	FAT	NUTRITIVE RATIO
Capons (9 to 12 pounds)......	2.30	0.06	0.30	1.74	0.20	I to 7.5
Hens (5 to 7 pounds)........	2.70	0.10	0.40	2.00	0.20	I to 6.2
Hens (3 to 5 pounds)........	3.90	0.15	0.50	2.95	0.30	I to 7.4

DIGESTIBLE NUTRIENTS REQUIRED PER DAY FOR EACH 100 POUNDS OF FOWLS IN FULL LAYING

Hens (5 to 8 pounds)........	3.30	0.20	0.65	2.25	0.20	I to 4.2
Hens (3 to 5 pounds)........	5.50	0.30	1.00	3.75	0.35	I to 4.6

DIGESTIBLE NUTRIENTS REQUIRED PER DAY FOR EACH 100 POUNDS OF GROWING CHICKS

First 2 weeks..............	10.1	0.5	2.0	7.2	0.4	I to 4.1
Second 2 weeks............	9.6	0.7	2.2	6.2	0.5	I to 3.4
From 4 to 6 weeks..........	8.6	0.6	2.0	5.6	0.4	I to 3.4
From 6 to 8 weeks..........	7.4	0.5	1.6	4.9	0.4	I to 3.7
From 8 to 10 weeks..........	6.4	0.5	1.2	4.4	0.3	I to 4.3
From 10 to 12 weeks........	5.4	0.4	1.0	3.7	0.3	I to 4.4

DIGESTIBLE NUTRIENTS REQUIRED PER DAY FOR EACH 100 POUNDS OF GROWING DUCKLINGS

First 2 weeks..............	17.2	1.1	4.0	11.2	1.4	I to 3.7
From 2 to 4 weeks..........	17.0	1.5	4.1	10.1	1.3	I to 3.2
From 4 to 6 weeks..........	11.2	0.8	2.7	7.0	0.7	I to 3.3
From 6 to 8 weeks..........	8.0	0.6	1.7	5.2	0.5	I to 3.8
From 8 to 10 weeks.........	7.0	0.5	1.4	4.7	0.4	I to 4.1
From 10 to 15 weeks........	4.6	0.3	0.9	3.2	0.2	I to 4.1

Throwing a pailful of food to a flock of birds at regular intervals is not feeding them in the strict sense of the term. To feed stock intelligently, which is to say economically and properly, means to cater to their bodily requirements in every particular. First of all we must have some knowledge of the demands of the body and its functions. We must have a similar knowledge of the composition of the different foodstuffs and their availability in meeting the demands of the body. We must consider the mechanical condition of the various foods, since some are more convenient to use than others. And we must pay attention to their digestibility and palatability. For example, some feeds possess most of the elements of a well-proportioned diet, but if fed exclusively or in large quantities, they are harmful in that they tax the fowl's digestion. Then again, some feeds are practically incapable of digestion by fowls.

Palatability and Digestibility.—On the other hand, no matter how digestible and nourishing a particular food may be, if it is not palatable to the fowls, they will not eat freely enough of it to sustain them in proper condition. It has been found in practice, however, that highly digestible foods are usually equally palatable, providing they are fed in sufficient variety, hence this point is a consideration. Moreover, it has been found that certain

(*Courtesy C. L. Opperman*)

Fig. 126.—Overhead rail system for carrying feed and other supplies.

foods are made more digestible when fed in conjunction with certain other foods.

The cost of the feed is another vital consideration, if not the most vital point. Economical feeding means to use those materials which render the greatest benefit at the least cost. This is saying a great deal, as we shall see. A food that costs very little money is not necessarily a cheap food. On the contrary the most expensive feeds to buy are sometimes the cheapest. In other words, price alone should not determine the cost of a food. Price plus results should form the decision.

No matter how cheap a food may be, if the fowl cannot secure substantial benefits from it, it is expensive to use in that it is largely wasted. See Table X. For example, rye, though similar to wheat, has little value as a poultry feed. The fat of rye is quite indigestible, and tends to cause organic disturbances. And there is something about rye grain which is distasteful to birds; they do not care for it, and seldom eat it unless compelled to do so. On the other hand, sprouted rye is exceedingly palatable, and fowls do well on it. This is characteristic of the knowledge which the successful poultryman must have.

Composition of Fowl.—Let us begin with the composition of the fowl's body. We should know its analysis, or how are we to supply the necessary elements? This is going to lead us into a technical discussion, though it is the writer's intention to devote as little space as possible to scientific expression. It is wearisome to the practical poultryman, and of little value outside of the laboratory.

The fowl's body, as with any form of life, consists of numerous substances or compounds. For convenience they are grouped under four general headings: water, protein, ash and fat. Table VIII shows the percentages of these elements.

TABLE VIII.—COMPOSITION OF FOWL AND EGG

	WATER	PROTEIN	ASH	FAT
	Per cent	Per cent	Per cent	Per cent
Leghorn hen...................	55.8	21.6	3.8	17.0
Plymouth Rock capon............	41.6	19.4	3.7	33.9
New-laid egg, complete..........	65.7	11.4	12.2	8.9
Egg, without shell..............	74.45	12.16	0.97	9.74
White of egg...................	86.2	12.3	0.2	0.6
Yolk of egg....................	49.5	15.7	1.1	33.3

It will be seen from Table VIII that the hen's body is over half water, and that the egg which she produces is nearly three-fourths water.   Obviously water is as essential to her welfare as food.   In fact, a fowl can go without food with fewer ill effects than it can go without water, especially during warm weather. Lack of water retards every function of the body, thickens the blood and raises the body temperature. Failure to supply a liberal quantity of clean fresh water is particularly hard on growing chicks.   It stunts their development quite as much as sickness.

TABLE IX.—COST OF PROTEIN FROM DIFFERENT FOODSTUFFS, ACCORDING TO CALIFORNIA EXPERIMENT STATION

	PRICE PER TON IN DOLLARS	PROTEIN IN POUNDS PER TON	DIGESTIBLE PROTEIN. PRICE PER POUND IN CENTS
Alfalfa hay	16	210	7.6
Alfalfa meal	21	210	10.0
Barley	30	168	17.8
Blood, dried	60	1216	4.9
Casein, dried	80	1500	5.3
Corn	22	140	15.7
Cottonseed meal	33	752	4.3
Linseed-oil cake meal (n. p.)	34	630	5.4
Linseed-oil cake meal (o. p.)	60	582	5.8
Meat, fresh	60	400	15.0
Meat scraps	60	1324	4.6
Milk, skim	4	66	6.0
Milk, granulated	80	1000	8.0
Oats	25	178	14.04
Peas	80	380	21.0
Rice	36	128	28.1
Rice bran	30	214	14.0
Rye	27	150	18.0
Soya-bean meal	45	600	7.5
Wheat bran	23	220	10.4
Wheat, plump	27	170	14.8
Wheat, shrunken	31	264	11.7
Wheat shorts	28	246	11.3

Protein, sometimes spoken of as nitrogenous matter, because nitrogen is its most distinguishing characteristic, is the most important group of materials found in the body.   In reality protein

is the foundation or base of all living tissue—the substance
through which life is manifested. The muscles, skin, tendons,
brain, feathers, blood and internal organs are composed chiefly
of protein. The white of the egg when dry is almost all protein,
while the dried yolk is one-third protein.

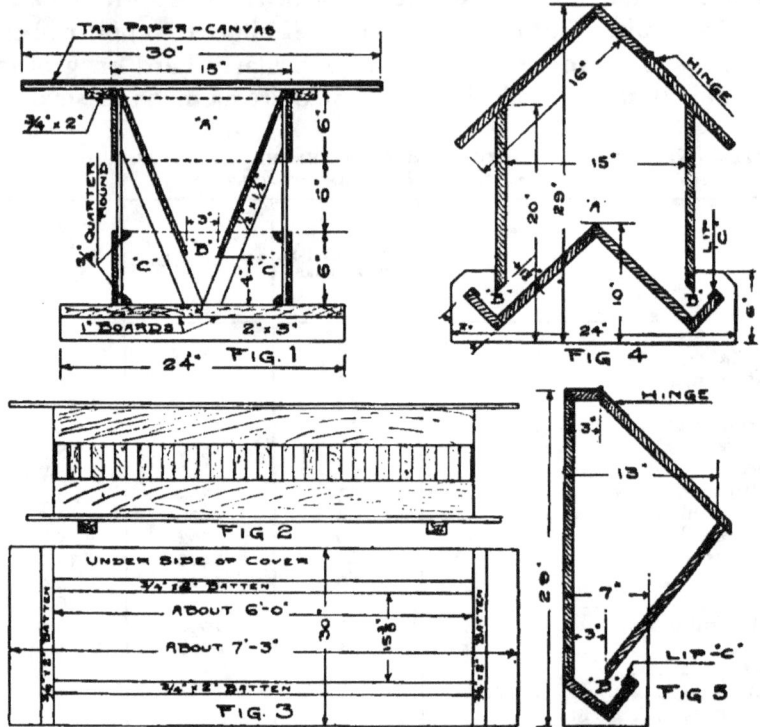

Fig. 127.—Designs for feed hoppers that are sanitary and non-wasteful.

Ash, or the mineral substances, such as lime, silicon, magnesium and so on, form the incombustible material of the body.
In short, ash is that part of the body which is left after the volatile and combustible elements are driven off by heat, the heat
being created by the fats. Considerable ash enters into the egg,
and it is an important element in the structure of bone and

feathers. Foods which supply ash are especially vital to the growth of young stock.

Fats comprise the fuel of the body. They are burned to maintain the body temperature, to produce the energy required in all muscular action, and where the supply is greater than the immediate needs of the fowl, they are stored up in fatty tissues under the skin or in globules distributed throughout the muscles, giving the fowl a full, plump appearance. This fat constitutes a reserve force, and will vary according to the health and condition of the fowl. Too much fat is as bad as too little, because it tends to make the fowl sluggish and non-productive.

TABLE X.—AVERAGE DIGESTION COEFFICIENTS OF DIFFERENT FOODS WITH CHICKENS, ACCORDING TO THE MAINE EXPERIMENT STATION

	NUMBER OF EXPERI- MENTS	ORGANIC MATTER	CRUDE PROTEIN	NITRO- GEN-FREE EXTRACT	ETHER EXTRACT (FAT)
Bran (wheat)..............	3	46.70	71.70	46.00	37.00
Beef scrap................	2	80.20	92.60	..	95.00
Beef (lean meat)...........	2	87.65	90.20	..	86.30
Barley....................	3	77.17	77.32	85.09	67.86
Buckwheat................	2	69.38	59.40	86.99	89.22
Corn (whole).............	16	86.87	81.58	91.32	88.11
Corn (cracked)............	2	83.30	72.20	88.10	87.60
Corn (meal)..............	2	83.10	74.60	86.00	87.60
Clover...................	3	27.70	70.60	14.30	35.50
India wheat...............	3	72.70	75.00	83.40	83.80
Millet....................	2	..	62.40	98.39	85.71
Oats (whole).............	13	62.69	71.31	90.10	87.89
Peas.....................	3	77.07	87.00	84.80	80.01
Wheat....................	10	82.26	75.05	87.04	53.00
Rye......................	2	79.20	66.90	86.70	22.60
Potatoes..................	6	78.33	46.94	84.46	..

Composition of Foods.—The same elements which are found in the fowl's body or in the egg are also found in different foods. But it should be understood that these elements are not transferred as such direct from the food to the body tissues. On the contrary, they are entirely remade. The food elements are taken into the body, digested and assimilated, and when they are in such form that the blood can absorb them, they are distributed throughout the body as required. Water is a great aid in this respect, in that it softens the food in the crop and prepares it for the grinding process in the gizzard. It also serves as a carrier in transporting the food through the digestive tract, and being a liquid it dilutes the digested elements so that they can be more readily absorbed.

(*Courtesy Cornell Experiment Station*)

Fig. 128.—Simple type of outdoor feed hopper. Cover is made water-tight by means of patent roofing material.

Nutrients.—Those portions of the food which yield heat or energy or serve the body in any way are termed nutrients. Nourishing elements would be another way of expressing the parts or constituents of food. They are classified in five groups, as follows: (1) water, (2) protein, (3) carbohydrates, (4) fats, (5) ash.

Water and ash are the inorganic nutrients; protein, carbohydrates and fats are the organic elements. For convenience carbohydrates and fats are sometimes grouped together. They contain the same compounds of carbon, hydrogen and oxygen, only in different proportions. The one great difference is that the fats have a greater power than the carbohydrates, being rated at two and a quarter times the power of carbohydrates.

Carbohydrates include the starches, sugars and gums, and like fats, they are burned to produce heat and energy. Carbohydrates as found in plants or grains are stored in a structure of cellulose, which constitutes the framework of the plant or the kernel of the grain. This framework or fibrous substance is mostly indigestible, and in livestock feeding, especially poultry feeding, it is spoken of as crude fiber. The digestible portions of the carbohydrates are called collectively NITROGEN-FREE EXTRACT.

(Courtesy Cornell Experiment Station)

Fig. 129.—Where fowls are grown in large numbers on free range it is economy to tend the houses with a team, preferably a low-wheeled truck, containing water-barrel, feed and so on.

Avoid Fiber.—The poultryman aims to feed materials which have the greatest quantity of nitrogen-free extract, with the least amount of crude fiber, since fiber is not only indigestible, but it is voided and therefore of no food value.

Protein nutrients are by far the most expensive portion of the fowl's diet. See Table IX. As a matter of fact, protein usually forms the basis upon which prices of feeds are determined. Wheat and its by-products rank first in popularity as sources of vegetable protein. Beef scrap, green cut bone and fish scrap are the most common sources of animal protein. Fowls require both animal and vegetable protein for best results, though chemically there is very little difference between the two kinds.

Most of the ash elements are found in sufficient quantities in the ordinary poultry feeds, except lime, which should be furnished in the form of oyster shells for laying hens, and in granulated bone for growing stock. Bran and alfalfa are both rich in ash, hence they are valuable feeds for chickens.

Wheat is the most efficient single feed for poultry, because it furnishes the nearest thing to a perfectly balanced ration. When fed in conjunction with corn and oats as a scratch feed, supplemented by a mash consisting of various meals, including a meat ingredient, properly compounded, wheat produces the best results at the least cost. When fed by itself corn is too fattening. Oats is a fattening feed, too, in addition to which it has a high percentage of fiber in the hulls, which is objectionable when fed in large quantities. Hulled oats or oatmeal, without this fiber, makes a splendid feed for young chicks.

CHAPTER XV

AVAILABLE GRAINS AND MEALS

Fowls are omnivorous feeders, in that they eat both animal and vegetable foods. They require both kinds of food to do their best. Under domestic conditions, they might be classed as granivorous, or grain eaters, because they subsist mainly on a grain diet. Their digestive tract is especially well adapted to the grinding and digesting of cereals, which the fowls relish.

Left to forage for themselves over a range that is abundant in plant life, the natural food of fowls would consist of grains, seeds, roots, grasses, worms and insects, tidbits of greens, berries and so on. The ration would comprise a wide variety of things, and in the main it would be wisely balanced. When kept in confinement the fowls must be given food that resembles this natural diet. It is impracticable to give them precisely the same articles which are found on free range, but these articles can be closely imitated, and the poultry keeper has a large list from which to choose.

How Foods are Grouped.—For convenience we will divide the available foodstuffs into four groups, about as they are grouped in everyday practice: (1) grains; (2) meals and prepared foods, or mash constituents; (3) green foods, or succulents; (4) mineral and condimental foods. We will analyze them in this order, confining ourselves more or less to the commonplace products.

Value of Wheat.—If it were necessary to limit the flock to a single article, wheat would probably give the best results, because it is the nearest thing to a well-balanced ration. It is not perfectly balanced, however, and gives best results only when fed in conjunction with other products. Wheat contains more protein and ash than corn, and less fiber than oats. At the same

TABLE XI.—COMPOSITION OF VARIOUS GRAINS AND THEIR BY-PRODUCTS

(The nutrients are given regardless of their digestibility. And it should be borne in mind that analyses of this kind vary somewhat according to different authorities.)

FEED	WATER	ASH	PROTEIN	FIBRE	CARBO-HY-DRATES	FAT
Corn group	Per cent	Per cent	Per cent	Per cent	Per cent	Per cent
Dent corn............	10.6	1.5	10.3	2.2	70.4	5.0
Flint corn...........	11.3	1.4	10.5	1.7	70.1	5.0
Sweet corn...........	8.8	1.9	11.6	2.8	66.8	8.1
Pop corn............	10.7	1.5	11.6	1.8	69.2	5.2
Corn meal...........	12.0	1.3	8.7	2.1	71.2	4.7
Corn meal (sifted)......	12.0	1.0	8.9	1.3	72.0	4.8
Corn-and-cob-meal.....	15.1	1.5	8.5	6.6	64.8	3.5
Gluten meal..........	8.2	.9	29.3	3.3	46.5	11.8
Gluten feed..........	8.5	1.7	26.2	7.2	53.3	3.1
Hominy meal.........	11.0	2.5	10.4	4.2	64.1	7.8
Wheat group						
Whole wheat.........	10.5	1.8	11.9	1.8	71.9	2.1
Wheat screenings......	11.6	2.9	12.5	4.9	65.1	3.0
Wheat bran..........	11.9	5.8	15.4	9.0	53.9	4.0
Wheat middlings.......	12.1	3.3	15.6	4.6	60.4	4.0
Low-grade flour.......	12.4	.6	10.0	.9	75.0	1.1
Old bread...........	31.2	Variable	6.9	Variable	44.2	.9
Oat group						
Oats, whole or ground....	11.0	3.0	11.8	9.5	60.7	5.0
Clipped oats.........	11.2	2.8	12.0	7.4	61.6	5.0
Oatmeal............	7.9	2.0	14.7	.9	67.4	7.1
Rolled oats..........	8.5	1.9	15.0	.6	66.0	8.0
Buckwheat group						
Buckwheat..........	12.6	2.0	10.0	8.7	64.5	2.2
Buckwheat bran.......	10.5	3.1	12.4	31.9	38.8	3.3
Buckwheat middlings....	13.2	4.9	28.9	4.1	41.9	7.0
Barley group						
Barley.............	10.9	2.4	12.4	2.7	69.8	1.8
Barley meal..........	11.9	2.6	10.5	6.5	66.3	2.2
Barley screenings......	12.4	3.6	12.2	7.6	61.6	2.6
Malt sprouts.........	10.2	5.7	23.2	10.7	48.5	1.7
Other grains						
Linseed meal (old process)	9.2	5.7	32.9	8.9	35.4	7.9
Cottonseed..........	10.2	3.5	18.4	23.2	24.7	19.9
Cottonseed meal.......	8.2	7.2	42.3	5.6	23.6	13.1
Rye..............	11.6	1.9	10.6	1.7	72.5	1.7
Rice..............	12.8	.7	7.5	.5	78.1	.4
Sorghum seed.........	12.8	2.1	9.1	2.6	69.8	3.6
Broom corn seed.......	12.7	3.4	10.2	7.1	63.6	3.0
Millet seed..........	14.0	3.3	11.8	9.5	57.4	4.0
Sunflower-seed........	8.6	2.6	16.3	29.9	21.4	21.2
Canada field peas......	13.4	2.4	22.4	6.4	52.6	3.0
Cowpeas...........	14.8	3.2	20.8	4.1	55.7	1.4
Soy beans...........	10.8	4.7	34.0	4.8	28.8	16.9

time it contains materially less digestible fat than corn, and slightly less nitrogen-free extract.

Fowls fed on a pure wheat diet are subject to digestive disturbances. When fed on a mixture of wheat, corn and oats, these troubles are minimized. This mixture constitutes the bulk of so-called *Scratch Feeds* on the majority of poultry farms in this country.

(Courtesy Kansas Experiment Station)

Fig. 130.—Growing stock on alfalfa pasture.

Cost of Wheat.—Considering the high cost of wheat, it is likely that too much importance has been laid upon it in respect to poultry feeding. Pound for pound, the by-products of wheat—bran and middlings—are more valuable than the whole grain, and they are very much less expensive. True, the whole grain of wheat has all the qualities—attractiveness of size, color, shape and texture—which go to make a palatable ration for fowls, and it should be used in the scratch feed by way of variety because of this palatability; but wheat need not be depended upon as a mainstay of the diet. Wheat is a better growing feed than a fattening feed, hence it is valuable in rearing young stock. It has been found that wheat gives a lighter color to the yolks of eggs and to fat, and according to many packers, it imparts a redder color to the flesh, the lean meat, than does corn.

TABLE XII.—FEEDING RATIONS FOR LAYING HENS RECOMMENDED BY POULTRY DEPARTMENT, CORNELL UNIVERSITY
WHOLE-GRAIN MIXTURE FED MORNING AND AFTERNOON IN LITTER

BY WEIGHT, WINTER	BY MEASURE, WINTER	BY WEIGHT, SUMMER	BY MEASURE, SUMMER
60 lbs. wheat	32 qts. wheat	60 lbs. wheat	32 qts. wheat
60 lbs. corn	36 qts. corn	60 lbs. corn	36 qts. corn
30 lbs. oats	30 qts. oats	30 lbs. oats	30 qts. oats
30 lbs. buckwheat	20 qts. buckwheat		

DRY MASH FED IN A HOPPER, OPEN AFTERNOONS ONLY

BY WEIGHT, WINTER AND SUMMER	BY MEASURE, WINTER AND SUMMER
60 lbs. corn meal	57 qts. corn meal
60 lbs. wheat middlings	71 qts. wheat middlings
30 lbs. wheat bran	57 qts. wheat bran
10 lbs. alfalfa meal	20 qts. alfalfa meal
10 lbs. oil meal	8 qts. oil meal
50 lbs. beef scrap	43 qts. beef scrap
1 lb. salt	½ qt. salt

"The fowls should eat about one-half as much mash by weight as whole grain. Regulate the proportion of grain and ground feed by giving a light feeding of grain in the morning and about all they will consume at the afternoon feeding (in time to find grain before dark). In the case of pullets or fowls in heavy laying, restrict both night and morning feeding to induce heavy eating of dry mash, especially in the case of hens. This ration should be supplemented with beets, cabbage, sprouted oats, green clover, or other succulent feed, unless running on grass-covered range. Grit, cracked oyster shell, and charcoal should be accessible at all times. Green feed should not be fed in a frozen condition. All feed and litter used should be strictly sweet, clean, and free from mustiness, mold, or decay. Serious losses frequently occur from disease, due to the fowls taking into their bodies, through their intestinal tract or lungs, the spores of molds."

No Need for Best Wheat.—It is a mistake to assume that the best grades of milling wheat are the best for poultry. Plump wheat is slightly more palatable, but it has been shown that shrunken wheat, caused by frost or drought, contains a higher percentage of protein, therefore it is more economical as a stock feed. Frequently the poorer grades of wheat, which are not suitable for the best grades of flour, can be had at prices which are so low that they are cheaper than corn and oats.

Musty, smutty and heated wheat should be carefully avoided, no matter how cheap they can be bought, because of the injurious effect they may have on the fowls' digestion. There is no objection to wheat screenings, however, which constitute the refuse or "screenings" from the better grades of wheat. The feeding value of screenings depends upon quality, and quality varies in each purchase. Screenings contain broken and shrunken wheat kernels, weed seeds, chaff and often considerable harmless trash. If the price

(Courtesy Wisconsin Experiment Station)

Fig. 131.—Fowls seldom derive sufficient exercise unless there are opportunities and incentives for scratching.

is low enough, screenings can be fed to an advantage. The point to consider is the price and quality of the screenings in relation to the price of sound feeding wheat.

Corn and its products are the principal sources of grain feed for poultry. It is the best relished of all the grains, it is easily digested, contains no poisonous substances, and except that it is too concentrated when fed alone, too fattening, it is an ideal ration. Corn has many advantages; it is attractive looking, free from shuck or husk, it is easily swallowed, and contains large amounts of oil and sugar which give it flavor and ease of digestion.   It satisfies the cravings of appetite and furnishes warmth. It is not the most desirable food for egg production and the development of lean meat, because it is deficient in protein.   Corn is a fattening food, and it is most valuable in this respect.   When fed in conjunction with foods which are rich in protein, and some bulky material, such as green stuff or vegetables, it makes an indispensable article of diet.

With the exception of flaxseed and sunflower seed, corn is the richest grain in carbohydrates and fat formers.   As previously mentioned, it is a little low in protein, and quite low in mineral substances.   To make up for this deficiency in mineral matter, oats are usually compounded with corn in poultry feeding. Barley is also used, though there is no better food for ash or mineral matter than wheat bran.

Cheapness of Corn.—Ordinarily, corn is the cheapest food for poultry, not alone because of its price per bushel, but because fowls obtain the greatest amount of digestible material from corn. With oats at 40 cents a bushel, corn at 60 cents and wheat at 80 cents, it has been found that the protein in wheat or oats costs practically fifty per cent more than the protein in corn.   Similarly, the cost of fat in wheat is about twice as much as the cost of fat in corn.   From every standpoint corn is the cheapest food, but since it does not contain the nutrients in the proportion in which they are required in the production of eggs and lean meat, it must be mixed with other products.

The majority of farm flocks of poultry are maintained on an exclusive corn diet, largely because the corn is grown at home, and because it is not sold so readily as wheat.   This practice is a great mistake, and cannot be condemned too harshly.   It invariably results in a small egg yield, and frequently no eggs at all.   Properly fed, corn is a most valuable feed—quite as valuable as wheat.   This is especially true in America, where corn is plentiful, and under normal circumstances cheap.

The value of oats as a poultry feed is determined largely by the percentage of fiber in the hulls.   This fiber is the greatest objection, because it is a tax on the fowls' digestion.   Light oats,

so called because of the large proportion of hull to kernel, which
makes a bushel of oats weigh light, is not a desirable poultry feed.
Heavy oats, with a smaller percentage of husk, are greatly relished by chickens, and should form an important part of the
diet. Hens are quick to recognize the quality of oats. They
will refuse to eat oats which contain a great deal of shuck, seeming
to know instinctively that it is as indigestible as straw.

No grain varies so much in weight per bushel as oats, which, of
course, is due to the variation in the proportion of husk to kernel.
It will pay to get heavy oats, though the price is more, because

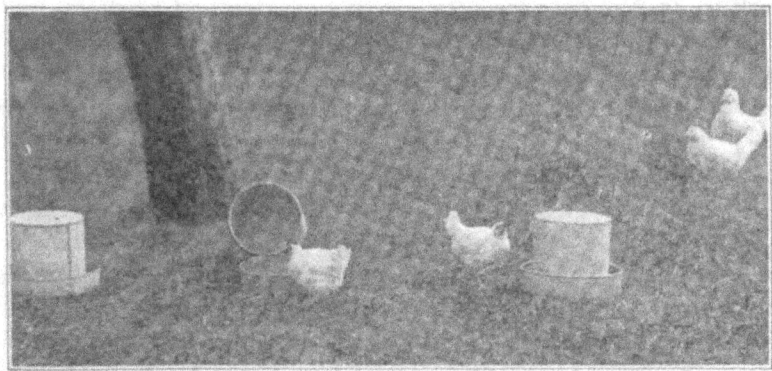

(*Courtesy Purdue Experiment Station*)

Fig. 132.—Chickens require an abundance of fresh drinking water at all times.

of the waste in the hulls. Hulled oats or oatmeal is an excellent
food for growing chicks. Some breeders maintain that oats contain an ingredient not found in other grains, which has a stimulating effect on the nerves and thereby promotes rapid growth.
Whether or not there is any truth in this belief, the fact remains
that there is no better food for baby chicks than pinhead oatmeal or hulled oats. It is expensive, but the results seem to
warrant the additional expense. Ground oats are used extensively in fattening poultry.

Barley ranks between oats and corn as a growing and fattening
feed. It contains less fat, fiber and ash than oats, but more

protein and carbohydrates.   It does not rate so high in digestible protein as wheat, and barley is not so palatable as either wheat, corn or oats; but it is a splendid ingredient for the scratch feed because of its variety.

The husk and beard of barley are somewhat against it, though not so much as in oats.   So long as the fowls will eat freely of it, barley can be fed in liberal quantities.   In Europe barley occupies the place filled by corn in America, in fact, it has much the same feeding value.   The great demand for barley for brewing purposes tends to keep the price high.   Malt sprouts and brewer's grains, which are the by-products of barley, are useful stock feeds, but because they are in such demand as dairy feed, little attention has been given to them as poultry feed, though they are a good source of succulent material.

Buckwheat.—Strictly speaking, buckwheat is a poor poultry feed.   It contains properties which supply heat and energy, but because of the black, woody hulls, which are both unappetizing and indigestible by reason of the fiber, fowls do not eat it readily. In fact, until the hens are accustomed to buckwheat, they are likely to ignore it completely.   Compared with other grains, buckwheat cannot be called a rich food, nor an economical food. It has a lower percentage of digestible organic matter than most grains, and is low in mineral matter.   The only excuse for feeding buckwheat is to add a little variety to the ration, but even this virtue is questionable because of its unattractiveness.

Rye is even less of a success as a poultry feed than buckwheat. Though virtually the same as wheat in composition, appearance, size and color, rye is not relished by fowls.   There is something about it which is distasteful, and if birds are compelled to eat it in grain form there is a tendency to create digestive trouble.   On the other hand, sprouted rye is exceedingly palatable, and fowls do well on it.   The fat of rye is said to be the indigestible portion of the grain, and apparently it has some medicinal quality or flavor which is objectionable.   In view of these facts, it is poor economy to try and use rye grain as a poultry feed.

Rice.—Though rice is reckoned among the available grains for poultry, it is used in very limited quantities. Even in the South, where it is grown, rice is not generally used, except as a chick feed, and then only in small quantities. The commercial chick feeds contain a small percentage of broken rice. Its value, however, is questioned by most breeders. Chicks do not seem to care for it, and for this reason it is not generally recommended. Boiled rice is sometimes fed to chicks as a wet mash, and the water

(*Courtesy U. S. Dep't Agriculture*)

Fig. 133.—Feeding time in a fattening station.

which is drained off the rice is cooled and given as drinking water. As a commercial proposition this is too troublesome.

Field peas, cowpeas and Soy beans are three nitrogenous seeds which may be used to vary the ration, though it is not advisable to feed them to poultry unless they can be bought at a reasonable price. Usually they are scarce and correspondingly high in price.

Kafir corn, milo maize, sorghum seeds and broom corn seeds may be added to the scratch grain ration to give variety. They can be made to take the place of corn, because they are pretty much the same in character. Fowls eat them with a relish. The one objection is cost; they are usually high in price, therefore prohibitive, except in small quantities.

Millet is fairly reasonable in price, and is found in most of the commercial chick feeds. It has practically the same composition as oats. Owing to an extremely hard shell, it is indigestible. There is a high percentage of fiber, which is against it as an economical feed. Because the seeds are small and brilliantly colored, little chicks are attracted to millet, and will eat it readily. This is accompanied by risks, since the bullet-like structure of the seeds resists the grinding action of the gizzard. Many breeders are opposed to the use of millet on this account. The writer discarded it long since.

Sunflower seeds are recommended for the molting season, and for birds intended for exhibitions, because of the vegetable oil which imparts a gloss to the plumage. They have a high fiber content (about 30 per cent) which renders them indigestible and wasteful. Then, too, sunflower seed is expensive. In the writer's opinion, it is not worthy of notice. The plumage can be stimulated by the use of cake meal (old process), which is the residue from linseed

(*Courtesy Wisconsin Experiment Station*)

Fig. 134.—Feeding battery used in packing houses. Note troughs on the outside of the cages.

after the oil has been extracted. This oil meal contains valuable nutrients, and is reasonable in price.

Every Process is Intensive.— Fowls have a higher temperature, increased respiration, and a more rapid digestion than most other animals. Their young is brought into being very quickly; they mature in from five to nine months; and their life is limited to about five years. Thus, every function, process and phase of the fowl's existence may be called intensive or rapid. In many respects the fowl may

(*Courtesy Wisconsin Experiment Station*)

Fig. 135.—Fowls with long, crow-like heads are usually poor feeders and do not make rapid gains.

be compared to a high-geared machine, which is delicate of operation, but thoroughly reliable, providing the adjustments are properly maintained.

We know that a low-geared machine can withstand more neglect than the fast-running, high-tensioned mechanism. In a similar way, the relation of food to bodily requirements is more exacting in fowls than in other animals of slower growth.

Do not misconstrue this statement: Fowls do not require greater care than other animals. On the contrary, they require much less attention. But fowls must be provided with the means by which they can take care of themselves, and of which they are fully capable. This

(*Courtesy Wisconsin Experiment Station*)

Fig. 136.—This bird's head is typical of a good feeder, one that can be depended upon to make rapid gains in flesh.

provision consists chiefly in giving the fowls access to the right sort of foodstuffs.

As explained in a preceding chapter, poultry must be fed those products which yield certain elements, such as protein, fats and mineral substances, which are necessary to meet the daily demands of the body. Moreover, these elements must be fed in a mechanical condition which will render the greatest good at the least expense. Expense in this instance embraces both the poultryman's pocketbook and the fowls' health.

(Courtesy Kansas Experiment Station)

Fig. 137.—The farm flock can be made to forage for the greater part of its keep.

Exclusive Grain Diet.—Time was when poultry was fed almost exclusively on whole grain. Such a diet was expensive and improperly balanced. Also, it was wasteful, because certain portions of whole grain were not necessary to a fowl's welfare. They were of greater value as human food. For example, wheat middlings and wheat bran are better poultry foods than wheat flour, whereas the latter is considered more desirable as human food. Therefore the logical thing is to mill and separate the whole grain. In recent years this has been done, with the result

TABLE XIII.—FEEDING RATIONS FOR LAYING HENS RECOMMENDED BY THE NEW JERSEY STATION

DRY MASH MIXTURE (WINTER)

KIND OF FEED	QUANTITY	
	Pounds	Quarts
Wheat bran............................	200	380
Wheat middlings......................	200	240
Ground oats..........................	200	200
Corn meal............................	100	95
Gluten meal..........................	100	80
Meat scrap (high grade)..............	100	86
Alfalfa (short cut)..................	100	200
Total.............................	1000	1281

Nutritive Ratio, 1:3.02

DRY MASH MIXTURE (SUMMER)

KIND OF FEED	QUANTITY	
	Pounds	Quarts
Wheat bran...........................	200	380
Wheat middlings......................	100	120
Ground oats..........................	100	100
Gluten meal..........................	50	40
Meat scrap (high grade)..............	25	21
Total.............................	475	661

Nutritive Ratio, 1:3.22

For Leghorns and other egg breeds the feed hoppers are kept open all day. For the heavier breeds, which are prone to put on too much fat, the hoppers are kept open in the afternoon only.

SCRATCH GRAIN MIXTURE (MORNING) FIVE POUNDS TO 100 BIRDS FED IN DEEP LITTER

KIND OF FEED	QUANTITY	
	Pounds	Quarts
Wheat................................	100	53
Oats.................................	100	98
Total.............................	200	151

Nutritive Ratio, 1:6.6

SCRATCH GRAIN MIXTURE (EVENING) TEN POUNDS TO 100 BIRDS

KIND OF FEED	QUANTITY	
	Pounds	Quarts
Cracked corn.........................	200	120
Wheat................................	100	53
Oats.................................	100	98
Buckwheat............................	100	66
Total.............................	500	337

Nutritive Ratio, 1:7.8

The above is intended for cold weather feeding. In the summer half the cracked corn is used, and barley is substituted for the buckwheat.

205

that we now have a valuable list of by-products, most of them meals, which are available as stock foods, at a much lower cost than the whole grain.

Merits of Mash Feeding.—To-day, instead of feeding their flocks exclusively on whole or broken grain, practical poultry raisers feed about half grain and half meals. The grain feed, which consists of a mixture of grains, such as corn, wheat and oats, is termed the "scratch feed"; the meals are mixed together in certain proportions and called a "mash." This mash is fed dry or moist, depending upon the object desired. Of late the dry mash has gained considerable favor over the wet mash, and largely because of the inconvenience in placing the wet mash before the birds. It must be fed at regular intervals and in specific quantities, because if moist feed is allowed to stand for any length of time, it will spoil in hot weather and freeze in cold weather. A dry mash is always available; it can be placed before the flocks in large quantities, providing the hoppers are suitable, and the mash remains sweet and wholesome until consumed.

TABLE XIV

At the Maine Experiment Station under Prof. Gowell the following quantities of feed were consumed per hen in one year by a flock of Barred Plymouth Rocks, averaging 144 eggs each:

	POUNDS
Grain and mash	90.0
Oyster shell	4.0
Granulated bone	2.4
Grit	2.0
Charcoal	2.4
Clover	10.0

Forced Feeding.—Excellent results have been secured through mash feeding. It is particularly valuable in fattening, and in forced feeding for increased egg production. This is due to the fact that meals are digested more quickly than the whole grains, therefore the nourishment which they contain is more readily available for absorption. Dry mash feeding is the only method by which fowls can be made to consume large quantities of food, which are so essential to rapid development.

The point to remember, however, is this, forced feeding must not be carried to the point where it affects and impairs stamina and health. In fattening poultry for market there is little need for caution in this respect, because the birds are killed before any ill effects are manifested. But in feeding laying stock, especially breeders, prolonged over-feeding is almost certain to result in weak offspring and a general deterioration in the vigor and productiveness of the flocks.

On first thought, ground feed or meals would seem to be more digestible than whole grains, because, as stated above, the meals are more quickly digested than the grains. It does not follow, however, that the meals are more thoroughly digested, as we shall see.

Gizzard.—Fowls are endowed with a powerful muscular organ —the gizzard, for the express purpose of grinding solid foods. The action of the gizzard is involuntary. If it is equipped with sufficient "molars" in the form of hard, sharp, angular grit, its grinding process is consistently thorough.

Gastric Juice.—The crushing and grinding process is not the only function of the gizzard. The food is partially digested in this organ, and to accomplish this digestion the food must remain in this part of the digestive tract for a certain length of time. It has been found that the gizzard secretes a gastric juice, which is one of the most powerful agents of digestion, especially in respect to digesting protein and in dissolving the mineral elements. Naturally, the harder the food the longer it is required for the gizzard to grind it, and the longer the food remains in the gizzard the greater is the action of the gastric secretion.

Sparing the Gizzard.—As might be supposed, meals pass through the gizzard much quicker than whole grains, and this spares the gizzard considerable work. In forced feeding it is found cheaper to do this work with power millstones. But the point to remember is this: If the gizzard is spared too much, it soon loses its efficiency, and thereafter the food will not be thoroughly digested, which is both a tax on the fowls' health and wasteful of food. In short, the fowls must be fed enough hard grain to maintain the gizzard and all other organs in their normal state. It is therefore considered good practice to feed from one-third to one-half ground feed.

TABLE XV

In a test at the Maine Experiment Station it was found that a flock of two thousand Barred Plymouth Rocks were raised to laying age on an average of the following quantities of feed per fowl:

	POUNDS
Grain and mash	28.0
Granulated bone	0.75
Oyster shells	0.5
Grit	2.25
Charcoal	0.5
Total	32.00

Controlling the Diet.—There is still another important point about mash feeding: The feeder is better able to control the flock's diet. This statement is easily explained. When whole or broken grain is fed, it is more good luck than good management if some hens do not eat more corn, others wheat, and others oats, or whatever are the ingredients. And since no one kind of grain is a complete ration, the error of this method is apparent.

In a properly compounded, well-mixed mash, which has the exact proportion of protein, carbohydrates and mineral substances required to preserve health and stimulate egg production, the hen has no selection. She simply gulps mouthfuls of the mash, and every mouthful has practically the same ingredients, consequently she is made to eat just what the feeder has prescribed. Furthermore, if the mash is accessible at all times, which is the generally accepted method of feeding it, the fowls soon learn that they can eat it at will, and this habit tends to eliminate gorging by the stronger birds, and under-feeding on the part of the weaker or more timid members of the flock. In other words, there is no mad rush for the mash hopper like there is for a periodical distribution of scratch feed.

Mash Constituents.—The feeder has a wide list from which to select the ingredients for a mash. We will commence with the by-products of wheat.

Bran is probably the most commonly used, all-round stock food. It is the outer layer of the wheat kernel and is rich in mineral matter. By reason of its flaky nature, bran gives the

(*Courtesy U. S. Dep't Agriculture*)

Fig. 138.—Chickens in feeding batteries, being fattened for market.

mash a light, bulky texture, and this condition is preferred by fowls. Bran is fibrous and slightly laxative. It is comparatively rich in nutrients, though its digestibility is rather low. Bran and corn meal comprise the bulk of most poultry mashes. As might be supposed, there is a wide variation in the quality of bran, which is due to the milling process—how thin the out-

14

side shuck of the wheat kernel is taken off and how much of the gluten is retained. The more gluten it contains the richer the bran is in protein.

TABLE XVI

Two Dry Mash Formulas Recommended by California Experiment Station

	QUARTS	POUNDS		QUARTS	POUNDS
Bran................	6.0	3.0	Bran..........	4.0	2.0
Middlings........	0.5	0.5	Alfalfa meal....	1.0	0.5
Linseed meal (o. p.)	0.5	0.5	Corn meal.....	1.0	1.5
Gluten feed......	0.8	1.0	Shorts.........	2.0	1.5
Ground oats......	1.0	0.75	Barley meal....	1.0	1.1
Corn meal........	1.5	2.25	Ground beans..	1.0	1.1
Beef scrap........	1.0	1.5	Beef scrap.....	1.0	0.5
Bone meal........	0.5	1.0	Bone meal.....	0.5	1.0

The above rations are calculated for 100 hens a day, to be fed in conjunction with from 9 to 12 pounds of scratch grains and sufficient green stuffs.

Middlings and Shorts.—In the center of the wheat kernel is the flour, which is largely starch. Between the flour and the outer shuck or bran is a layer of gluten, called middlings or shorts. Shorts are practically the same as middlings, except that a larger percentage of fine bran may be present in the former. As a rule middlings are ground finer than shorts and resemble a low-grade flour. Middlings are richer in carbohydrates and fats than bran; and lower in mineral matter and fiber, therefore more digestible. Middlings are heavy and form a compact mass, and for best results they should be mixed with a bulky feed, such as bran, mealed alfalfa or ground oats.

Feeding Flour.—Some feed markets do not distinguish between middlings and shorts. Then again, there is only a slight difference between middlings and low-grade feeding flour. The feeding flour resembles the patent grades of white flour intended for human consumption, and is richer in gluten than middlings.

Corn meal, sometimes spoken of as regular feed meal, as the name implies, is simply the whole corn kernel ground fine, or it may be the siftings from cracked corn. It possesses virtually the same merits as the whole grain, and is used extensively for fattening purposes.

Corn-and-Cob Meal.—When the cob is ground with the corn the product is called corn-and-cob meal. No matter how fine the cob is reduced to a meal, it consists mainly of fiber, with low feeding value, which is objectionable in feeding poultry. A little corn-and-cob meal can be used in the mash if bulk is desired, but it should not be made to take the place of straight corn meal.

Gluten Meal.—In the manufacture of starch the kernels of corn are separated into hulls or bran, gluten and starch. Gluten meal is therefore corn less the bran and starch. It is rich in fat and protein and highly concentrated. When fed in large quantities it produces ill effects. To overcome this concentration, and to provide a market for the corn bran, manufacturers now mix the gluten meal with the bran and sell the mixture as gluten feed.

TABLE XVII
WAR TIME RATIONS RECOMMENDED BY HARE, OF THE SOUTH CAROLINA
EXPERIMENT STATION

DRY MASH FOR SOUTHERN BREEDERS		DRY MASH FOR WESTERN BREEDERS	
	Pounds		Pounds
Corn meal	100	Corn meal	250
Wheat middlings	100	Ground oats	200
Rice bran	100	Wheat middlings	200
Ground oats	200	Wheat bran	200
Velvet bean meal	100	Meat scrap	100
Cottonseed meal	100	Alfalfa meal	50
Salt	3.5	Salt	5

Gluten feed contains less fat and protein than gluten meal, and greater bulk, therefore it should be bought on a guaranteed analysis, depending upon its content of protein. Gluten feed and gluten meal are both palatable and digestible. They are valuable sources of protein, as will be observed, and are excellent egg producers. It is a mistake, however, to depend entirely upon this vegetable protein. Some animal protein is necessary in the mash, such as bone meal or meat scrap.

Hominy meal or hominy chop is still another valuable by-product of corn, useful as a poultry feed, but not so generally used because of the limited supply. It consists of the hull, germ and portions of the starch cells—the residue from the process of making hulled corn or hominy. It is similar to gluten and may be used in about the same way.

Ground oats, as the name suggests, is simply the whole oats ground fine. The fiber is not reduced in any way, though it is rendered more palatable. In buying ground oats care should be taken that there is not an excess of hulls, since unreliable manufacturers sometimes use a very light or poor grade of oats for this feed. Ground oats are useful in all mashes, but because of the high percentage of fiber, they must be used judiciously. For example, it would be a mistake to compound a mash having large quantities of mealed alfalfa, ground oats and bran, because all three have large percentages of fiber, whereas the fiber content in poultry feeds should be kept as low as possible.

Ground barley, sometimes called barley meal, has pretty much the same value as the whole grain. Like ground oats, it must be used carefully because of the hulls. It is not a common feed in this country. Dried brewers' grains, a by-product of barley from the breweries, is a valuable addition to the mash, rich in fat and protein, and used about the same as corn meal and wheat middlings. It is more widely used as a dairy feed than for poultry.

Buckwheat bran and middlings, which are the by-products of buckwheat flour, are not particularly desirable as a poultry feed because of the large percentage of indigestible matter. The middlings are sometimes used in fattening poultry for slaughter, especially where white flesh is desired.

Cottonseed meal, a by-product from the manufacture of cottonseed oil from cotton seed, is thought by some to be a substitute for meat scrap. Fowls do not take kindly to it, however, and the general opinion among poultry feeders is that it is not a desirable feed. The consensus of opinion is that cottonseed meal should be used with extreme caution, or not at all.

Oil meal, or linseed meal, a by-product of the manufacture of linseed oil from flaxseed, is rich in protein and fat, and a valu-

Fig. 139.—Grading poultry from the hanging racks in a refrigerated packing room.

able feed for molting fowls.   It is exceedingly laxative, however, and because of this tendency it must be fed sparingly. Oil meal contains about three times as much protein as wheat or oats, and about five times the protein of rice.   It is one of the richest foods available for poultry, fairly reasonable in price, and it should find a place in all mashes for mature stock; but because of its concentrated nature, oil meal should not form over five per cent of the ration.

Old process oil meal, or O. P. cake meal, are other names given to this product.   In the old process of manufacturing linseed oil the flaxseed is crushed and the oil extracted by great pressure. In the newer methods the oil is dissolved by means of naptha. The old process meal is best, because it contains more oil, which stimulates the growth of feathers, and gives the plumage a glossy appearance, so much desired by fanciers.   When moist, oil meal is very sticky, and this quality tends to thicken or gum the mash. It is therefore objectionable in wet mashes.

CHAPTER XVI

ANIMAL AND GREEN FOODS

Animal Protein.—A controversy has existed for some time over the relative merits of vegetable protein and animal protein. From the chemist's standpoint they may be practically the same; but not so from the fowls' point of view, which is the all-important angle for consideration. We cannot escape the fact that chickens are scavengers. They are meat eaters, naturally. Witness their eagerness for worms and insects, or for dead animal matter, if they find it. Apparently they crave meat in some form, hence it behooves the feeder to furnish it.

Important feeding experiments have shown that protein and fats are more digestible when supplied from animal sources than from vegetable sources. Chicks and ducklings fed on rations containing animal food consume greater quantities of food and make faster gains in growth. Pullets are brought to maturity, and older hens lay more eggs when fed animal protein. It might be said that animal food is essential to economical production—an absolute necessity. No other food constituent stimulates the ovaries into activity in such a brief space of time. Because of this stimulation, however, animal foods must be used judiciously or ill effects will result.

Meat Scrap.—There are several sources of animal food, though meat scrap, otherwise known as beef scrap, poultry meat, animal meal and similar names, is unquestionably the most widely used. It consists of meat trimmings from slaughter houses and butcher shops, including considerable bone, which are cooked under steam pressure to render out most of the fat, and then dried and ground to different degrees of fineness. The cooking and curing process, if properly done, also sterilizes the product,

215

thereby destroying any bacteria which might prove harmful to the fowls.

Guaranteed Analysis.—In the cheaper grades of meat scrap the product is not so carefully purified, and in some instances it is treated with acid, which is injurious. These grades should be avoided. In no other form of food is it so essential to use the best grades as in meat scrap. And usually the best grades are the cheapest in the long run, because they are almost certain to

Fig. 140.—Spading and turning over the soil not only keeps the yards sanitary, but it affords worms and insect life for the fowls.

contain higher percentages of protein, which is the sole reason for buying this sort of food. High-grade scrap will contain from 50 to 60 per cent protein, the low grades from 25 to 40 per cent. Obviously the price should be based on a guaranteed analysis.

Testing Meat.—The appearance of meat scrap will go far in convincing the purchaser as to its quality. To test it, pour scalding water over a sample, and if it continues to smell sweet, the chances are it is all right. If it gives off a putrid odor—beware. Then examine it in a dry state: If it seems to have large quantities of hair, horn and hoof, it is not desirable. Particles of horn and hoof resemble bits of brown glass.

Tankage, dried blood and kindred packing-house by-products, though suitable for hog feeding, are not relished by poultry. They are rich in nitrogen, but not capable of digestion by fowls. Shun them.

Green cut bone is one of the most palatable foods of animal origin, and also one of the most forcing foods for egg production. It must be fed sweet by all means, which renders it a difficult food to handle on a large scale, especially in the summer time. The practice is to collect the bones, together with meat trimmings, from nearby butcher shops, and grind them at home by means of a power bone cutter. The ground bone and mangled bits of meat which cling to the bone are then fed separately or mixed with a little dry mash.

Green cut bone heats and spoils very quickly, even in cold weather, unless it is spread out so that the air can circulate freely through it. If fed in a heated condition it is almost certain to develop diarrhea and kindred troubles. It is laxative at best, therefore it is not a good food for young chickens. When given to laying stock to assist egg production, green bone should not comprise more than ten per cent of the ration. Excessive use of such stimulating food is a serious tax on the liver, and brings about symptoms similar to gout in man. Very few egg farms of any size have the facilities to feed green bone, consequently they depend upon the commercial forms of meat scrap, which can be kept indefinitely in any climate, providing they are stored in a dry place, and are not allowed to heat.

Granulated bone, or bone meal, which is a finer grinding of the bones, are both excellent foods for poultry, and especially for young stock, since they are valuable sources of both protein and mineral substances. This product is free from meat. The raw bones are dried, then ground to a meal. The best way to feed it is in the mash. It has no laxative properties to speak of, and fowls relish it. From five to ten per cent of bone meal should enter all mashes for chicks of all ages.

Fish scrap corresponds to meat scrap in many respects, and in some localities it is extensively used. The great difficulty with it is that it is likely to impart a fishy flavor to the eggs and flesh. This is due to the oil. In the best grades, wherein this oil is practically all removed, fish scrap makes a valuable substitute for meat scrap, and in most cases it it much cheaper than the latter.

TABLE XVIII.—COMPOSITION OF DIFFERENT ANIMAL FOODS

FEED	WATER	ASH	PROTEIN	FIBER	CARBO-HY-DRATES	FAT
	Per cent	Per cent	Per cent	Per cent	Per cent	Per cent
Meat scrap (high grade)......	10.70	4.10	60.20	..	..	25.00
Meat scrap (low grade)	15.40	2.50	45.00	..	..	37.10
Pork scrap................	11.0	2.2	55.0	..	..	31.8
Ground bone (dry).........	8.19	56.95	31.36	..	..	3.50
Green cut bone............	38.94	26.12	20.37	..	..	11.67
Animal meal...............	4.90	42.65	30.45	3.30	10.32	8.38
Blood meal................	10.61	4.65	75.69	1.28	1.46	7.11
Dried blood...............	6.70	6.60	65.10	..	5.30	16.30
Fresh fish................	44.0	1.00	10.50	42.00	..	2.50
Fish scrap (variable)........	5–10	2–8	34.0–50	..	..	17.0
Clams and other shell fish....	34.10	2.30	6.00	55.00	2.00	.60
Whole milk................	87.2	.6	3.6	..	4.9	3.7
Skim milk................	90.6	.7	3.1	..	5.3	.3
Buttermilk................	90.1	.7	4.0	..	4.0	1.2
Whey.....................	93.8	.4	.6	..	5.1	.1
Cheese....................	40.6	3.4	23.7	..	1.7	30.6
Granulated milk...........	28.5	3.6	13.7	..	51.1	3.1

Milk albumen, a by-product from the manufacture of milk sugar, is a good source of animal protein, but in the main it is too expensive for general use as a poultry food. Some breeders use it for their chicks. It is largely casein and lime, ground to a meal, and contains no crude fiber.

Granulated milk is another product of this kind, except that it is whole milk evaporated and crystallized. Sometimes it is spoken of as powdered milk. It is useful for chick feeding for the first few weeks, but its cost is high, therefore prohibitive for general use.

Milk in any form—fresh, sour, skimmed or buttermilk—is a wholesome feed for all forms of livestock, and should be used wherever possible in regular quantities. See Fig. 141. It is

Chicks fed corn and green clover

Chicks fed wheat and green clover

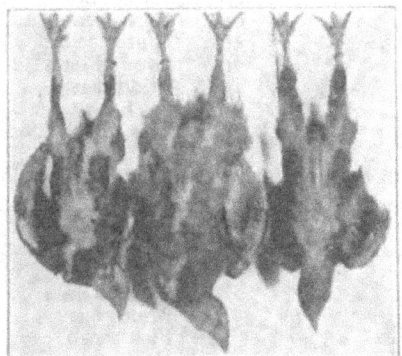

Chicks fed corn, green clover and milk

Fig. 141.—Experiment illustrating the benefit of milk in the chicks' diet. All three lots are the same age. (Wisconsin Experiment Station.)

palatable, aids digestion, carries a high nitrogenous content, and the presence of lactic acid tends to correct bowel troubles and reduces the spread of white diarrhea. Fowls will consume large quantities of it with the greatest relish.

Clean Fountains.—As previously mentioned, milk may be fed sweet or sour, but wherever possible it should be given one way or the other, and not alternated. To change it about constantly sometimes results in digestive disorders. Milk is given to fowls in drinking fountains, or it may be used to moisten the mash. When given to drink, the vessels must be scalded at regular periods, otherwise the milk is likely to become putrefied, especially in warm weather, which is equivalent to putrid meat, and is injurious, resulting in limberneck and similar troubles.

Milk whey, a by-product from the manufacture of cheese, is not so valuable as the other forms of milk, but it may be used whenever it can be obtained conveniently. It is relished by poultry.

TABLE XIX.—THE FOLLOWING FEEDING FORMULAS WERE USED IN FOUR LAYING COMPETITIONS AT STORRS, CONNECTICUT

SCRATCH GRAINS	Pounds	DRY MASH	Pounds
Cracked corn	60	Wheat bran	200
Wheat	60	Corn meal	100
Heavy oats	40	Gluten feed	100
Barley	20	Ground oats	100
Kafir corn	10	Middlings	75
Buckwheat	10	Beef scrap	60
Coarse beef scrap	10	Low grade flour	25

In the fifth contest the formulas were simplified, as follows:

SCRATCH GRAINS	Pounds	DRY MASH	Pounds
Wheat	100	Wheat bran	100
Cracked corn	100	Corn meal	100
		Ground oats	100
		Middlings	100
		Fish scrap	50
		Beef scrap	50

Salt adds to the palatability of the fowl's diet the same as it does for human food, also it assists digestion; but it must be used with more or less caution. Too much salt will result in diarrhea, while excessive quantities will prove fatal in very short order. Authorities recommend the use of four ounces of salt to each hundred mature fowls, and none to stock under ten weeks of age.

Ground salt, or the kind used in the preparation of human food, is best. It should be well distributed through the dry mash. If wet mashes are used, the salt should be dissolved in the water used to moisten the mash. Avoid rock or coarsely granulated salt, because the fowls are likely to mistake it for grit, in which event they will eat excessive quantities, and perhaps die. In most cases where fowls have died from salt it was attributed to this mistake. The writer was once called in by a neighbor who lost several hens suddenly, and on examination it was found that an ice-cream tub had been dumped near the kitchen door, and the hens had devoured most of the rock salt used in connection with the ice around the cream container.

Condiments, such as Cayenne pepper, should be used very sparingly, if used at all. By some they are severely condemned; others claim increased production with no ill effects. Opinion is divided on the subject, though one's common sense dictates that if certain elements are highly stimulating, there must follow a corresponding reaction.

Condiments should be given in moderation, for the purpose of seasoning the food about as would be desired by the human taste. This adds palatability to the ration and increases the fowls' appetites. With increased appetite more food is consumed, which is the sort of stimulation to be sought, not the direct action of the condiments themselves. Ground mustard or mustard bran and some of the reliable poultry regulators are to be recommended.

Charcoal is a very necessary part of the bill-of-fare. It is not a condiment, and strictly speaking, it is not a food. It is given for its effect as an absorbent and intestinal corrective. It has a marked influence on gases, impurities and acids, relieves sour stomach and indigestion, and generally sweetens the whole of the *inner workings*. It is thought to be a laxative when eaten in large quantities, and for this reason many breeders are opposed to giving charcoal to little chicks. The writer has never experienced such trouble. Some fowls do not take kindly to charcoal. The best plan is to keep it before the birds at all times in a separate hopper, the same as grit and oyster shells, or to mix it with the dry mash. Granulated charcoal is a commercial article, usually sold in three grades, coarse, medium and fine.

Shredded alfalfa, or mealed alfalfa, which is finely ground alfalfa hay, makes an excellent mash constituent, providing the quality is good. It is high in protein, ash and fat, resembling wheat bran in this respect. Moreover, it contains a great deal

(Courtesy Purdue Experiment Station)

Fig. 142.—Complete equipment for chicks on range: colony house, feed hopper and water-barrel.

of fiber, like bran, which is the worst feature about it as a poultry food.

If manufacturers would only use the leaves of the alfalfa in the preparation of mealed alfalfa for poultry, instead of including the stalks, which have very little feeding value for fowls, this product would find much greater favor. Another objection is found in the susceptibility to adulteration. Timothy and other grasses are sometimes used as fillers.

It is not advisable to use more than fifteen per cent of alfalfa in the mash, and then only as a substitute for bran or ground oats. If bran, ground oats and alfalfa are to be included, their combined total should not exceed twenty-five per cent of the ration.

Short-cut clover and clover meal are very similar to the alfalfa products, and are used in about the same way. If anything, the clover is richer in ash, protein and carbohydrates, with a trifle less fiber; but it is not so attractive looking as alfalfa in a dry state. Alfalfa is a bright green and has the true alfalfa smell, whereas the clover is brown and not so volatile.

Cut clover or finely shredded alfalfa soaked in boiling water or steamed over night, and fed separately or with a little mash, makes a good substitute for green food in the winter months. They could hardly be termed succulent food, however, any more than they contain considerable moisture.

Dried beet pulp, shredded, a by-product of the sugar-beet industry, and dried green cracked peas are two other substitutes for green food. Either one may be fed to advantage, and a combination of the two makes an ideal ration. It is appetizing in appearance, highly nutritious, a bulky food, comparatively inexpensive, and one that has given satisfactory results for a number of years. Most feed dealers carry these products in stock, or they will order them for a customer. The best pulp is used extensively as a dairy food, and is rich in carbohydrates. It contains considerable fiber.

To prepare this mixture, put equal parts of the pulp and peas into a tub or other receptacle, and add about three times their volume of water. Allow them to soak over night, and they will absorb all the water and swell to three or four times their original bulk. Before feeding next morning, loosen the mass from its compact state by turning it over a couple of times with a fork, and then place it before the fowls in troughs or hoppers.

If the fowls do not eat freely of the mixture at first, it is because they are not accustomed to it. To overcome this hesitancy mix some middlings or the regular dry mash with the pulp. This will give it a more familiar smell and appearance, and in a short time the birds will take to it readily.

Importance of Succulence.—Green foods are likely to be regarded more as an accessory to the fowl's bill-of-fare than an essential requirement. This notion is a mistake. Greenstuffs, sometimes spoken of as succulents, are just as important as grains, meals and animal foods—not so much from the standpoint of nutrition, because in this respect they are far below grain,

(Courtesy Million Egg Farm)

Fig. 143.—Gathering greenstuffs for the fowls.

rather because of their medicinal and hygienic value. Greens act as a tonic. They tone the system; counteract the evils of the more concentrated foods, and assist in their assimilation. Besides, they do furnish some nutrients, especially the mineral elements; also they contribute water.

The human taste craves lettuce, celery, water-cress and other crisp greens, not so much for their food value, but because of their beneficial effect on the system. If we tried to sustain ourselves on lettuce and celery alone, we would fail, because they do not contain sufficient nourishment. On the contrary, if we tried to sustain ourselves without greens of any kind, we would fail also, because the more concentrated foods would soon reduce our systems to a break-down. The same condition applies to poultry.

TABLE XX.—COMPOSITION OF GREEN FOODS

FEED	WATER	ASH	PROTEIN	FIBER	CARBO-HY-DRATES	FAT
Grasses	Per cent	Per cent	Per cent	Per cent	Per cent	Per cent
Alfalfa (green)...........	80.00	1.80	4.90	4.70	7.90	.07
Alfalfa (dry).............	11.90	7.13	14.12	27.09	37.34	2.42
Clover (green)..........	70.80	2.10	4.40	8.10	13.50	1.10
Clover (dry)............	10.00	8.10	16.32	17.84	45.99	1.75
Lawn clippings (green)...	76.40	2.40	2.30	4.10	13.80	1.00
Lawn clippings (dry)	15.30	5.50	7.40	27.20	42.10	2.50
Barley (green)..........	76.00	7.30	2.71	6.90	7.00	.09
Peas and oats (green)....	80.50	1.74	2.90	6.00	8.80	.06
Roots						
Potatoes (white)........	78.9	1.0	2.1	.6	17.3	0.1
Potatoes (sweet)........	71.1	1.0	1.6	1.3	24.6	.4
Beets (mangel)..........	90.9	1.1	1.4	.9	5.5	.2
Beets (sugar)...........	86.4	.9	1.8	.9	9.8	.1
Beet pulp (dry).........	8.0	5.4	9.5	15.4	61.3	.4
Beet pulp (wet).........	89.8	.6	.9	2.4	6.3	..
Onions.................	87.6	.6	1.4	.7	9.4	.3
Turnips................	90.5	.8	1.1	1.2	6.2	.2
Carrots................	88.6	1.0	1.1	1.3	7.6	.4
Artichokes.............	80.0	1.0	2.5	.8	15.5	.2
Leaves						
Cabbage................	90.5	1.4	3.8	1.5	2.4	.4
Lettuce................	95.9	.8	1.6	.5	1.0	.2
Beet tops..............	88.0	2.4	4.4	2.2	2.6	2.2
Rape..................	89.2	2.0	3.4	2.6	2.3	.5
Onion tops............	91.0	1.1	3.9	3.0	.8	.2
Chard (Swiss)..........	87.8	2.4	4.4	2.9	2.5	.4

Fowls are vegetable eaters, and they are grazers, though not to the same extent as geese and ducks. They like to forage, which affords much needed exercise as well as tidbits of greens. On the majority of egg farms, and the same thing is true of most of the backyard flocks, there is not sufficient range to maintain

15

abundant pasture. It takes a large acreage of land to support any considerable number of birds on the natural vegetation, because the fowls soon destroy it. Result, bare yards and runs, in which event it becomes necessary to supply the greens in a more or less artificial manner.

When the poultry raiser's yards become bare, and this erstwhile plant food is not supplemented from the outside, his venture is in a fair way to quit the business. He may not be aware of this fact, but it is true nevertheless. Fowls will not thrive pro-

(Courtesy Million Egg Farm)

Fig. 144.—Relation of ordinary feed oats to sprouted oats. The pile on the left was sprouted from the same quantity as shown in the right-hand pile.

fitably without greenstuffs of some kind. Their eggs are likely to be weak and to lack fertility. Weak eggs mean poor hatches and chicks with weakened constitutions, than which nothing is more difficult to combat in the brooder. It is unreasonable to expect strong, productive pullets from chicks with impaired stamina, hence in a few generations the vigor and productiveness of the flock is reduced to an unprofitable level. The next step is failure.

Failure to appreciate the importance of green food is not so much ignorance as indifference. Most poultry keepers are aware of the benefits of succulence, but because it is the most troublesome part of the ration to supply, particularly in winter, there is a tendency to side-step it—to take chances on getting by without it. As cold weather approaches the fowls receive less and less green food, finally none at all, until the advent of spring, when the problem is temporarily solved by the sprouting of a new crop of grass and weeds.

There can be no doubt as to the trouble involved in supplying greenstuffs, especially for large flocks. You can buy grains and meals and most of the other supplies in sacks, and there is very little care required to store them. But not so with greenstuffs. Plant food involves farming; the crops must be sowed, cultivated and harvested, and then carefully stored for the winter so that they will not freeze. The alternatives are to sprout oats in rotation, or to purchase plant food in the form of vegetables, such as cabbage, turnips and small potatoes, from nearby growers.

TABLE XXI.—FEEDING RATIONS FOR LAYING HENS RECOMMENDED BY IN-
DIANA EXPERIMENT STATION

SCRATCH GRAINS	Pounds	DRY MASH	Pounds
Cracked corn	10	Wheat bran	5
Wheat	10	Middlings	5
Oats	5	Meat scrap	3.5

A light feeding of the scratch grains is given in the morning, and all the birds will clean up at night. Dry mash is kept before the light breeds all the time, and for the heavier breeds from noon on. Succulent food is supplied in the form of sprouted oats, cabbage or mangels. The following variations are suggested: (1) Replace the beef scrap with 62 pounds of skim milk; (2) drop the wheat and increase the corn and oats; (3) give fowls abundant range and cut down a portion of the mash.

The poultryman has a wide list of plants from which to derive greenstuffs, and a great number of ways of placing them before the fowls. Practically speaking, almost anything that has tender, juicy foliage, fruit or roots is suitable, from the grasses to beets—*mangel wurtzels*. The point to bear in mind is, that the product must be appetizing and palatable, or made so.

Clover and alfalfa pasturage are considered to be the finest sources of greenstuffs. Rye is another excellent crop, especially for a fall planting, because it will germinate in cold weather, following which the sprouts are available in the early spring, weeks ahead of any other crop, except wheat, which can be used for the same purpose.

Oats and peas sown together, with a thin sprinkling of clover and rape seed, make a good planting for poultry pasture. The oats and peas furnish the first growth of greens, and serve to protect the more delicate shoots of the clover and rape. The latter will grow late into the fall, and so long as the crown of the rape plants is not destroyed, they will continue to bear foliage in abundance, which is much relished by fowls of all ages.

(*Courtesy U. S. Dep't Agriculture*)
Fig. 145.—Commercial fattening plant, well lighted and ventilated.

If fowls are not accustomed to unlimited quantities of green food, it must be fed with more or less caution at first, especially rye pasture, or the birds are likely to develop a mild form of diarrhea.

Turning fowls out on pasture is the easiest way to furnish greens, but unless the pasture is given an opportunity to obtain a strong growth, the birds will quickly destroy it. To overcome this, the practice is to have two or more yards, and to use them successively. Or the greenstuffs can be cut with a scythe or mower, raked up and fed to the birds inside their houses. If this method is followed, the greens must be fed in small quantities, once or twice a day, to keep them fresh. As soon as they wilt they are not eaten with any degree of relish.

The best grasses to sow for permanent poultry pasture are blue grass, red top, low, Dutch or white clover. Other seeds, such as red clover, timothy, alsike and alfalfa, make very good hay, but they are not durable enough for poultry. They are soon killed off.

Swiss chard and lettuce are useful for feeding baby chicks in the early spring. They should be cut into short lengths and fed in small quantities to keep them fresh.

Onion tops and sliced onions are both excellent for chicks. If you have a brood which is inclined to mope around, out of sorts, so to speak, and you want to put a little "pep" into the chicks, try a few onions sliced fine. In short order the chicks will be tussling and tugging at the slices of onion as though they were bugs or worms.

Onions are very good for mature stock, too, except that when fed in large quantities to laying hens they are apt to impart the flavor of the onion to the eggs.

Beets—mangel wurtzels—is the best all-round vegetable for poultry. They are easy to grow, and keep well for winter feeding. From twenty to twenty-five pounds per day per hundred hens is about the correct ration. Shredding the beets by means of a root cutter is the best way to feed them; or they may be cut into large pieces and spiked on nails in the poultry houses. Suspending the halved beets in a fish net is another way to place this sort of food before fowls. The idea is to keep the beets from being tracked around in the dirt and litter. In cold weather the middle of the day is the only time to feed succulent food, so that it will not freeze and become unpalatable.

TABLE XXII.—WAR-TIME RATIONS FOR LAYING HENS RECOMMENDED BY THE AMERICAN EGG LAYING CONTEST

SCRATCH GRAINS	Pounds	DRY MASH	Pounds
Cracked corn	400	Wheat bran	150
		Middlings	150
		Beef scrap	100
		Charcoal	4
		Fine salt	3

Cabbage is relished by fowls, though it should be fed in moderation, lest it impart an objectionable flavor to the eggs. It is somewhat difficult to store in winter, and not always available at an economical price. If the cabbage is grown at home, a good plan for its storage is to take up the heads with the roots, then turn a fairly deep furrow, place the cabbages head downward in

(Courtesy Cornell Experiment Station)

Fig. 146.—Home-made rack for sprouting oats. Note the seven trays, one for each day in the week, grown in rotation.

this furrow, cover them over with soil, allowing the roots to stick above ground, whence they can be taken up as desired.

So much of the cabbage grown for market is not saleable, due to small, misshapen and loosely headed specimens, that it is possible for the poultry keeper to bargain with farmers for this condemned cabbage at a low price. The writer has bought large quantities of cabbage for fifty cents a load, and, of course, I furnished the labor of picking and hauling. In this way cabbage is a very cheap food.

Undersized potatoes and similar vegetables are often procurable at low prices, and make good succulent food for winter use. Hard vegetables, like potatoes and turnips, should be boiled first, or they will not be eaten in large quantities. Cook them in their skins, and when soft crush them slightly and mix with a little dry mash. Fed warm, this is a splendid food for cold mornings.

Waste fruits, pumpkins, melons and other garden products are available for poultry. Fowls eat them with the greatest avidity. The main consideration is to try and feed them in regular quantities, rather than in large doses some days and none at other times. And always feed them in troughs or hoppers to keep them as clean as possible.

Sprouted oats make a convenient form of raising greenstuff in limited quarters, or during the winter months when plants cannot be grown outdoors. Chickens eat them greedily. They are commonly used and possess unquestioned merits. Some writers refer to them as a cheap food, and they have been widely advertised as the secret of feed at fifteen cents per bushel. This idea is a delusion. Oats are oats, whether you feed them as dry hard grains or in the form of sprouts. Sprouting changes the form of the feed and increases the bulk by means of water, but it does not add nutriment. According to some authorities, sprouted oats contain about 76 per cent water, 3.2 per cent protein, 0.8 per cent ash, 16.3 per cent carbohydrates, 1.3 per cent fat and 2.5 per cent fiber.

Warmth and moisture are the essential conditions for sprouting oats. Use a good grade of heavy feed oats, natural oats, not bleached or clipped oats. Soak them in a pail of warm water for about forty-eight hours, and then spread them out on trays to a thickness of about an inch. At this time the excess water is drained off, but each day the oats must be sprinkled and kept as moist as possible without having the oats actually immersed in water.

A good plan is to cover the oats with pieces of burlap, something to act as a blanket and conserve the moisture. For the first couple of days, or until there is danger of breaking the roots and sprouts, the oats are stirred about on the trays. In about a week they are ready for feeding; the sprouts will be two or three inches long, and the roots will be so closely interwoven that the whole tray of oats can be picked up in a mass or cake.

Mould.—The one difficulty in sprouting oats is mould. To prevent it the oats should be treated with a solution of formalin. In the pail of water in which the oats are first soaked, add about ten drops of formalin. All trays and other fixtures should be sprayed with a solution of formalin at regular intervals.

There are numerous types of oat-sprouting cabinets on the market, heated by kerosene lamps, which are convenient of operation. Similar appliances are easily contrived by anyone who is mechanically inclined.

Mineral Matter.—As will be noted from the analyses of different grains and feeds, practically all foodstuffs contain a certain percentage of mineral substances or ash. Mineral matter may be said to determine the efficiency of a ration, in that it increases the digestibility of all the other nutrients, especially protein. Besides, it is required for the upbuilding of bone and in the formation of egg shells, not to mention many other uses throughout the body.

Mineral matter should be supplied in two forms: that which is quickly available as such; and in a hard form, not so easily assimilable, but for use in the gizzard for the purpose of crushing and grinding the other feeds. A diet consisting of a well-balanced mash, scratch feed and an abundance of green food will supply the first-named sort of mineral matter, especially if the mash contains either granulated bone, bone meal or meat scrap. The second kind of mineral matter, the hard variety, is supplied

(Courtesy U. S. Dep't Agriculture)

Fig. 147.— Killing room in a western packing house.

233

by grit and oyster shells.   No pen of poultry should be without these two products.

Oyster shells, also other kinds of sea shells, crushed to small particles, are rich in carbonate of lime, so necessary for the formation of egg shells.   They are quickly dissolved by the action of the gizzard, hence they will not answer the purpose of grit.

Grit, to aid the gizzard in grinding solids, corresponds to teeth in other animals, consequently it must be hard, sharp and angular, and preferably bright, so that fowls will be attracted to it. The commercial forms of poultry grit are usually made from crushed quartz, granite, phosphate rock and feldspar.   They come in various sizes for poultry of all ages.

TABLE XXIII.—COMPOSITION OF OYSTER SHELLS

	PER CENT
Carbonate of lime	93.71
Carbonate of magnesia	1.39
Phosphate of lime	0.76
Organic matter	4.24

Grit contains very little lime, and very little of the other mineral substances in soluble form.   What nutrients it may supply are but incidental; its chief function is a mechanical one—grinding.

Grit and oyster shells should be kept before fowls in separate hoppers at all times.   It is surprising the quantities they will consume.   If fowls have been deprived of grit for a considerable period, they will eat it as ravenously as though it were grain. Lack of grit, especially among flocks kept away from the soil, has had much to do with the failures in the poultry industry. Apparently, the soil furnishes some material which the system of the fowl craves, and which is deficient in most of our foods. Witness how hens revel in scratching in free earth.

CHAPTER XVII

ARTIFICIAL INCUBATION

Simplicity of Operation.—To-day few people question the value and convenience of the incubator, though there are some who suspect that hatching machines might be difficult to operate, or that they require special training and considerable experience before good results are to be expected. This is a mistaken idea. Incubators are not difficult to run, and previous experience is altogether unnecessary. Naturally, however, the more practice one has with a machine, whether it be an incubator, gas engine, cream separator or other implement, the more proficiency one develops. On the other hand, there is a kind of beginner's luck in the matter of these things. The writer has seen some unusually large hatches brought off by novices.

Regularity and Care.—Perhaps it is a discredit to call this success *luck*. Let us call it the *beginner's watchfulness*, or *beginner's faithfulness* in observing the necessary rules, for such it really is. Later, when we become fully experienced, we are apt to develop over-confidence, or a sort of carelessness, which may cause difficulties. Regularity and carefulness, both in the management of the heating device and in the handling of the eggs, go far toward insuring success. Reliable incubators are made almost automatic and fool-proof these days. They are so simple that children can operate them. Therefore when poor hatches result, it is well to look for other causes beside the machine.

Condition of the Eggs.—No incubator can vitalize eggs with weak germs, or overcome conditions which may have had a deleterious effect on the eggs before they were even taken to the incubator cellar. The first and one of the most important steps in the process of either natural or artificial incubation begins in

235

.

the condition of the eggs. See chapter on selecting eggs for incubation. If the eggs are in every way strong and hatchable, the incubator will hatch them almost as well as the hen, and a great deal more economically, because the machine can handle such large quantities.

The prospective purchaser of an incubator is sometimes per-

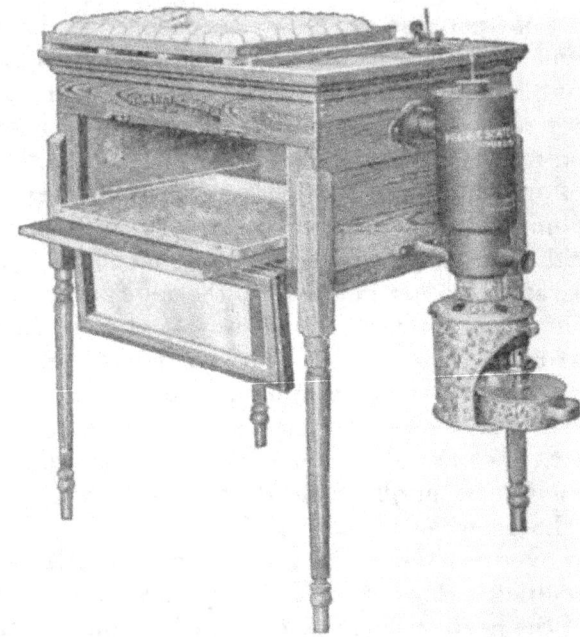

Fig. 148.—Small single tray incubator suitable for the farmer and backyard enterprise.

plexed as to which is the best type of machine for his purpose. There are many different makes of incubators on the market, which may be divided into two general classes: hot-air machines and hot-water machines, all of which are constructed along *moisture* and *non-moisture* designs. In principle they are essentially the same. The reason for this is clear: the whole theory of incubation is based upon the fact that, if a fertile egg is kept for a sufficient period of time under certain conditions of heat, ventilation, moisture, and position, it will be transformed into a healthy fowl.

Small incubators are heated by gas or kerosene, though most of them use the latter. Gas burners require less attention and are desirable in every way, but it is unfortunate that gas is not

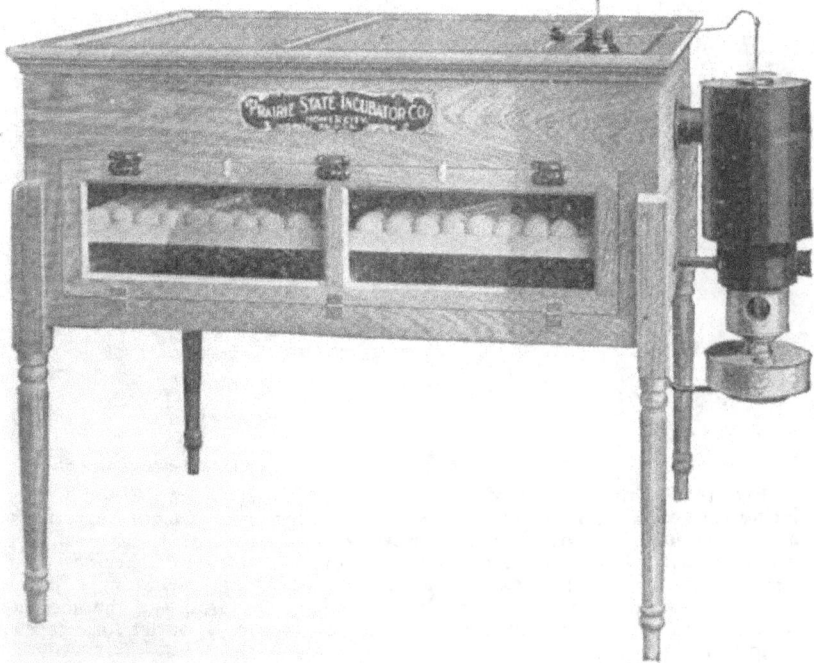

Fig. 149.—Double tray lamp incubator with capacity of from 200 to 400 eggs.

available in all localities. Electricity is also used for heating incubators, and has proved satisfactory, but it requires a totally different method of radiation. The large incubators—those which have a capacity of many thousands of eggs, popularly known as mammoth machines, are mostly heated by a coal-burning stove, though gas is used to some extent.

The hot-air-heated machines are those in which fresh air is taken in at the lamp, heated as it passes around the drum, which corresponds to the chimney of a lamp, and passed through the egg chamber by means of a diaphragm in the ceiling of the machine, and then out through variously arranged outlets. See Fig. 150. In some machines the heated air only passes over the

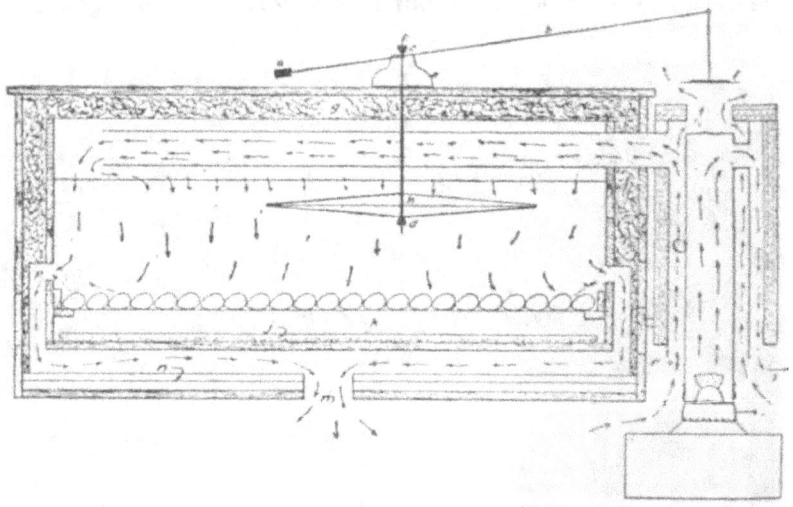

(*California Experiment Station*)

Fig. 150.—Cross-section of a hot-air incubator, showing method of regulating the temperature and ventilation. *a*, Counterpoise weight; *b*, regulator arm; *c*, connecting rod; *d*, thumb-screw; *e*, pivot casting; *f*, heater disc; *g*, cotton batting filling between inside and outside cases; *h*, thermostat; *i*, egg chamber; *j*, moisture pan; *k*, nursery; *m*, bottom ventilator; *n*, insulation in bottom of incubator; *p*, outlet to discharge air from egg chamber into false bottom beneath moisture pan; *r*, fresh air intake; *s*, outlet for escape of lamp fumes.

radiator above the egg trays, and never actually enters the egg chambers.

Hot-water machines are heated by tanks or a system of pipes above the egg trays. In both types the heat is supplied with a regulator which, acting upon a valve or damper, controls the admission of heat to the egg chamber. Such a device is called the thermostat.

Moisture.—Whether moisture should or should not be supplied has never been definitely decided; both principles have their advocates. Some machines are built with pans to hold moist sand or water, others have none. Some machines are built with a solid bottom, the idea being to conserve the moisture within the eggs, and others are built with slatted bottoms, through which there is a constant circulation of air. All types are in general use, and all give equally good results. The question of moisture is one that must be solved by the individual operator. Common sense should tell us, however, to conform in a general way to the instructions of the manufacturer. Incubators which are designed to use moisture pans seldom give best results when operated without the pans. Similarly, the non-moisture machines are seldom improved by the addition of moisture pans.

In the opinion of the writer there can be no set rule advanced on the subject of moisture. That excessive evaporation is bad for a hatch, there can be no doubt. And too much moisture will ruin a hatch, also. It seems to be better to err upon the side of too little than too much. But everything depends upon the incubator, its location, the season of the year, climate and the external atmosphere at the time of the hatch. If the weather is extremely dry and hot, more evaporation will take place than when the weather is cool and damp; this is only natural. An examination of the air cell in the egg by means of a candle will reveal the exact condition of the contents. See chapter on testing hatching eggs.

Temperature.—An absolutely uniform temperature is not at all necessary to success, and in fact, it is seldom obtainable. That statement does not mean the operator should disregard temperature or cease his vigilance. On the contrary he should aim to keep the thermometer as near a certain degree as possible. The point is this, variations of a degree or so either up or down, if corrected within a reasonable length of time, will cause no damage. The operator should aim to maintain a temperature of 102½ degrees for the entire hatch. If the temperature drops to 100 degrees or rises to 104½ degrees, no ill effects will result, providing either extreme is discovered and corrected. These temperatures might be termed the limits of natural incubation. Some operators prefer to start a hatch at 102 degrees, and gradually allow it to creep up to 103½ degrees at the close.

Hen's Temperature.—Personally, I think there is no better authority than an imitation of the hen's ways. If we insert a thermometer under a sitting hen we will find a temperature of 104½ degrees against her bare breast; in her plumage in about the center of the nest the temperature is a trifle more than 103 degrees; while the temperature inside the sitter's wings, a position that corresponds to the outside row of eggs, is about 100 degrees, sometimes a

(*Courtesy Buckeye Incubator Company*)

Fig. 151.—Incubator tray showing the relation of thermometer to the eggs.

little less. Thus, the eggs are exposed to a constant variation in temperature, with an average of 102½ degrees. The hen turns her eggs about eight times daily in order to overcome this variation, but it cannot be doubted that some eggs will remain in the warmest part of the nest for a greater length of time, no matter how faithful and careful the hen may be.

Thermometer.—Notwithstanding the existence of this variation, it should not be construed to mean that the office of the thermometer is unimportant. The thermometer should be absolutely accurate, since it is only by this instrument that the operator can attempt an average temperature. We have shown where a variation of a degree or so will not seriously affect a hatch, if it is shortly corrected, but if the temperature varies a degree or even a half degree for the entire period, it will make a big change in the hatch.

There are two general styles of incubator thermometers, while each style has numerous modifications. One is mounted on a metal frame and placed on the egg tray, with the bulb located at the level of the centers of the eggs; the other is hung directly above the eggs, with the bulb as close to the eggs as is practicable, but not touching them. See Fig. 151. It is a mistake to have the thermometer in contact with the eggs, because the temperature will vary with the vitality of the egg touched by it. Touching good eggs it will be high; next to weak or dead germs it will be low, thus never registering the real temperature of the egg chamber.

Unless there is a very good reason for changing the position of the thermometer, it should be used in strict accordance with the manufacturer's directions. In some egg chambers a difference of an inch in the height of the bulb will make a difference of a degree in the temperature.

Testing Thermometer.—It is a good plan to test the thermometer once a year, even with a new machine. This can be done by comparing it with a clinical thermometer, which has about the same scale and range; place both instruments in warm water, heated to about 103 degrees, which should be stirred, and the bulbs kept in about the same position. Or a thermometer can be certified by an optical goods store or a reliable drug store.

The thermometer is placed near the front of the egg chamber so that it can be easily read without opening the door to the machine. Even so some thermometers are particularly difficult to read, and it is surprising that so little improvement has been made in this respect. An electric flash lamp is very helpful in reading a thermometer. By all means learn to read the thermometer without opening the door; to so do will cause the temperature to fall.

Set Machine Level.—The incubator should be set up level by all means. Heat rises to the highest point, and if the machine is not plumb, one part of the egg chamber will be warmer than the other. Do not guess at the level of the machine, but test it with a spirit level, both ways, for length and breadth, and correct any irregularity with thin blocks of wood under the legs of the machine.

Some manufacturers are careless about the fit of the door, or

16

the tracks for the egg trays. The door should fit snugly, but it must open and close easily, without jarring, which will tend to throw the regulator out of adjustment, and the machine out of plumb. If the doors stick, which they are likely to do with the first hatch, plane them down a bit, or rub the edges with sandpaper. If the trays do not run smoothly, plane them also, and then rub the bearing edges with soapstone, or some other substance that will reduce friction. Do not use oil.

Fig. 152.—9,600-egg double deck coal-burning incubator. Note handles at end of machine which operate the egg-turning mechanism.

In selecting a machine here are a few pointers: The size to buy will depend upon circumstances, of course, but it should be borne in mind that it does not take any more time to care for the heating device on a 400-egg machine than it does for a 60-egg machine. And in most cases the fuel cost per egg is reduced with the increased capacity of the machine. Furthermore, the larger the machine the less likelihood there is for a variation in the temperature.

In regard to price, it is well to consider that the value of the machine is small compared to the value of the eggs placed in it during the lifetime of the incubator; hence it is poor economy to purchase a machine just because it is cheap. Buy one that is reliable. The good hatches that it will produce will soon return its initial cost.

Where to locate the machine?—This question is important. It has much to do with the convenience of running the machine, and when a device is conveniently run, it goes a long way towards successful operation. One needs room to work about an incu-

(Courtesy Niagara Farm)

Fig. 153.—Interior incubator cellar.

bator, and sufficient light to perform the work properly. Then there must be plenty of fresh air, yet without direct drafts, and the place must be clean, free from shocks and vibrations, such as are transmitted by a weak floor and which would disturb the adjustment of the heating apparatus, and the location should be permanent and easy of access for the attendant.

Cellar is Best.—Incubators are run in a great variety of places and under the most varying conditions; I have seen them operated in the barn, attic, loft, cellar, shed, spring-house, parlor and in a tent. Be that as it may, authorities agree that the best

Fig. 154.—Erection of a mammoth incubator. Upper picture shows the sections and parts ready to be assembled. Middle picture shows the heater set up and the foundations temporarily braced to receive the body of the machine. Lower picture shows the first section in place.

Fig. 155.—Erection of a mammoth incubator, continued from opposite page. Upper picture shows the second section in place. Lower picture illustrates the completed machine, with temporary bracing removed.

location is where the atmosphere is more or less moist, where the temperature is practically uniform, and cool, and where at the same time it is possible to obtain ventilation without draft. The air in the incubator apartment must be sweet and fresh or the eggs will not receive the oxygen which they absolutely require at all times.

These conditions are seldom found in buildings above ground, especially in frame structures, consequently it has become customary to run incubators in cellars and basements. In fact, there is a generally accepted view that they must be run in cellars for best results.

If only a few machines are in use, they may be operated in the cellar of the poultryman's house, which is perfectly practicable, except that it is advisable to partition off a tight room for the incubators, so that the heat from a furnace or other objectionable influence will not affect the hatches. Where the incubating equipment is extensive, particularly if mammoth machines are contemplated, it is better to have the apparatus housed in a cellar specially built for the purpose. See Fig. 155. Common sense teaches us that if equipment of any kind is not afforded suitable quarters, we cannot hope for its fullest efficiency.

For convenience in shipping and to save freight incubators are packed and crated within the smallest compass. With each shipment the manufacturer sends out a book or card of directions, telling how to unpack and set up the machine, and how to operate it. These instructions should be carefully followed. Read them carefully before attempting to set up the machine, count and inspect the parts, and then put them together in a workmanlike manner. See Fig. 154.

In the case of a second-hand machine, it should be thoroughly cleaned and disinfected, and the heating apparatus carefully examined—taken down and re-assembled, if necessary.

Care of Lamp.—In oil-burning machines be particularly careful of the lamp. Always start with a new wick, which should be slightly trimmed at the corners, so as to produce a nicely rounded flame. Avoid the fish-tailed flame, for it is apt to smoke and start trouble. Fill the bowl of the lamp with a good grade of oil, not too full, because as the oil warms it will expand and may overflow. This is not dangerous, of course, any more than it is disagreeable to have oil spilled on things. It may cause smoke, or the oil may find its way to the eggs by means of the operator's hands, which will endanger the hatch. For this reason it is a good plan to turn and handle the eggs before caring for the lamp.

Regulating Flame. — If the machine is of the hot-water type, then, of course, the tank must be filled with water before the lamp is lighted. After lighting the lamp and placing it firmly in position, which should be accomplished without jolting the machine, regulate the burner so that the flame is about half as high as it should be finally. The reason for this is, as soon as the chimney or heater warms, it will draw the flame upward and increase its height. It is, therefore, a safe practice to return about a half-hour after the lamp is lighted, to see that the flame is all right.

Skilled operators visit their incubators at frequent intervals, including an inspection the last

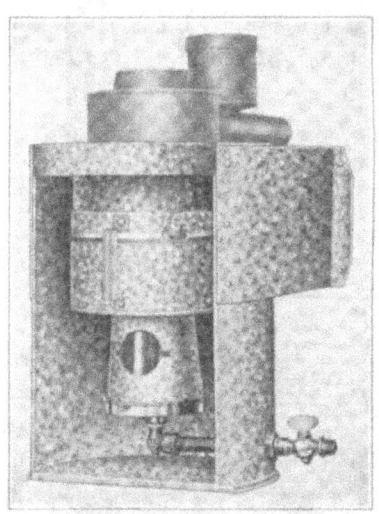

(*Courtesy Buckeye Incubator Company*)

Fig. 156. — Well-built incubator heater protected by metal casing. Heaters of this kind meet the approval of fire underwriters.

thing at night. The machines seldom need attention, but it is gratifying to assure oneself that everything is running properly. In hot weather, when the temperature of the cellar is likely to rise in the middle of the day, especially toward the close of a hatch, when the eggs themselves radiate considerable heat, it is often necessary to lower the burner of an oil-burning machine almost to the point of extinguishing it. In some cases I have found it expedient to turn the lamp out completely, sometimes for several hours, re-lighting it again toward evening.

Always test an incubator before filling it with eggs. This advice applies to both old and new machines. A few hours spent in adjustment will be time and money saved later on, perhaps. Just because a machine worked satisfactorily the last time it was used the preceding season, is no guarantee that it has remained in perfect adjustment. During the six months or more in which it was idle a great many things could happen to it. Dust or dirt may have clogged some of the parts, dampness may have caused swelling or warping, a child, stray fowl, cat, rat or mouse may have interfered with the regulator or some other vital part.

Heat Control.—It is impossible, of course, to attempt to explain the individual peculiarities of the heating devices of different makes of machines. The card or book of directions which accompanies the incubator should be followed for

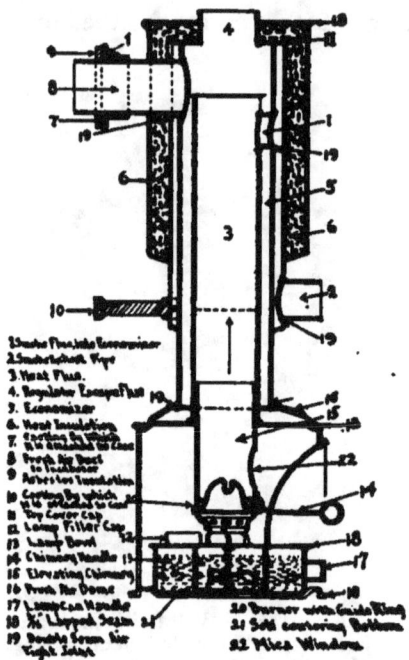

(Courtesy Prairie State Incubator Company)

Fig. 157.—Diagram of an incubator heater and lamp.

this advice. The general principle, however, is the same in all machines. Inside the egg chamber there is a thermostat, an instrument made of a combination of metals, such as steel, zinc and aluminum, which contract and expand, and thereby operate a damper. Wafer or disc thermostats, which contain some fluid used for expanding the disc, are also used. The point is this, they are connected by a thin rod to a bar or regulating arm, which is nicely balanced by means of a counterpoise weight. On the other end of this arm there is a damper, suspended over the top of the heater, which opens and closes, thereby permitting the heat to enter the machine or to escape around the damper. Some thermostats regulate the size of the flame as well as the position of the damper.

Heater without chimney

The position of the counterpoise weight is usually located at the factory, and seldom needs attention; but it should be so placed in point of balance that it will develop sufficient leverage to raise the damper which is hung at the other end, as soon as there is the slightest pull on the connecting rod which leads to the thermostat. In other words, the damper or tin wafer over the heater should be a trifle heavier than the counterpoise weight, consequently lie flat over the vent in the heater when the machine is cold.

We are cautioned not to tamper with the heat-regulating appliance, which is good advice in a way, except that intelligent tampering, which amounts to adjusting, is sometimes necessary. Occasionally the regulator arm is

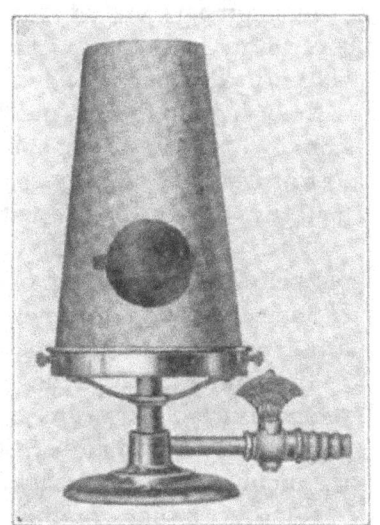

(*Courtesy Buckeye Incubator Company*)
Fig. 158.—Complete gas heater for incubator or brooder.

found bent out of line, if it is metal, or perhaps warped, if it is wood, in which case the alignment will have to be corrected, or the damper will not fall directly over the vent in the top of the heater.   Then again, the tin wafer which acts as the damper may be bent or twisted in some trifling manner, or thrown out of poise, which will not allow it to settle evenly over the vent.   Note these things carefully, and your common sense will tell you if they seem to function correctly.

Adjusting Regulator.—As soon as the thermometer registers 102 degrees, the connecting rod between the regulator arm and the thermostat should be adjusted by means of a thumb-screw so that sufficient tension is placed on the regulator to raise the damper about an eighth of an inch, perhaps a sixteenth of an inch, over the vent in the heater.   Then, if the heat increases and the thermostat expands, this expansion will transfer greater tension to the connecting rod, which in turn lifts the damper and permits the excess heat to escape.

The temperature of 102 degrees should be maintained for several hours, preferably twenty-four hours, before the eggs are placed in the machine, and this temperature should be attained without having to turn the flame up so high as to be in danger of smoking.   Once the machine has been adjusted to "blow off," so to speak, at 102 degrees, and it has been found to work satisfactorily, do not meddle with it.

When the eggs are first placed in the chamber, though the temperature was correct previously, the heat will fall instantly. In fact, the thermometer will probably fall so low as not to read at all; but do not be alarmed, this is to be expected, and is due to the temperature of the eggs.   It is likely that the eggs had a temperature of 50 degrees, and if there are several hundred of them in the one chamber, it will take twelve hours or more for the incubator to warm them to the correct temperature.

Remember that you have this latitude to depend upon in the operation of a machine: Several hours are required to affect the interior of the egg a single degree.   That is, if you should suddenly find something wrong with the heating apparatus and the thermometer registering 98 degrees or 106 degrees, it does not necessarily follow that the interior of the eggs is that temperature. And the more advanced the hatch, the greater the increase in this reserve so far as a deficiency in heat is concerned, by reason of the heat generated by the embryos. From the fourteenth day this natural heat is quite noticeable.

Cases have come to my attention in which chickens have hatched from eggs left out of the incubator all night, due to forgetfulness. Therefore, if you meet with accidents, do not assume that the hatch is ruined, and destroy the eggs; but correct the error, and then in a day or so test the eggs to see what dam-

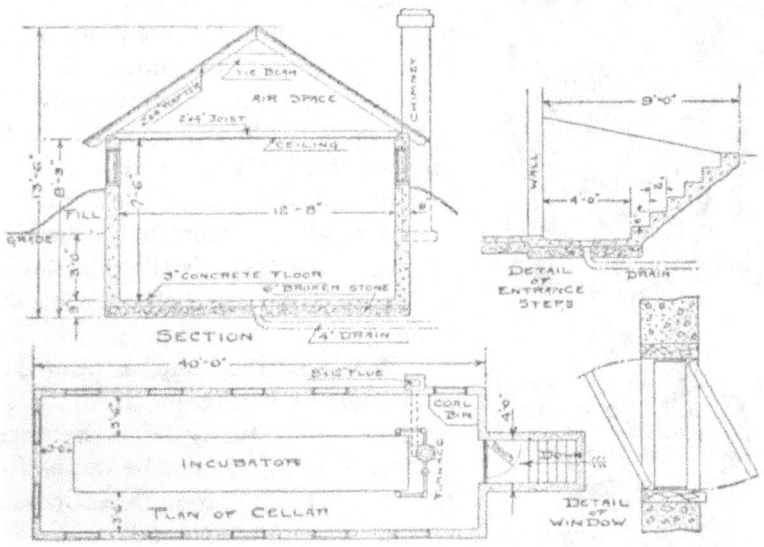

Fig. 159.—Design for concrete incubator cellar.

age, if any, has been done. It is surprising to learn the extent of improper usage and handling to which eggs can be subjected at times, and still develop into perfectly formed chicks. I have thrown presumably unhatchable eggs into a manure pit, and had them hatch chicks.

Loading the Trays.—If a machine can be operated for a couple of days with no appreciable variation in temperature, one can feel reasonably sure there is no fault in its mechanism, and that it is safe to commence hatching. Place the eggs on the trays on their sides, never on end, and never on top of one another. Fill the trays comfortably full, but avoid crowding them. This is most unwise and poor economy; it will make turning very diffi-

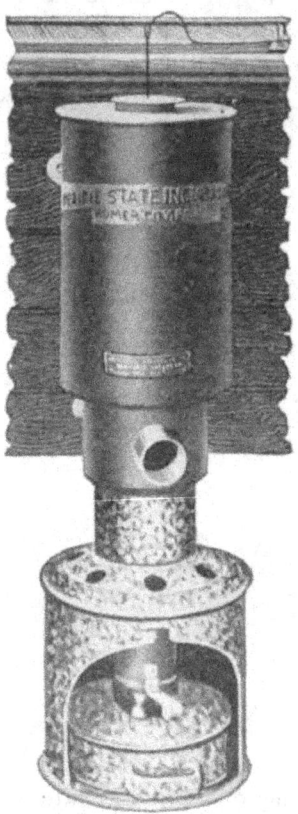

Fig. 160.—Type of oil-burning incubator lamp which is approved by the fire underwriters.

cult, and breakage is likely to occur. A good plan is to allow room for two or three more eggs at each end of the tray.

For the first three days after the eggs are placed in the machine, little attention is required save the daily filling and trimming of the lamp. This, of course, must be done punctiliously. It is advisable to have a special time to do this work, and never deviate from it, any more than you would neglect your breakfast. Morning is the best time, because there is then all day in which to note the behavior of the lamp.

In trimming the lamp, a term that really means caring for it, it is not necessary to actually trim the wick. In fact, this should not be done after the wick is first lighted, unless it be to cut off a pointed thread or corner. Simply rub the charred portion of the wick, using a rag for this purpose, taking care to remove all dirt from the perforated screen through which the air passes to feed the flame, or the lamp will smoke. Try to make a practice of rubbing the wick in the one direction. In this way there will be no trouble in keeping the wick perfectly level.

Turning the Eggs.—On the evening of the third day, and twice daily thereafter, the eggs should be turned. By turning it is not meant that each egg should be turned over an exact 180 degrees, simply that the position of the egg should be sufficiently altered, so that the germ will not gravitate and adhere to the membranous lining of the shell. Some operators turn their eggs three or four times a day. It has never been satisfactorily demonstrated that these extra turnings insure better hatches,

(*Courtesy Buckeye Incubator Company*)

Fig. 161.—Phantom view of mammoth incubator in which the egg trays are arranged in tiers to save space. An even distribution of heat is maintained by means of electric fans inverted in the top of the machine.

though it is reasonable to suppose that they do. The hen is known to turn her eggs eight or ten times a day; but in her case this is necessary to overcome the variation in temperature in the different parts of the nest.

Best results are obtained when the eggs are turned by rolling them with the hands. The trays of the larger sized machines are built with inclined bottoms, divided in the center with a low partition, therefore it is a simple matter to remove a few of the eggs from the middle rows, gently roll the others toward the center, replacing the middle eggs at the ends of the trays. Lay the palms of the hands flat on the eggs and endeavor to move them with a slight rotary motion. Avoid jarring them or handling them roughly, for the delicate membranes and blood-vessels are apt to be ruptured.

When the tray is replaced in the machine, it should be reversed end for end, and if there are two trays to a compartment, which is the customary arrangement in large incubators, the trays should be alternated from side to side. This is done to equalize any irregularity in the temperature of the egg chamber; the eggs are constantly moved to every point in the interior. In performing this work the hands should be clean, and particularly free from any kerosene that might have been acquired in handling the lamp. For this reason, it is well to handle the eggs first, and then trim the lamp.

Cooling the Eggs.—Some operators declare that cooling the eggs is unnecessary. I do not agree with this idea. I believe that cooling is just as necessary as turning the eggs, perhaps more so. I have found it to be a decided help in hatching chicks of strong vitality, whereas lack of cooling often produced weak chicks. The amount of cooling depends almost entirely upon the season of the year and the temperature of the cellar. During the first week of incubation sufficient cooling is obtained in turning the eggs. After the seventh day leave the trays on top of the machine or on tables until the eggs are almost cool, or until an egg when it is placed against the eyelid feels neither warm nor cool. From fifteen minutes to a half-hour is about the correct time.

While the eggs are cooling in this fashion see that they are not in a draft, and always keep the doors to the egg chambers closed. The temperature in the machine should be maintained, though, of course, it will fall as soon as the trays are replaced, by reason of the lower temperature of the eggs. It is customary to cool the eggs at the morning turning only, not in the evening, unless the hatch is pretty well advanced and the weather is exceedingly warm. On the eighteenth day the eggs are turned and cooled for the last time, after which the machine is closed and not opened until the hatch is off.

This cooling process not only exposes the eggs to fresh air, but it causes the contents of the eggs to contract, and thereby draw through the pores of the shell a fresh supply of oxygen, without which the embryo could not thrive.

The above methods are really in imitation of the hen's ways. For the first few days the sitter remains close to her nest, after which she leaves it once or twice a day, sometimes staying away for an hour or more. Cooling can be overdone, like anything else. But in most cases I have found too little cooling rather than too much, to be in practice. After all, there can be no set rules for an operation of this kind. The rules must be flexible, and governed largely by the operator's judgment. For example, in addition to weather conditions and the temperature of the cellar, the amount of cooling should be regulated according to the temperature of the machine for the past twenty-four hours. If the temperature has been low, little or no cooling is advisable; if the temperature has been high, then extra cooling is in order. Moderation—common sense—that is the watchword in running an incubator. Live up to the reason for a rule, rather than to the letter of it.

CHAPTER XVIII

SELECTING HATCHING EGGS

Composition of Eggs.—Chemically, the composition of an egg remains practically constant, notwithstanding serious changes may take place in the hen's bill-of-fare, her environment and general care, which are in turn reflected in her state of health. In other words, we find in this condition, which has been established by careful scientific experiments, one of the highest laws in nature—that the animal will sacrifice its own bodily strength and health in a supreme effort to produce a perfect offspring.

Comparisons have been made between the eggs of fowls in robust health and those in a more or less anemic, sickly condition, and it has been found that they are almost identical in composition. Also, there is little or no difference in the composition of eggs from different breeds. This is an interesting idea, especially so when we consider the controversy that sometimes breaks forth between the advocates of brown-shelled and white-shelled eggs.

This does not mean, however, and should not be so construed, that there is no difference in the flavor of eggs, because there is a vast difference in this quality—a very great difference. Furthermore, there is a great difference in the fertility of eggs and in their hatchability, also in the vitality of the chicks, which are manifestations of the differences between the well-fed, vigorous fowl, and the bird which is impoverished by disease or improperly nourished. In short, though the chemist may not detect any difference in the make-up of the egg, from a reproductive point of view it is there, nevertheless.

The most logical way to account for this is due to the fact that egg making is in reality a double process—it is a reproductive

256

and secretory process combined. It differs from milk-making in this respect, which is purely secretory, consequently egg-making is a lot more exhaustive and vital. The perfect egg not only contains the embryo from which the chick is developed, but it also contains the materials for the embryo to draw upon, a complete shelter for the chick during its development, and last of all it provides the chick with several days' food after it is hatched, and until it is strong enough to learn to eat. We are

Fig. 162.—"Doing the work of four hens with a small incubator."

speaking of the yolk, which is absorbed by the chick before it pips the shell.

Thus, if we consider that the profitable hen is expected to produce at least ten dozen eggs a year, and in some cases specimens have laid three hundred and over, which is many times the weight of the hen, it is easy to understand what a terrific strain this must be on the fowl's constitution, and illustrates as nothing else can, what a really intensive organism we have in the hen.

17

Some idea of the effort and strain may be gleaned from the fact that the temperature of the hen, which under normal conditions burns with much greater vitality than the temperature of man, rises two or three degrees above normal at the time the egg is being laid.

The first step in the important work of incubation lies in the selection of the eggs. Hatching eggs of prime quality must be laid by hens which are intelligently bred, carefully fed and quartered, and from blood lines of known reliability as to stamina and prolificness.

(Courtesy Newtown Giant Incubator Co.)

Fig. 163.—Sectional view of coal-burning incubator stove, with automatic fuel feeder.

Poor hatches are often blamed on the incubator, on the eccentricities of a perverse hen, or the poultry raiser will rail against weather conditions, or what not, as the cause of his failure, when in reality the quality of the eggs is entirely at fault. The first essential being the character of the parent stock, the next step is the selection of eggs suitable for incubation, for all eggs, even though they are laid by strong, healthy birds, are not equally hatchable. A third step might be called the care of the eggs prior to hatching them, since they are extremely susceptible to surrounding conditions.

It is useless to attempt to obtain strong, vigorous, livable chicks, the kind that are calculated to perpetuate one's stock with profitable, productive fowls, from a sickly, poorly fed, listless, degenerate, in-bred, dwarfed and anemic flock of breeders, for the reason that it is impossible to produce any first-class article from inferior materials. This is a natural law, and it is immutable.

Eggs from over-fat breeding stock seldom produce a large percentage of chicks. Hens that are closely confined, without sufficient exercise, and where they have little sunshine and fresh air, are almost certain to lay eggs which are low in fertility, and even lower in vitality. Fowls that are fed on highly concentrated foods and forced to the limit for egg production usually lay eggs which are so nearly devoid of the life-giving principles that they are virtually useless as breeders.

To get fertile eggs, with strong, hatchable germs, plenty of fresh air, sunshine, exercise, green food and a well-balanced rational diet throughout are necessary. For this reason the breeding stock should be kept on free range as much as possible. Abundant range is the greatest panacea in the world for chick and grown fowl alike.

The importance of selecting none but well-formed eggs for hatching purposes cannot be overestimated. Not only is this essential to the actual hatching capacity of the eggs, but it goes a long way toward improvement generally. Eggs which are well shaped and normal are almost certain to produce chickens which will later mature and lay well-shaped eggs, consequently the poultryman is enabled to secure better prices for his products by reason of their uniformity and superior quality.

Select the eggs of a medium size and an average as to color and shape. By that I mean, let the selection be governed by the average product of the hen or breed, rather than by some arbitrary standard. Eggs that would be considered abnormally large on one farm, might be considered medium-sized on another plant, or even small on a third. Discard all eggs which are unusually long, too round, flattened on one side, elliptical, wedge-shaped, and those which have any excrescence or ridge. Monstrosities cannot be expected to hatch.

The ideal hatching egg is the real *egg-shaped* specimen, nicely rounded at one end, with a gradual taper to the other, and having a firm shell of good texture, free from bumps, corrugations and other imperfections. Avoid eggs with very thin shells or exceedingly thick shells, or those with invisible cracks. A good plan is to sound each egg as it is selected, by tapping it gently with a lead-pencil or with one's finger-nail. I have seen eggs with cracked shells hatch, providing the cracks were sealed with adhesive plaster, but as a rule it does not pay to bother with them. Cracks mean excessive evaporation or ruptured blood-vessels.

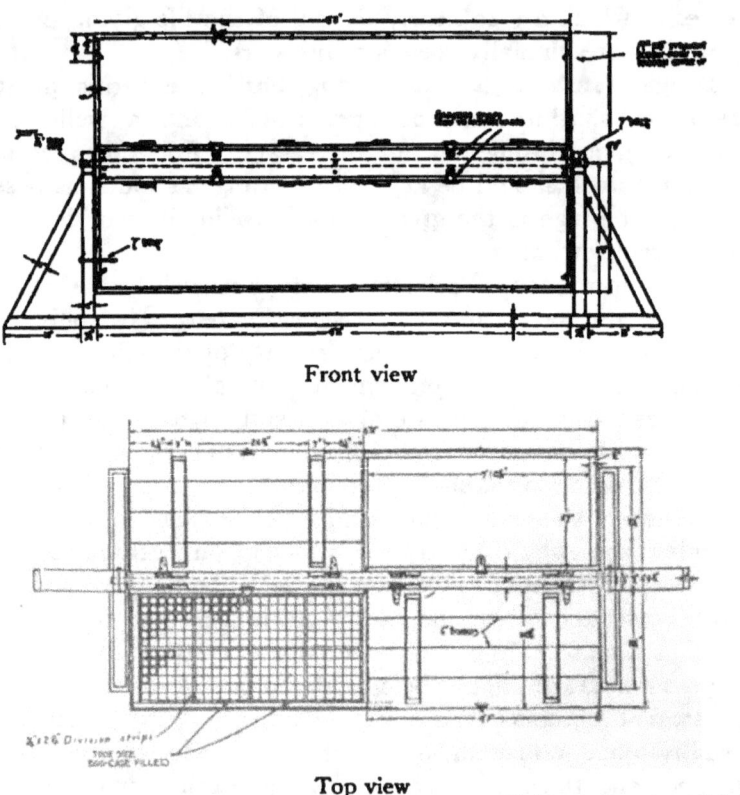

Front view

Top view

(*California Experiment Station*)

Fig. 164.—Plans for a revolving hatching egg cabinet.

There is a commonplace that round eggs will hatch pullets and long eggs cockerels. This is a notion and should not be taken seriously. The shape of an egg is influenced entirely by the contour of the oviduct in which it is cast, and has nothing to do with the sex of its embryo. For further proof of this, we observe that the hen laying a long egg or round egg will continue to lay the form peculiar to her with very little variation, providing she is not frightened or injured in any way.

Freshness.—Common sense teaches us that freshness is a prime necessity in hatching eggs. Successful poultrymen aim to set eggs as soon as possible after they are laid. Not only because eggs a day or two old hatch from 12 to 24 hours earlier than those kept a couple of weeks, but because the longer an egg is kept the more evaporation takes place, which weakens its vitality. If eggs are held in too low a temperature, the chilling is likely to injure them. If they are stored where it is too warm

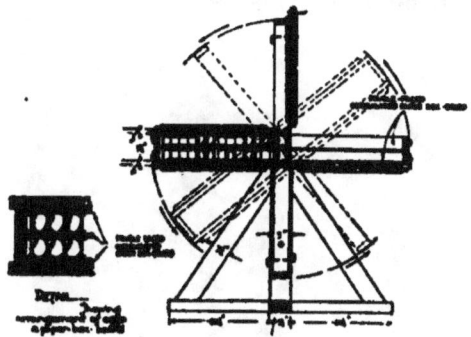

Fig. 165.—End view and section of egg cabinet as shown in Fig. 164.

the development of the germ is apt to start, and later die. A temperature of about 50 degrees F. seems to be best.

The eggs should not be permitted to stand in a direct draft, or exposed to steam, vapor or fumes of a deleterious nature. The shells of eggs are exceedingly porous, therefore they are predisposed to outside influences. It is a bad plan to wash eggs intended for incubation. If they are so dirty as to make this necessary, they had better be discarded altogether. Moisture only helps to convey any soiled matter into the interior of the egg through the pores in the shell.

In cold weather eggs intended for hatching should be gathered several times a day to prevent chilling, and in hot weather they should be collected frequently to avoid heating.   If they are to be kept more than two or three days before being set, it is best to turn them once a day.   There are revolving egg cabinets made for this purpose (see Fig. 164), but equally satisfactory results can be had by packing the eggs in an ordinary egg crate and turning it over gently from day to day.   This is done to prevent the yolk, also the germ, from gravitating to the membranous lining of the shell and adhering to it.

A little attention paid to the foregoing simple directions will work wonders in the possibilities of hatching eggs, and will insure much better results in the brooder.   It is another application of the doctrine of preparedness.

CHAPTER XIX

TESTING EGGS DURING INCUBATION

Inasmuch as all eggs are not fertile, and because all fertile eggs do not contain embryos that develop properly, it has become a rule to test the eggs during the period of incubation, first for fertility, and later for the strength of the embryo. It will be readily understood why this information is desirable: In the first place, by removing the infertile or clear eggs, which may be used for culinary purposes or hard boiled and fed to little chicks, additional room will be created on the egg trays, which means greater ease and convenience in the operation of the incubator. Then again, if the dead germs are removed the egg chamber is more easily kept free from bad odors, and this is a worthy factor.

Since there is no way to determine if an egg is fertile before placing it in the incubator, we must endeavor to gain this information while the eggs are being hatched. Claims have been made that fertility can be told before incubation. These *theories* have no basis in scientific fact. They are usually offered for some pecuniary gain, and should be discarded.

Construction of an Egg.—The yolk of a fresh egg floats in a dense mass of albumen, popularly known as the *white*, which is in the form of layers. These layers are particularly noticeable in the hard-boiled egg. Attached to the yolk are two cords, called the *chalaza*. The office of these cords is to suspend the yolk in the *white* and keep it from injury, and to keep the life germ which is attached to the vitelline membrane of the yolk in a certain position where it will receive the fullest effect of heat during incubation. See chapter on the development of an egg.

Experiment and you will soon see that no matter how you twist or turn an egg, the yolk will always return to a definite position.

263

That is—it will do so providing the egg is fresh.   These cords, or chalaza, lose strength with age, the albumen becomes thin and watery, and the yolk, instead of being supported in the center of the *white*, settles or gravitates to the surface, where it finally adheres to the membranous lining of the shell.   In time the vitelline membrane which surrounds the yolk loses its strength and ruptures, and thereby allows the escape of the yellow substance—called the vitellus—into the albumen.

(*Courtesy Million Egg Farm*)

Fig. 166.—Turning and cooling eggs for hatching.

There are several fallacies in connection with the production of eggs which all breeders should aim to clarify.   One is that a pen of fowls must be headed by a male bird for the production of eggs.   Another idea is that only fertile eggs contain life germs, sometimes called germinal spots.   Still another supposition is that the germ contains an infinitesimally small chick, and that the process of development is simply enlargement.   All of these ideas are logical enough in their conception · nevertheless all are wrong.

The presence of the male is not essential to the production of eggs. Hens will lay just as well without his society as with him. Some breeders claim that their hens do better without males. I do not think there is anything of importance to this idea, except in rare cases where it might be found that a large number of males were annoying the hens. If anything, I am inclined to think that egg production is slightly improved by the presence of a few males. They are very attentive in the matter of locating food, and in escorting the flock afield, which induces exercise and encourages foraging. This value, however, is more

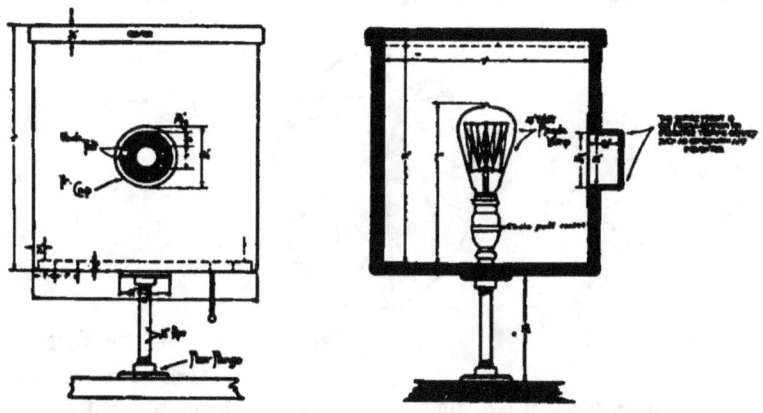

(*California Experiment Station*)

Fig. 167.—Plans for an electric egg-candler or tester.

than offset by the disadvantage in having fertile eggs at times when they are not required for hatching purposes, because they are so easily affected by heat. These days the slogan is—"swat the rooster," and in the long run it is a good rule.

Now for the second fallacy: An egg, whether fertile or not, has a small grayish spot the size of a pin-head on the surface of the yolk, known as the life germ, because it is the vital principle of the egg. If an egg is broken into a saucer, this germ is usually plainly visible to the naked eye; sometimes it is quite conspicuous. Examine it closely; if it has a clear outer rim or circle with little white dots in the center, it is fertile. The infertile germ is whitish in appearance and lacks the clear outer ring.

The activity of this life germ is temporarily suspended as soon as the egg is laid. But as soon as the egg becomes heated to the proper temperature, either by contact with the hen's body or by other means, its development is resumed. It has been found that this germ contains no definite organs, but that its function is to reproduce other cells like itself, each enlarging and reproducing more cells with the same functions, which ultimately establish the form and body of the chick.

Signs of Life.—After about 24 hours' incubation life is per-

| Dead Germ | Fertile Egg | Infertile Egg |

(California Experiment Station)

Fig. 168.—Appearance of hatching eggs through candle on seventh day of incubation.

ceptible, but only if the egg is broken open. Blood-vessels may be seen, and on the third day the heart appears. On the fourth day the eye can be distinguished, and from the eye and heart blood-vessels radiate in all directions, which, to the mind of the candler, resemble a sort of spider. See Fig. 168. This network of blood-vessels continues to grow until it completely surrounds the shell membrane. Its function is to take up oxygen penetrating the shell, and act as a respiratory apparatus until the lungs are completed on or about the nineteenth day.

The lungs start to take form on the fifth day, and on the seventh day the bill is noticeable. The bones are pretty well shaped by the ninth day, later the muscles of the wings are visible, and by the eleventh day the arteries are quite distinct.   Up to the end of the first week the embryo lies very still, after which it shows unmistakable signs of voluntary motion.   The yolk is now perceptibly thinner, for the growing embryo draws heavily upon it

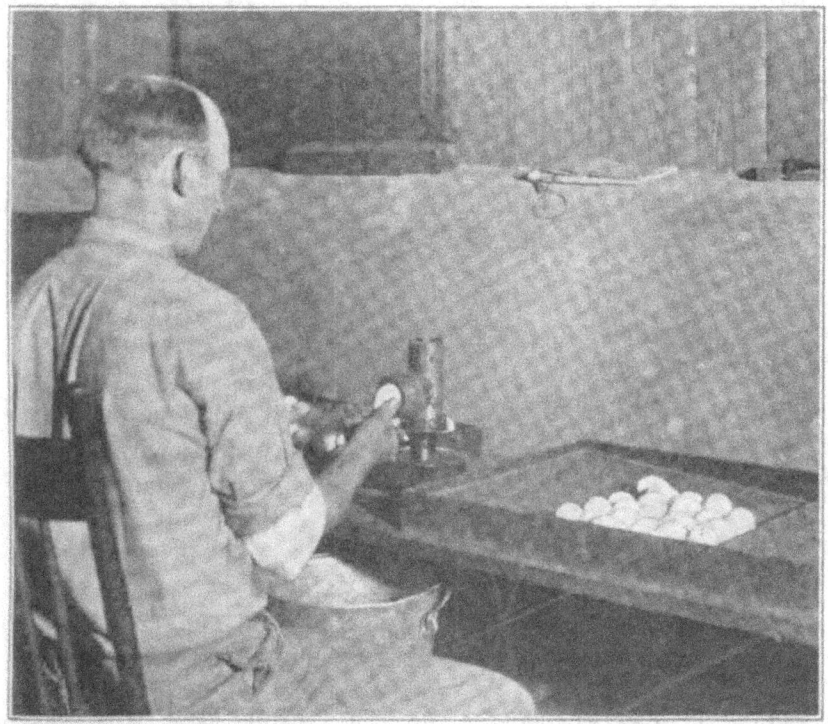

(Courtesy Million Egg Farm)

Fig. 169.—Testing hatching eggs.

for nourishment.   The feathers are well developed by the fourteenth day; on the eighteenth day the first cry is usually heard. On the day following the yolk should be nearly all taken into the body.   The beak of the chick then breaks through the membrane into the air cell, after which it soon pips the shell and extricates itself.

Testing Eggs.—It is customary to test hatching eggs twice, the first time on or about the seventh day, for fertility, in which the clear eggs are removed; and the second time at the end of the second week, though some operators, if they make their first test on the seventh day, which is pretty sure to remove most of the dead germs, prefer to make the last test on the evening of the eighteenth day, when the eggs are turned and cooled for the last time, and the machine is to be closed until the hatch is completed.

The shells of white eggs are so translucent that they can be tested on the third day, which is a big convenience. The shells of brown are so dense and thick that it is difficult to see the contents with any degree of accuracy until the seventh day.

The process of testing eggs is simple enough after a little experience is acquired. It consists mainly in holding the egg between the eye and a strong light, which illuminates the interior of the egg. In fact, it is precisely the same as candling eggs. See Fig. 169.

Fig. 170.—Simple egg candling outfit—ordinary lamp and cardboard box.

Egg Testers.—There are many types of egg testers on the market, and the manufacturers of incubators usually furnish one with each machine. The simplest device is a tube or chimney of tin to fit over an oil lamp, and on one side of the tube there is a small opening, against which the egg is held for an examination. See Fig. 170. Eggs may also be tested by sunlight, using a shutter or curtain with a small hole in it for the light to shine through.

Evening is the best time to make a test by artificial light. Arrange all the details as conveniently as possible before the eggs are removed from the machine, especially for the first test, as the work should be done rapidly to avoid prolonged exposure. Have a low table, large enough to accommodate two egg trays, and locate them on either side of the tester.   Two baskets should be placed near at hand, one for the clear eggs, and the other for the dead germs.   One by one the eggs are held before the spot of light from the tester. If fertile, they are placed on the empty tray, and if infertile or dead, they go into one of the two baskets.

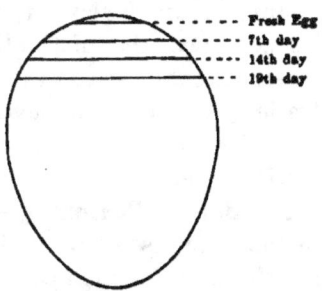

(*California Experiment Station*)

Fig. 171.—Increase in size of air cell, due to evaporation during period of incubation.

Experience soon teaches the most unfamiliar operator how to distinguish between the spoiled egg and the egg which is developing properly. And there is no difficulty at all in detecting the clear egg. If the formation appears as a black, stationary spot, the egg is one that was fertile, but the germ of which is now dead.   Other manifestations of a dead germ are blood rings, which indicate a hemorrhage, or, if the albumen appears cloudy and watery, it is a sign that life has started and then died. See Fig. 168. Some dead germs show only a streak of blood.

Eggs from the Mediterranean breeds, such as the Leghorns, often run so high in fertility that some breeders do not bother to test them for fertility at all, preferring to test them but once— about the fourteenth day—for dead germs. Or the eggs may be given their first test without handling, by passing an electric torch under the eggs as they lie on the wire bottom of the incubator trays, which quickly discloses the clear ones.

CHAPTER XX

DAY-OLD-CHICK INDUSTRY

Unique Development.—The day-old-chick industry probably constitutes the most remarkable development in the history of animal husbandry. Chick producers themselves do not boast about this achievement, and the industry is still too young to have gained the distinction to which it is rightfully entitled. In sharp contrast to other discoveries of its kind, the baby chick trade is one of those evolutions which arrived unheralded and unnoticed, but which instantly secured recognition through sheer merit alone.

Accidental Beginning.—The baby chick business is unique in that the discoverer of it did not know that he had discovered anything. The idea came into being by accident, one might say. And from this accidental beginning it has progressed by leaps and bounds that are almost bewildering even to those who have made a practice of following the enormous strides taken by other branches of the poultry industry in the past twenty years.

Old Idea.—We speak of the baby chick trade as a new idea, whereas it is little more than the adaptation of a very ancient practice. Artificial incubation is an old custom. It was practiced by the Chinese and Egyptians centuries before the Christian Era. Tradition credits the invention to the priests of the ancient Temple of Isis. The Egyptian hatcheries, which were little more than brick ovens heated by wood fires, were public institutions, operated on a toll basis. The farmers brought their eggs to the hatcheries, and later they returned for the chicks. Practically the same idea is carried on in this country, and it is called *custom hatching*.

It remained for an American farmer to expand this custom

270

hatching into the actual sale of baby chicks. The story goes that a poultryman agreed to do some hatching for a neighbor as an accommodation. This neighbor died suddenly while the hatch was in progress, whereupon the poultryman was at a loss to know what to do with the five hundred chicks which he had incubated. He had no facilities for brooding the extra chicks, and he could not afford to kill them. It occurred to him that some of the townspeople might buy them to place under their sitting hens.

Accordingly this poultryman placed some of the chicks in a

(Courtesy Watson Mfg. Company)

Fig. 172.—Series of double-deck mammoths in a large Eastern hatchery.

basket, covered them over carefully to keep them warm, and set out to peddle what he conceived to be a very strange assortment of wares.

Sold Out.—To make a long story short, this poultryman not only sold all of his extra chicks without any trouble, but he discovered that he could have disposed of hundreds more. He found that farmers and backyard poultry keepers were only too eager to avail themselves of the opportunity, and to pay a fair price for the chicks. It was an inspiration to the poultryman. He reasoned that if farmers in his own community wanted to buy chicks already hatched, in preference to going to the trouble of incubating their own eggs, it was likely that people in other communities would want to buy them for the same reasons.

Forthwith this poultryman increased his incubator capacity, and for the remainder of the season he found a ready sale for all the chicks he could hatch. The following year he added more incubators, and to be sure of a sale for his increased output, the poultryman inserted a small advertisement in a local paper. The notice was to the effect that he had little chicks for sale, and that he would deliver or ship them when one day old. Orders came in thick and fast, and in a comparatively short time the season's output was sold or reserved.

The next year this progressive poultryman's incubator capacity was trebled, also he did more advertising. He sold out again. Most of the buyers to whom he sold the first season wanted more and more chicks. They were satisfied customers and told their friends about the scheme. Whereas the first year's chicks were sold to farmers living in nearby sections, now orders began to come in from more distant points and from other states. The poultryman soon found that he had more business than he could possibly handle.

Idea Spreads.—Then it was that others engaged in the enterprise, until the baby chick industry quickly spread throughout the entire country. In less than five years hatcheries were dotted in almost every state from the Atlantic seaboard to the Pacific, and almost all reported a thriving business.

To-day the chick industry has assumed operations on a gigantic scale. It has capital invested in plants and equipment running into millions of dollars. Millions of chicks are produced each year, and the demand is constantly growing. In the opinion of many, the business is still in its infancy.

Capacities of Hatcheries.—Apparently the size of a hatchery has no limitation. There is one in Ohio which has an incubator capacity of over 600,000 eggs—about forty tons of eggs—at each hatching. See Fig. 174. During the spring of 1917 this establishment produced 1,500,000 chicks. Numerous other

Fig. 173.—California hatchery; capacity, 125,000 eggs.

hatcheries ship from a quarter to a half million chicks in a season, while scores of smaller plants turn out numbers varying from twenty thousand to a hundred thousand chicks.

It is doubtful if any enterprise could make the enormous growth of the day-old-chick trade within such a short space of time, unless it possessed some peculiar advantages. That is the secret of the success of the chick trade—it has unique advantages, and natural ones.

(*Courtesy Smith Standard Company*)
Fig. 174.—One of the incubating rooms in a hatchery of 600,000 egg capacity.

The first poultryman stumbled over them, but even he did not see the commercial possibilities at the time. His awakening was not so much in the baby chick, as in the realization of the extent to which farmers and poultry keepers generally were disgusted with their own efforts in trying to hatch eggs. The opportunity to purchase chicks already hatched filled a long-felt want, unquestionably the greatest want in the poultry industry. It ran counter to the familiar proverb—"Never count your chickens before they are hatched," of which everyone who had struggled with the mechanics of an incubator or the eccentricities of a perverse hen, had the fullest appreciation.

Specialization.—What is even of greater importance, the chick trade is an application of the principle of specialization. It is generally admitted that the man who devotes all his time, energy and thought to one thing exclusively is likely to become more skillful in that particular line than the man who must do the same thing and a dozen others besides. The poultryman who produces baby chicks is a specialist. He has trained himself for the work.   He has experimented and put into operation methods and appliances best suited to secure maximum results at the minimum cost.

The beginner with poultry, who is the chick producer's best customer, as a rule, is especially benefitted by the merits of specialization.   While the beginner's attention is occupied with a hundred and one details incident to the commencement of operations, and before he has had time to learn his own lessons in

(*Courtesy Watson Mfg. Company*)

Fig. 175.—10,000 chicks ready for shipment.

the economical operation of an incubator, he has at his command for a very nominal charge the services of a veteran operator.

Still another advantage: If a poultryman meets with a misfortune of some kind, either with sitting hens, artificial incubation or brooding, and there is no time in which to make a fresh start, his project sustains a severe setback, maybe for a year, except for the baby chick operator.

To-day it is comforting to think that if you are caught in a tight place through unforeseen circumstances, you can turn to the baby chick man.   He takes the responsibility of counting your chickens before they are hatched. He assumes the risk of your not having broody hens early enough to produce chicks for autumn layers. He can furnish you with broods without your having hens at all, and in whatever quantity and at whatever time or times you want them. Hence a start can be made on short notice, and the only equipment required is a brooder.

The baby chick man, perhaps, has forgotten more about the

Fig. 176.—Incubator building on a large hatchery in Ohio.

operation of an incubator than the average operator will ever take the time to learn. Therefore its eccentricities, if it has any, do not keep him awake nights. He is operating on a very large scale, consequently he can afford to hire the services of trained assistants. And the scale on which his business is conducted enables him to incubate an egg for maybe one-third the cost others could do it. Briefly, the baby chick man is one of the greatest assets to the poultry industry. He is a sort of fly-wheel, with sufficient impetus to carry other breeders over their fluctuations in power.   As the baseball fan would say—He's a good pinch hitter.

Many were prejudiced against the day-old-chick trade at one time, and a few are still opposed to it.   They feel that baby chicks, above all creatures, seem so tiny and delicate as to require the utmost care and attention for the first hours of their existence, and that to ship them hundreds of miles at the mercy of a cardboard box is little short of barbarous.   Others ask: "Will the little fellows survive the shipment in express cars without being chilled, or without permanent injury to their vitality and productiveness?"

Natural Provision.—Let us consider these fears: Chicks require neither food nor drink for the first couple of days of their life.   Practically the only attention needed consists of rest, warmth and air.   During the period immediately following incubation the chick is sustained by the assimilation of the yolk of the egg.   The general practice is to allow the chicks to remain in the incubator for about twenty-four hours after the hatch is completed, then to place them in the brooder for another day before giving them food.   In fact, some poultrymen do not give their first feeding until after the third day, believing that this much time is required for the proper assimilation of the yolk.

In any event it is this provision of nature which gave rise to the possibility of shipping chicks long distances while they were in this dormant state.   This natural provision was greatly aided by the invention and perfection of special boxes or carriers, which were designed with the idea of conserving the warmth radiated by the chicks themselves.   See Fig. 175.

The principle of this shipping box for chicks is much the same as the fireless cooker or the vacuum bottle, both of which were regarded rather skeptically at first.   The principle is based on a very simple law, that of retaining heat or cold by non-conducting enclosures, and corresponds to the insulation in a refrigerator.

One of the best packages yet devised for the shipment of baby chicks is that made from corrugated fibre-board, than which there is no greater non-conductor of heat or cold, weight, strength and other qualities being considered. Carriers of this description are made expressly for the purpose. They are strong,—capable of withstanding the weight of a man,—durable, easy to assemble and handle, and represent a great saving in transportation charges by reason of their light weight. They are usually made in three sizes, for shipments of 25, 50 or 100 chicks; the small size containing but one compartment, the 50-chick size two compartments, and the large size four compartments.

Early in the season, when the weather is quite cold, more chicks

may be shipped in each box than is stated above, because the chicks huddle closely together and require very little space. Similarly, in very warm weather the number should be reduced to about twenty chicks to a compartment, to prevent overheating.

(Courtesy Smith Standard Company)

Fig. 177.—Hatchery of 600,000 egg capacity; it resembles a warehouse or refrigerating plant.

Ventilation is obtained by cutting small holes in the sides of the box near the lid, and in such a way that drafts are prevented. In the winter very few holes will supply all the air necessary, while as the season advances, depending also upon the climate to which the shipment is made, more holes should be cut.

To prevent the chicks from slipping around over the smooth surface of the bottom of the box, cut hay, straw, alfalfa or other material is placed in the bottom to give them a foothold, and to absorb any manure. If the weather is extremely cold, feathers may be substituted, which will make the little fellows as warm and comfortable as though in a brooder. As a matter of fact, these shipping boxes are precisely the same as fireless brooders, in which the heat from the chicks' bodies is bound to keep them sufficiently warm. The point to remember is to place the correct number of chicks in a given size box. With this done, there is no question as to their security.

Mortality.—A visit to any of the large hatcheries will show that it is now a common thing to ship chicks a thousand miles and have them arrive at their destination in as healthy and active a condition as the day they were removed from the incubator. The average mortality during shipment is two per cent, which is a negligible factor, and would probably exist anyhow, in placing the chicks under the hover in a brooder for the first twenty-four hours. Most hatcheries make a practice of including three or four extra chicks, to allow for this mortality. In addition to this, they guarantee safe arrival and full count, and will make good any losses, providing these are reported at the time the packages are delivered by the carrier's agent.

(Courtesy Million Egg Farm)

Fig. 178.—500 chicks ready for express shipment. Corrugated boxes are slipped into a wooden crate for extra protection.

A poultry plant in Maine shipped a box of fifty chicks to a town in Wyoming, a distance of 2600 miles, and only four of the birds perished. Another shipment was made to New Orleans, in which the chicks were two days and three nights on the road, and they arrived none the worse for their long journey. Express messengers and other railroad employees are now so accustomed to handling baby chick shipments, and they seem to have such a great deal of sympathy for them, that the shipper is usually assured of the very best treatment.

Pet Shop Trade.—For the past couple of years the hatcheries have found a big outlet for chicks in pet shops and 5-and-10-cent stores. Hundreds of thousands have been sold to eager buyers in this way, mostly to women and children who find it difficult to resist the attractiveness of little chicks. Seemingly nothing arouses the human interest so quickly as a flock of chicks, which fact constitutes a potent advantage, in that the chicks advertise themselves. Every child and most women have an impulse to love and fondle a tiny chick, though generally to the discomfort and injury of the latter.

It is doubtful if this phase of the business should be encouraged. The chicks are sold almost exclusively as pets, and the stores handle them largely as a drawing card, often at prices less than they have to pay, simply to attract buyers for other goods—an advertising scheme. Very few of the chicks outlive the pet stage or serve a useful purpose, thus the practice amounts to a waste. If stores insist upon retailing chicks, they should distribute a pamphlet on the care of the birds, or make some attempt to enlighten inexperienced persons not to kill their pets with kindness.

Many of the objections against the baby chick trade were raised by fanciers who claimed that the increased sale of chicks had seriously impaired their business in breeding stock and hatching eggs, for which they were accustomed to receiving good prices. If there is any truth in this belief, which is doubtful, to complain about it is working on the wrong tack. Poultrymen must sell what the buyers want, and not what the breeders choose to offer them. It is quite evident that customers want chicks, because they represent the most convenient form of acquiring stock, in which event it is up to the fanciers to get into the chick game, as many are now doing. To oppose the progress of the chick trade is a policy that is almost certain to result in a reaction against the fancier.

Appeal to Farmers.—For years State Experiment Stations and agricultural organs have been endeavoring to induce the farmers to substitute standard-bred poultry for their flocks of mongrels. But the farmer has never been a heavy buyer of hatching eggs from thoroughbred stock, chiefly because of the uncertainty of success with his hatches. Now that he can buy well-bred poultry in the form of baby chicks, ready for the brooder, his interest is aroused, and the farmer is fast becoming a regular customer.

Chick producers will do well to direct their appeal more directly to this class of trade, and not so much to the beginners. The farmers produce the bulk of our poultry, and the chances are they always will. Furthermore, they produce it more or less as a side line, at the least possible outlay for grain, labor, housing and range.

(*Courtesy Watson Mfg. Company*)

Fig. 179.—Double-deck mammoth incubators of large capacity.

Few Complaints.—It is curious, perhaps, but poultrymen declare they receive fewer complaints over the sale of baby chicks than with hatching eggs, therefore they prefer to sell chicks. This anomaly is attributed to the fact that the customer may secure unsatisfactory results from eggs, for which he is practically certain to blame the poultryman, when in reality the customer's ignorance or carelessness, or the integrity of his hens or incubator, is entirely at fault.

It must be admitted that any one of a dozen things can befall a shipment of eggs, which may weaken or destroy their hatchability, and over which the poultryman has no control. Nevertheless he is held responsible. He seldom has a chance to defend himself. He must either endure a dissatisfied customer, which is a poor business associate, or he must make good the losses, which wipes out his profit on the transaction.

Most of this obscurity is eliminated with baby chicks. The purchaser sees at a glance what he is getting, and thereafter if he mismanages the chicks in the brooder, he cannot blame his incompetence on the shipper.

Of course, there will always be some doubt as to the quality and productiveness of a flock of fowls raised from a shipment of chicks. It is a matter of dealing with responsible hatcheries. The reliable chick man is in business, not for a season, but for an indefinite time; he has made a considerable investment on plant and equipment, and to hope to derive profit from this investment it is absolutely necessary for him to render satisfaction. As in all enterprises, satisfied customers are his chief assets.

Satisfaction is more than landing baby chicks alive at the express station of the customer. It is giving the customer chicks hatched from strong, vigorous, well-bred productive parent stock, chicks which were properly incubated, and those which should make rapid growth in the brooder, mature early and become prolific layers. In other words, the success of the hatchery depends largely upon what becomes of its products under the care and management of its customers. It behooves the chick man to aid his customers wherever possible, and to give them a little more value than what was promised.

On the other hand, the customer should not expect unreasonable things. He must not anticipate exhibition specimens from utility chicks, which were sold at utility stock prices. The average hatchery chicks, which sell for about ten dollars a hundred, should be from well-bred, standard stock, but not show birds. They cannot be expected to have the refinements in shape and plumage of specially mated pens, whose eggs are seen advertised at from five to twenty-five dollars per setting.

CHAPTER XXI

NATURAL INCUBATION

Artificial incubation and brooding are to be recommended because of their economy. They save the hen's time, and in so doing the hen is enabled to produce more eggs. Very often, however, it is more important to save the poultry-keeper's time, as in the case of farmers and backyard growers, in which event the business of rearing young stock is left entirely to Mistress Biddy.

Though incubators are widely used on farms, it is not likely that they will entirely replace hens, because the hens are capable of looking after hatching and brooding details with practically no outside attention. Unfortunately, the hatching season comes at a time when farm work—plowing and planting—is most pressing. To escape the responsibility of looking after an incubator, the farmer prefers to depend upon his hens, knowing that they are fully competent to secure results, if not so economically, at least as thoroughly, as the machine.

It is a mistake, however, to ignore the hens completely. Unless the quarters intended for the sitting hens are convenient, sanitary and comfortable, not alone for the hens, but for the person who feeds them, it is likely that the results will be disappointing. The hens will do their part, providing they are given the opportunity, and it is this opportunity which is so often neglected.

Avoid Stolen Nests.—On farms where little or no attention is paid to the chickens, it is customary to allow the hens to steal their nests in out-of-the-way corners of the buildings, sometimes in the hen house, under the barn, in the loft, or in barrels or boxes scattered about the barnyard. This hit or miss plan neither gives comfort to the hen nor security to her brood; it is wrong.

The sitting hen is entitled to just as much consideration as the brood mare or cow. Aside from the humanity involved, to treat animals decently is the only way to obtain the full benefit of their efficiency.

Give the Sitters Privacy.—It is a mistake to allow the sitters to bring off their hatches in the regular poultry house along with the rest of the flock. In the first place, the sitting hens are almost certain to be pitilessly tormented by the other fowls. The layers will fight for possession of the sitters' nests, lay in them, sometimes drive the rightful owners off the nests entirely, or break their eggs, which is not only a loss in itself, but the presence of broken eggs seriously endangers the safety of the rest of the hatch. There is also the risk of allowing fresh laid eggs to become heated and spoiled by the sitters, or of removing the hatching eggs in mistake for fresh eggs. Or, if the nests are entirely ignored, it is likely that the layers will fill them to over-flowing with their eggs, making it impossible for the sitting hens to cover the real hatching eggs, in which case they are chilled and the hatch is a failure.

Fig. 180.—A farm brood.

Vermin.—Then, again, hens set in the poultry house are more apt to be troubled with vermin than if they are given a clean, new nest of their own somewhere else. This is an important consideration.

We like to think that our flocks and houses are free from lice and mites; but free from these pests they seldom are; make no mistake on this point. To delude ourselves on this score is to invite trouble and losses. The brood is no sooner hatched than it becomes infested with vermin, both from the mother hen and from the nesting material. Nothing is more devastating. The chicks are weakened, their growth and development are dwarfed, they fall easy prey to disease, and those that survive are finally reckoned as unprofitable.

Remote corners in the outbuildings, in mows and under sheds, are objectionable places for sitting hens because they are so remote. With nests scattered or stolen in this way the hens are

Fig. 181.—Simple devices for sitting hens.

troublesome to feed and water; frequently the hens are neglected, either through ignorance of their whereabouts or because it is too inconvenient to reach them.

Stolen nests may be free from vermin and free from the disturbances of other layers, but as a rule they are not so secure as the hens suppose. Rats are likely to abound in obscure corners, and these pests are a constant menace, or the nests may be visited by an inquisitive cat or dog. Then, again, hens sometimes choose locations which are insanitary or wet, and both are detrimental to good hatches.

Fowls are accredited with astounding judgment in some respects, yet for all this show of intuition they do stupid things. This point is mentioned to emphasize the fact that no matter how attentive a hen may be to her eggs or to her brood, in a general way her efforts must be supervised by the owner of the

Fig. 182.—Give the hens and their broods a grassy range and they will thrive like weeds.

flock. Your labor will be amply repaid by the additional chicks reared.

To set eggs successfully the first step is to get the hen—the right hen, because they are not all good hatchers. Chickens have a certain amount of individuality; some, indeed, might be said to be temperamental. Some hens are quiet and long-suffering; in spite of everything you do to them they will remain on the nest. Others are wild and nervous, and take flight at the approach of the attendant. These excitable birds are not to be intrusted with eggs. Usually they make poor work of hatch-

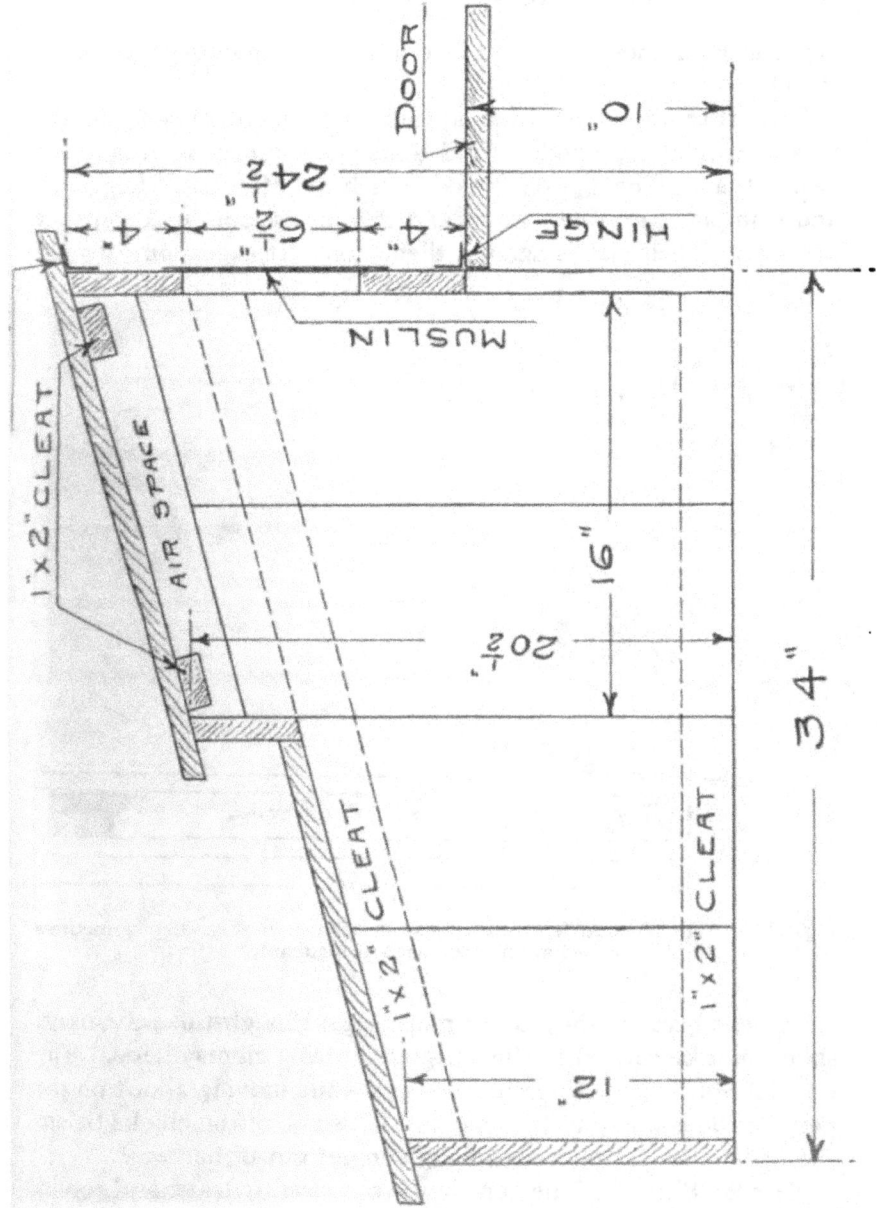

Fig. 183.—Cross-section of coop intended to house sitting hens in units of five or more.

287

ing, and later they are found incompetent to manage a brood of chicks.

Hens from the heavy breeds, such as Plymouth Rocks, Brahmas, Wyandottes, Rhode Island Reds and Orpingtons, make the best sitters. The lighter breeds, such as Minorcas, Leghorns and Campines, are too flighty, and they are seldom used on that account. Their size is against them, too. It is economy to use

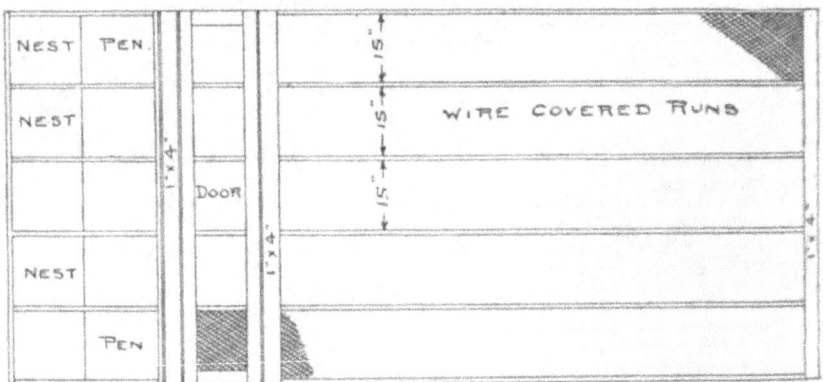

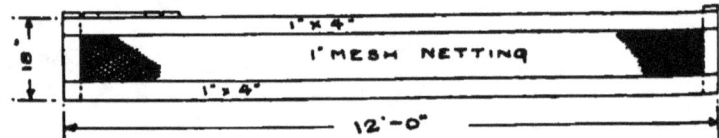

Fig. 184.—Plan of coop for sitting hens, as shown in Fig. 183, sometimes called an outdoor natural incubator.

large hens because they cover more eggs, though this advantage should not be carried to the extreme of using clumsy hens. The ungainly bird is apt to break the eggs while moving about on the nest, or she is likely to trample and kill some of the chicks before the little fellows are strong enough to get out of her way.

Test the Sitters.—The hen that is observed to leave and return to the nest with care and precision and to step lightly is the bird

to select for setting.   But do not be misled into thinking that
all hens found on the nest after nightfall are really and truly
sitters.   When broody hens are removed from the laying nests
to the place where it is desired to have them sit, some may go on
a strike and not sit at all.   It is therefore best to first test the
hens.   Give them some dummy eggs for a couple of days, and
if they show unmistakable signs of settling down to business,
give them real eggs.

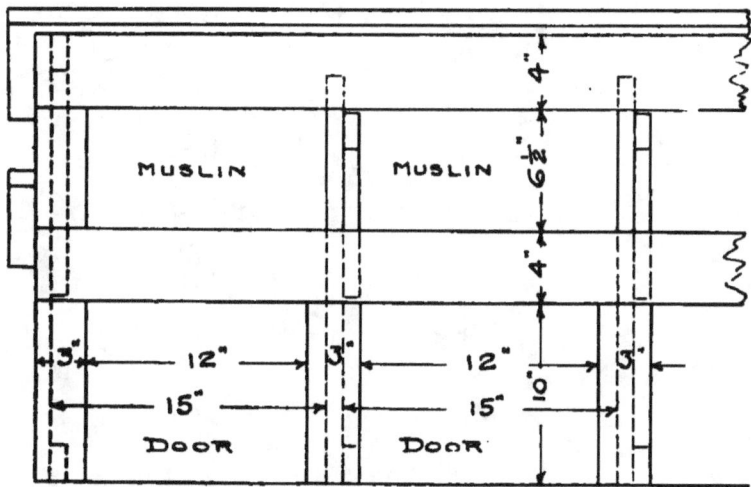

Fig. 185.—Front elevation of coop for sitting hens, as shown in Fig. 183.

Quarters.—A clean, cool, well-ventilated room or coop is the
best place for the sitters.   If it can be darkened after feeding
time, so much the better, as the hens will be quieter.   Arrange
the nests along the walls and in such a manner that the hens will
not have to fly or jump into them, a practice which is likely to
break the eggs.   If the coop has a dirt floor, the earth will serve
as a dust wallow, otherwise the building should be fitted with
a special dust bath.   It is customary for sitting hens to take a
dust wallow about once a day, which they seem to realize is

necessary to rid themselves of vermin. Therefore encourage this habit.

Nests.—A nest fifteen inches square is none too big for the sitter, and six inches is about the right depth. Place some clean earth or sand—better still, a piece of sod—in the bottom of the nest, about two inches deep, nicely hollowed to receive the litter and finally the eggs. Avoid corners into which the eggs can be rolled and left to chill; yet do not have the nest so much like a

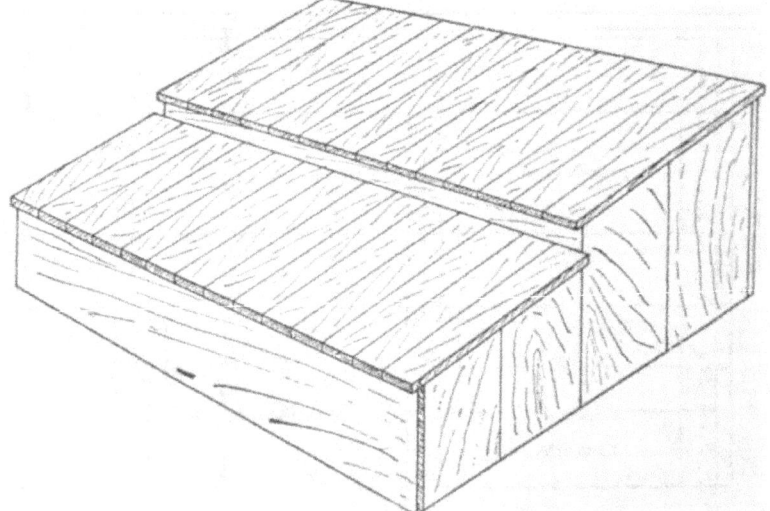

Fig. 186.—Rear perspective of coop for sitting hens, as shown in Fig. 183.

hole that there is a tendency for the eggs to pile on top of each other.

Litter.—Straw, cut hay, excelsior, shavings or fresh pine needles make excellent nesting material, and shape the stuff so that it will conform to the body of the hen. Never use old nesting material, because of the germs or vermin which it might contain. If the nest boxes have been used for previous hatches, it is advisable to give them a thorough cleaning with some disinfectant, or to whitewash them.

The number of eggs to allow a hen will depend upon the bird's ability to cover them properly. Never put so many eggs under a hen that they are even slightly visible from beneath her plumage. This is especially important in cold weather; the outer row of eggs is almost certain to be chilled, and in view of the fact that the hen changes the position of the eggs from six to ten times a day, it means that all of the eggs are likely to be chilled at intervals. Even if this does not spoil the hatch, it is sure to retard it, which is not good for the chicks.

From ten to eighteen eggs is correct; thirteen eggs are reckoned as a setting. When a number of hens are set, it is well to start them in pairs or trios, then at hatching time the chicks from three hens may be divided between two hens, or the chicks from two hens may be given to one hen, thereby saving the toil of extra hens.

On farms it is quite common to see a number of hens with their broods, each brood having perhaps six chicks. If these broods were doubled up, the efficiency of the hens would be greatly increased, since it is no trouble for a good biddy to care for fifteen chicks. The work of transferring the chicks must be done at the start, however, or the hens are likely to refuse to accept the extra chicks. A good plan is to make the transfer on the evening of the day the hatches are completed, before the hens have left their nests. Slip the chicks under the hens when it is dark, and on the day following the chicks will not be able to distinguish their foster mothers, neither will the hens be able to detect their adopted offspring.

Dusting the Sitters.—Special efforts must be made to render the sitters free from vermin. When the hatches are started, the hens should be thoroughly dusted with a good insect powder. See Fig. 187. Repeat the operation at the end of the first and second weeks of the incubating period. Watch for mites, and in case any are discovered, move the hen and eggs to a clean nest. Lice and mites are not only a danger to the health of the chicks, but they annoy the sitters to such an extent that sometimes the hens are driven from their nests to escape torment from the pests.

It is unwise to dust the hens at hatching time, unless one is familiar with the nature of the insect powder, because some of them are of such strength that they may injure or kill the newly hatched chicks. As soon as the chicks are a week old it is safe to dust the mother, and this dusting should be repeated weekly until the chicks are weaned. If, despite these efforts, the chicks are bothered with head lice, which sap their vitality and stunt their growth, it is necessary to grease the heads of the chicks with lard or carbolated vaseline. This method will drive the lice away, and tend to prevent others from coming.

(Courtesy Wisconsin Experiment Station)

Fig. 187.—Sitting hens should be carefully dusted to eradicate vermin.

Feeding.—It is a good plan to feed the sitters at regular times each day, for this teaches them to expect it at certain intervals, and they will come off the nests, eat and return, promptly. Whole grain should be fed together with a dry mash, but nothing in the shape of sloppy feed should be given, because moist food tends to loosen the bowels. Provide clean water in abundance, grit and charcoal. In warm weather it is a good plan to place a can of water beside each nest, so that the hen can drink at will without leaving the nest. Whole corn is an excellent food for sitters; it is fattening, which is a desirable effect, because the broody hen is prone to become very thin and poor. Should undue looseness of the bowels occur, the addition of a little sulphate of iron in the drinking water will usually correct the trouble.

Disturb the hens as little as possible while they are sitting. If the nests need attention, because they are sometimes soiled from one cause or another, clean them while the hens are off eating.

Fig. 188.—The makeshift coop is all right, providing it is weather-proof, vermin-proof and sanitary.

See to it that cats, dogs or other animals are unable to gain entrance to the room or coop, also other poultry.

Testing Eggs.—Few farmers bother to test the hatching eggs, though this is advisable, because the infertile eggs may be used as food for other broods. Then, too, if the fertility is poor, let us say if half of the eggs are clear, the eggs from two hens may be placed under one sitter, providing the hatches were started at the same time, and fresh eggs started under the hen from whom the eggs were removed. Here again we add to the efficiency of the hens.

By all means darken the nests at hatching time, and do not

disturb the hens unless they step around on the nests a great deal, in which event they are likely to trample the chicks, or if they pick at the chicks, then the chicks should be removed as soon as they are hatched and placed in a basket lined with flannel or some other warm material, and kept near a stove until the balance of the hatch is completed. Occasionally a hen will manifest a vicious attitude toward the brood from the start, and nothing that the attendant can do will alter the situation, in which case it is best to take the chicks away from the vicious mother and give them to a quieter bird.

(*Courtesy Kansas Experiment Station*)

Fig. 189.—Rectangular brood coop and run.

Fig. 190.—V-shaped brood coop and run.

If the sitters are well fed immediately before hatching time, they are not so likely to leave the nests in search of food, therefore the brood is not chilled before the chicks have had a chance to dry off. Confine the hens for a few days after the hatches come off, or they will take their broods too far afield and tire them. It is well to provide separate coops for the broods, and wherever possible place the coops on a grassy range, preferably where there is some shade. For the first two weeks confine the hens to the coops in the early morning, or until the grass has had time to dry off; otherwise the hens will stalk their young through the dew-laden undergrowth and get them soaking wet.

CHAPTER XXII

ARTIFICIAL BROODING

 If artificial incubation is practiced it necessitates artificial brooding, unless the hatches are small, in which case the chicks can be given to hens, though this practice really defeats the idea of the incubator, which is to conserve the hen's time. It takes at least four weeks to wean a brood of chicks, sometimes six weeks, and during this period the mother hen is a non-layer. In fact, she seldom starts to lay for a month after she has weaned her brood, due to the fact that she is usually so run down and out of condition, as the result of her maternal efforts and responsibilities, that she must first rebuild her vitality. This represents a great deal of lost time so far as egg production is concerned, and the time lost is usually during April, May and June, the months of heaviest laying.

Farmers are the greatest patrons of this combination method, and their idea is to escape the care of the brooder, which they regard more or less with suspicion. It must be admitted, of course, that no brooder is equal to the hen as a mother; we cannot improve on nature in this respect; but we can do the work a whole lot cheaper with the brooder, and this is an important consideration.

Were it not for artificial incubation and brooding it would be impossible for commercial poultry plants to conduct their operations on such a vast scale. It is not practicable to raise large numbers of chicks by hens. In the first place, it is virtually im-

possible to secure enough sitting hens at the right time—early
enough to hatch pullets which will mature as fall layers. In the
second place, a large number of sitting hens and their broods
require a great deal of equipment and range, not to mention care
in feeding; and in the third place, the expense of maintaining the
hens, without egg production, would wipe out the profits which
might be made from their broods.

(Courtesy Candee Incubator Company)

Fig. 191.—Double hot-water brooding system. A row of hovers located on
either side of a central alleyway.

These factors have always been of importance: to-day they
are vital to success. The hen must be kept on the job of laying.
Her work of rearing young must be left to the machine.

Makes of Brooders.—There are a number of brooding systems
in vogue, and a wide variety of makes from which to select or
evolve a particular scheme, one that is adaptable to given circumstances. There is as much, if not more, choice with brooders

Fig. 192.—Brooder house with outdoor runs on both sides; capacity, 3000 chicks.

297

than with incubators.   One might say, there is a brooder for any or all circumstances.

It goes without saying that artificial brooding should be in imitation of natural brooding, except that the hen's shortcomings should be eliminated.   And the hen has her faults, make no mistake on this point.   For example, she will take her brood afield in wet weather, or lead them through wet grass in the early morning, where the chicks may become chilled and die.

The chief requirements of artificial brooding are these: (1) a compartment in which the temperature is equal to the warmth of the hen's body, which is accessible to the chicks at all times; (2) an abundance of fresh air as well as warmth, because if heat is obtained at the expense of ventilation, the chicks will not thrive; (3) a well-lighted, moderately warm compartment in connection with the heated hover, which will provide a place for exercise, feeding and everyday activities; (4) a protected run or yard where the chicks can be given outdoor freedom in nice weather, and an opportunity to pick up greens and grits; (5) the interior of the brooder must be dry, capable of being flooded with sunlight, and safe from fire risks; (6) every compartment and all appliances must provide means for convenience in cleaning and disinfecting.

Common Defects.—The absence of one or more of the foregoing requirements is quite common in many of the brooding systems in use.   The greatest difficulty seems to be that the installation of the average brooding system is without sufficient latitude—a margin of safety which will automatically take up the slack resulting from sudden changes in temperature and so forth.   Many brooders are run too hot or too cold, some have insufficient ventilation, and others are over-crowded.   Heavy losses are likely to occur from these conditions, for which perhaps the operator condemns his apparatus, when as a matter of fact the trouble probably exists solely in its management.

Careless Operation.—Nine times out of ten losses in the brooder are due to carelessness or mismanagement, and the most conspicuous blunders are made at the beginning of a new brood, in that the brood is not properly broken to the brooder. Briefly, the brooder will provide warmth and so on, but the operator must teach the brood of chicks how to avail themselves of its care.

No one make of brooder or system of brooding is superior to the others. The poultry raiser must choose one that is best suited to particular requirements, such as the size of the flock, size and

(*Courtesy Buckeye Incubator Company*)

Fig. 193.—Brooder stoves are great labor-savers inasmuch as they can be made to care for chicks in large flocks.

type of houses, climate and so on. Also the funds available for permanent equipment of this sort.

There are two principal ways of brooding: one is in small units, consisting of about fifty chicks, and the other is in large flocks, ranging from 200 to 1500 chicks. Then again, the small unit plan may be carried out in two ways: either by single hovers in small coops (see **Fig. 194**), usually spoken of as colony brooder coops, or by a long, continuous brooder house in which the hovers are arranged side by side and heated from a central plant, which is generally a hot-water system. See Fig. 191.

Hot-Water System.—The earliest method of heating the hovers in a long brooder house was by means of a series of hot-water pipes arranged about eight inches above the floor of the brooder. The chicks huddled together under these pipes, and ventilation was controlled by means of apertures in the tops of the hovers. A development of this idea was found by heating a compartment

Fig. 194.—Colony coop brooders on a Government experiment station.

or duct under the brooder floor with a hot-water system of piping, and then conveying the warmed air up through a vent in each hover.

Fireless Brooders.—Another method is to heat the brooder house to a moderate temperature by the use of a few coils of hot-water pipes, but to heat the hovers themselves by means of individual kerosene lamps. An adaptation of this method is to use fireless hovers, so constructed as to conserve the heat thrown off from the chicks' bodies. See Fig. 195. These fireless brooders have never been widely used; the whole principle is against them. Adequate ventilation is impossible, since to ventilate the hovers means to lose the warmth created by the chicks.

As a matter of fact, all long brooder house systems are rapidly disappearing in favor of large flock systems reared with brooder stoves. The long brooder house usually represented the most expensive building on a poultry farm, and because of its equipment it could not be used for any other purpose, consequently for six months each year it was idle. See Fig. 192. And capital invested in idle equipment is unprofitable.

Individual Hovers.— For farmers and backyard flocks, where but a hundred or so chicks are raised each year, the individual hover, heated by hot air or hot water, with a kerosene lamp as the source of heat, seems to be the most popular device. See Fig. 196. Most of these hovers can be installed in any sort of

(*Courtesy Wisconsin Experiment Station*)

Fig. 195.—Homemade fireless brooder. The principle is that of conserving the warmth given off by the chicks' bodies.

a coop or building without alterations, or with some minor preparation, such as cutting a hole for the exhaust pipe from the lamp. Some makes have the lamp in the center and are entirely portable, others have the lamp on one side, which is housed in a separate box fastened to the outside of the house. They are called universal or adaptable hovers, and practically all makes can be depended upon to give satisfactory results.

Colony Brooders.— Some manufacturers of portable hovers make a brooder coop in connection with the hover. The coop is built in sections, screwed together and easily handled. See Fig. 197. The coop is about six feet long, three feet wide and about three feet high at the front; it has a shed roof, which is removable, and the interior of the coop is divided by a removable partition.   One compartment is fitted for the hover; the other is intended for an exercising pen or nursery.   Both compartments are easily ventilated, comfortable and convenient for the chicks, and the entire coop is readily cleaned.

This type of brooder is commonly known as the outdoor colony brooder.   It can be purchased complete at a reasonable price, and there is no better outfit for the beginner or for one who in-

(*Courtesy Prairie State Incubator Co.*)

Fig. 196.—50-chick size lamp brooding hover which may be adapted to any type of house.

tends to raise but a few chickens.   The average capacity is fifty chicks.   When artificial heat is no longer required, the hover, lamp and central partition can be removed and the coop used for a growing coop.   It is also convenient for housing extra male birds at odd times, as a conditioning coop for show specimens, or as an isolation pen for sick birds.   When not in use, it can be taken apart and stored under a shed.   Paint it occasionally and it will last for many years.

Brooder Stoves.—Whereas the outdoor colony brooder is a splendid outfit for the farmer, backyard flock and small poultry plant, the colony brooder stove is to be recommended for larger operations. In reality the brooder stove is a development or enlargement of the colony brooder idea. It is of recent origin, yet for all it has gained greater popularity than all the other appliances. It is the one device which permits of a gradation of heat and a reserve heat, and it is by far the most economical system of brooding, both as to labor and fuel.

Fig. 197.—Outdoor colony brooder. The front and top are removed to show interior equipment, which consists of a hover similar to that shown in Fig. 196. The lamp box is on the outside, and all fumes from the lamp are carried off by means of the T-shaped duct leading from the top of the hover.

Works on a Large Scale.—The brooder stove had its inception in the need for performing its work on a large scale, at the least possible cost for special buildings, for fuel, for operating costs and for the care of the chicks, and this it does. These stoves are made in various sizes, with capacities ranging from 200 to 1500 chicks. They have passed the experimental stage; they are giving results, though until one has had considerable experience in raising chicks in fairly large numbers, it is not advisable to attempt a brood greater than, say, 300 chicks. Then, as skill is developed, the flocks can be enlarged. Not that brooder stoves will fail to perform the functions required of them, but the operator must become familiar with ways of feeding large flocks of chicks running together, so that they can be induced to exercise sufficiently, and not get into bad habits, such as toe-pecking and feather-pulling.

Gradations of Heat.—In many of the earlier brooding appliances there were, generally speaking, two distinct temperatures and no gradations of heat; the interior temperature of the hover, so frequently stuffy and hot, and the outside air, which

(Courtesy Prairie State Incubator Co.)

Fig. 198.—Outdoor colony brooder—a complete outfit.

was very apt to be too cold. Either of these the chick had to accept, and both were weakening—to be chilled or partly suffocated.

The brooder stove is a high-power furnace capable of radiating a great deal of heat, which, by means of a wide-spread sheet-iron deflector, is distributed downward over the backs of the chicks, where it is most needed. See Fig. 199. When taken from the incubator direct to the brooder, the chicks instinctively learn to form a circle around the stove. In the majority of times they will gauge their distance from the base of the stove entirely by the intensity of the heat most comfortable to them. See Fig.

200.   In short, the brooder stove provides a gradual decrease in temperature, from the base of the stove, which is very hot, to the farthest corners of the room, which can be kept as cool as desired by means of ventilation.

This arrangement of providing warmth is thoroughly practical, for it permits every chick to seek the degree of warmth best suited to its individual comfort.   The plan is in imitation of natural brooding, which is best, except that it is not economical.

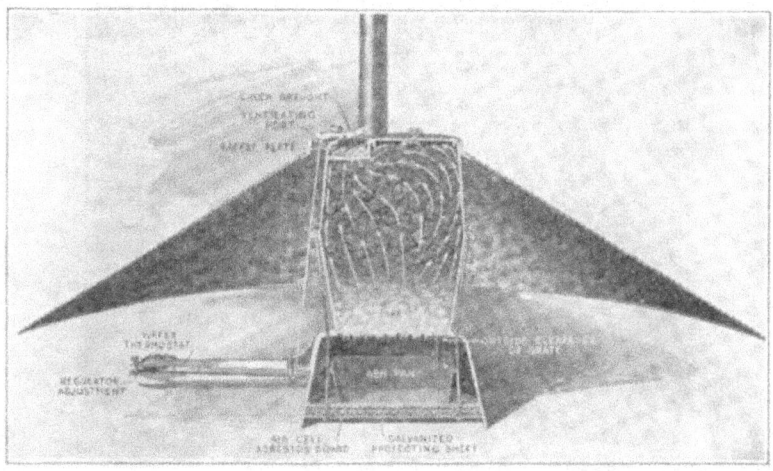

(*Courtesy Buckeye Incubator Company*)

Fig. 199.—Sectional view of a coal-burning brooder stove. All the warmth is radiated downward over the backs of the chicks, where it does the greatest good.

The bare breast of a sitting hen has a temperature of about 105 degrees, which is greater heat than the chicks require under ordinary circumstances.   The temperature of the hen's plumage ranges from about 100 degrees to 90 degrees.   But the chicks are not compelled to accept any one of these temperatures for all time.   They may seek that which is most comfortable at a particular time.   If they have been exposed and are cold, the chicks can be quickly warmed by nestling close to the bare breast of the hen.   Later they can seek a lower temperature within the

20

confines of the wing feathers; or, if they are too warm, they can push out their heads and cool off. Thus it will be seen that the natural condition is a very flexible one, and we should aim to imitate this flexibility as much as possible in artificial brooding.

No Special Buildings.—A brooder stove does not require a special type of building, and this factor is one of its greatest virtues. It can be installed anywhere, providing the room or

(Courtesy *Newtown Giant Incubator Co.*)

Fig. 200.—Coal-burning brooder stove. Note the circle formed by the chicks.

building is fairly well constructed, weather-proof, dry, and capable of being well ventilated without direct drafts.

Stoves may be set up in colony houses, and later, when the brood no longer require artificial heat, the stoves may be removed and the same quarters used for rearing the young stock. See Fig. 201. Stoves may be erected in laying houses, and when the broods are weaned the stoves are taken down and stored elsewhere, or removed to other quarters to take care of new broods. The idea is simplicity itself; it is flexible and economical.

Fuel.—There are various types of brooder stoves on the market; some designed for crude oil, gasoline, kerosene, distillate and coal; but, since no form of combustion is safer than that confined within the iron castings of a coal stove, where coal is obtainable for anything like a reasonable price, this fuel should become the

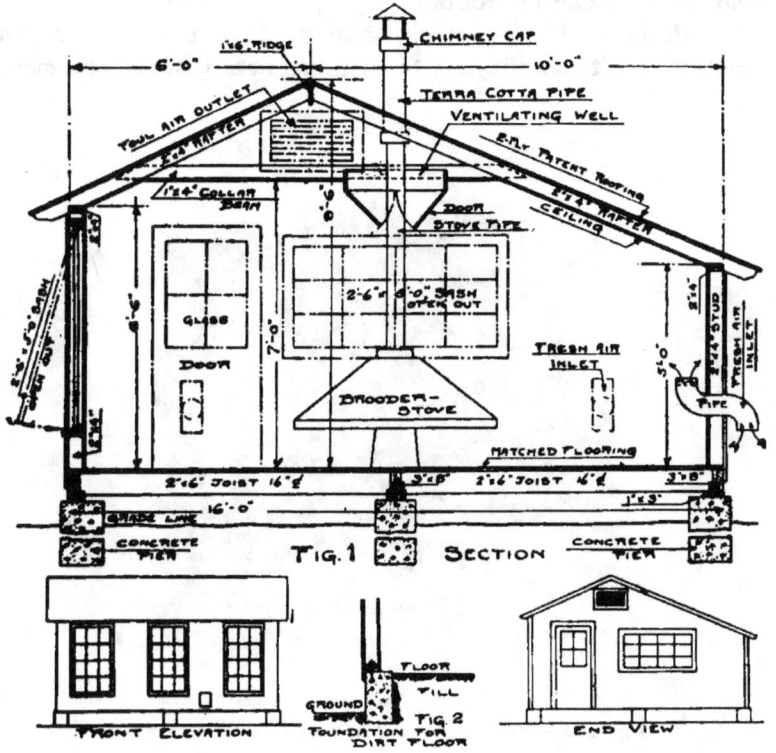

Fig. 201.—Combination brooder house, colony house and laying house.

most popular. One scuttle of coal will run a good-sized brooder stove for twenty-four hours.

The ideal house for an 800-chick stove is a building about fifteen by thirty feet, divided in the middle by a solid partition. This affords two rooms, one to be heated by the stove, and the other without heat, to be utilized as a scratching pen after the chicks are about a week old. There should be large windows on at least three sides, and so arranged as to flood the interior with sunlight.   Except in very warm weather, these windows should not be relied upon for ventilation, for it is almost impossible to ventilate in this manner without creating drafts.   It is the draft, not fresh air, that causes trouble.

Ventilation.—The following system of ventilation has given excellent results, and it may be installed in any house at a moder-

(*Courtesy Prairie State Incubator Co.*)

Fig. 202.—Filling the coal hopper of a brooder stove.

ate cost: In the four corners of the brooder house, and about one foot above the floor line, cut circular openings in the walls about six inches in diameter, and with ordinary stove-pipe and elbows construct an S-shape ventilator, pointing downward on the outside and upward on the inside.   Over the inside opening place a screen to prevent chicks from flying into it, also a damper to control the intake of air.   In the center of the roof install an exhaust vent or cupola having an area twice as great as the combined areas of the fresh air intakes. Cold air expands when heated, hence the necessity for doubling the size of the exhaust ventilator.

Ventilating cupolas of galvanized iron may be purchased in varying sizes from sheet-iron workers, or a similar device may be made of wood which will answer the purpose nicely. If one does not wish to cut an opening in the roof, the highest point in the

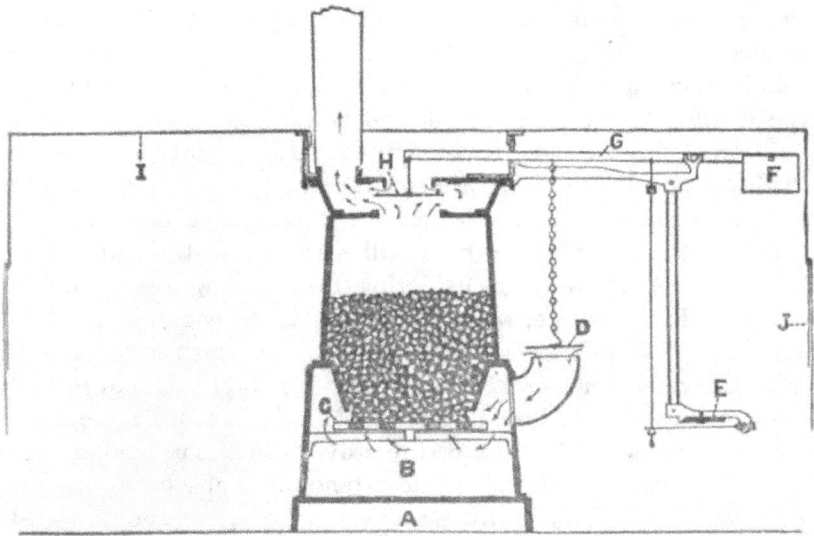

(*Courtesy Prairie State Incubator Co.*)

Fig. 203.—Sectional view of coal-burning brooder stove. A, Base; B, ash-pit; C, grate; D, fire draft; E, thermostat wafer; F, counterpoise weight; G, regulator arm; H, escape vent; I, metal hover; J, curtain.

front wall of the building may be converted into a ventilator, and fitted with a shield or baffling plate to prevent back currents of air or driving rains from entering the house.

Test the System.—When the brooder stove and ventilation system are installed, it is advisable to try out the scheme for a few days before the chicks are brought out, in order that the operator may familiarize himself with every phase of the work. Ascertain the exact heating capacity of the stove under severe weather conditions, how often the fire requires stoking, if it is a coal fire, and how to bank the fire at night.

A thermometer should be hung three feet from the base of the stove and one inch above the litter, for that is where the chicks sleep and rest. The stove should be hot enough to make the thermometer in that position register 90 degrees. No harm will ensue if it registers more than 90 degrees, but try not to have it register less particularly if the night is coming on cold. When the brood has settled down comfortably for the night, the nearest chicks should be about two feet from the base of the stove, and this intervening distance amounts to a reserve heat, so that as the night advances and the fire cools off somewhat, the chicks may draw closer to the stove for the desired warmth.

Hover Breaking.—Late afternoon or early evening is the best time to remove the chicks from the incubator and place them in the brooder, for the darkness will serve to restrain the more venturesome fellows from exploring the building and becoming chilled. Furthermore, when they are quiet they can be made to obtain their first impression of where the warmth is to be had. This is termed *hover breaking*, and it is the first and paramount issue in the artificial brooding of chicks; they must be taught to seek the hover or the base of the stove, which corresponds to a hover, for warmth, and until the attendant is absolutely assured that this instinct is firmly established, he must adopt special means of confining them within or very close to the stove.

Wire Partition.—A good plan is to erect a small mesh-wire partition, poultry netting will do, in a circle around what is considered to be a safe distance from the stove, removing it in the daytime, or when it is found the chicks no longer require it as a safeguard. See Fig. 193. If the chicks are not confined close to the stove in this manner for the first few nights, it is quite likely that some of them will stray from the warmth and be unable to find their way back, whereupon they will soon become chilled and die.

With proper training, two or three days is usually sufficient to break a brood of chicks to the hover, and the attendant must be as patient as possible, for the success of the brood later on largely depends upon this early discipline. It sometimes happens that a brood will be particularly obstinate, and insist upon spending the night in every place but the right one, all of which is very exasperating and likely to tax the patience of the most careful operator; yet the will of the operator must dominate.

The expert brooder operator realizes the importance of this, and aims to *break* his brood into the ways of a particular apparatus, much as a dog is trained to follow a scent, or a horse is broken to the harness. Poor results are often blamed on the brooding appliances, when in reality the fault rests entirely on the persons running them. A brooder stove, no matter what make or type, is seldom better than the operator in charge of it.

CHAPTER XXIII

FEEDS FOR BABY CHICKS

Baby chicks seem like such helpless, frail little creatures that the first month of their life is likely to be a period of much concern to their keeper, especially the beginner. A certain mystery seems to attach itself to the undertaking, which forebodes distressing, troublesome times ahead, maybe serious losses, even failure. We speak of it as the brooding period, and it is well named. According to the dictionary, brooding means "to dwell upon with anxiety."

Rearing young stock is the most difficult operation, though largely because it requires the closest attention to details. Other than this it need hold no fears, even for the novice, because the work is really quite simple, and almost certain of success, providing a few elemental facts are borne in mind. Moreover, there is comfort in the fact that once a chick is weaned from the brooding period, which lasts from four to six weeks, depending upon the season and weather conditions, it is practically as hardy as a mature fowl.

In its fullest sense, brooding means to provide shelter, warmth and comfort, a quiet retreat for the chicks, a resting place and a place to sleep, under the most healthful conditions, such as sanitation and ventilation, which will promote rapid development and a strong constitution.

Hatch all the chicks you can during March, April and May and the problems of brooding are greatly simplified, because conditions at this time are naturally favorable. The weather is on your side, also plant life, because it is fresh and appetizing. Then, too, March, April and May hatched pullets will start laying in the fall and furnish winter eggs, which is the goal of every poultry raiser.

Remarkable Growth.—When one considers the remarkable growth made by chicks during the first month or two, it is easy to understand why they require a watchful attendant. At hatching time the chick weighs about one and a half ounces. It doubles this weight in six days, and under normal circumstances it can be made to weigh two and a half pounds in three months, which is more than twenty-five times its original weight.

If we humans grew that fast, we would weigh about two hundred pounds at the age of three months. This comparison furnished some idea of the naturally intensive, high-powered capacity of the chick, and of the need for suitable nourishment.

(Courtesy Wisconsin Experiment Station)

Fig. 204.—Summer-hatched chicks require plenty of shade. Inexpensive coops like these are easily moved from place to place, which moving gives the chicks fresh greens and clean soil.

Proper nourishment is the crux of the whole business. Chicks require a great deal of food in relation to their size, but do not mistake this to mean that you should simply pile the food in front of them. On the contrary, they must be fed in just the right quantities. The feeding program is the most important consideration; it is paramount; it makes for success or failure.

Overfeeding.—Oddly enough, perhaps, more chicks die from overfeeding than from underfeeding. And the trouble usually starts by feeding the brood too soon after it has left the incubator.

The last development in incubation prior to breaking the shell is the embryo chick's absorption of the yolk. This highly nutritive material is capable of sustaining life for two or three days, or until the chick is strong enough to walk about freely and pick up food. Investigations have shown that this absorbed yolk contains almost half of the original energy of the egg. It is a wise provision of nature, with which it is folly to interfere. There is no need for additional nourishment at this time, and to supply any will invariably work more harm than good, in that it interferes with the proper assimilation of the yolk. We might say

Fig. 205.—Feeding frame for young chicks. It is wire-covered, and none but the small chicks can crawl under the lower rail.

that the chick's system is not ready for food until the third day after incubation.

This condition makes it possible to ship chicks hundreds of miles without imposing any hardships through lack of feeding and watering. It is a good plan, however, to place water before the chicks as soon as they are removed to the brooder, preferably water with the chill taken off, if the weather is cold. Care should be taken that the little fellows do not wallow in the water and get wet; therefore it is best to use shallow vessels protected in some way so that the chicks cannot fall inside.

As a rule, chicks learn to eat as soon as they are able to stand. They will even commence picking at things while they are in the

(Courtesy Buckeye Incubator Company)

Fig. 206.—Brooder stove with the hover or heat deflector raised to permit sweeping and cleaning around the base of the stove. The deflector is suspended by means of rope and pulleys attached to a counterpoise weight.

incubator. The idea seems to come to them as instinctively as walking. In the case of drinking it is sometimes necessary to teach them, or at least to point out the presence of water. This is easily done by dipping the bills of two or three chicks in the water. They will catch on to the idea in a few minutes, whereupon the rest of the flock will quickly imitate them.

Feed little but often, is the slogan to be adopted in the brooder. This is especially important with flocks which are confined indoors, or leg weakness is apt to result. Chicks on free range are not susceptible to this ailment. Leg weakness comes from excessive feeding without sufficient exercise; the bodies of the chicks become too heavy for the muscles and bones of the legs, consequently the chicks are unable to walk or stand. They present a pitiable appearance and are likely to be trampled and killed by the rest of the flock. ▶ **On the morning of the third day,** which is to say when the chicks are forty-eight hours old, give them a light meal of easily digested food, preferably soft food,

(Courtesy Prairie State Incubator Company)

Fig. 207.—Small lamp hovers are easily moved from place to place.

which has been sprinkled with fine grit. Repeat this feeding every two or three hours, so that the chicks receive four or five meals a day.

The first feed may consist of bread crumbs moistened with milk, bread crumbs and hard-boiled eggs ground fine, shells and all, johnnycake, or pinhead oatmeal. Food moistened with milk is of great value in giving the chicks a good start, but the mixtures must be fed in a crumbly state, never in a sloppy condition. In the case of bread crumbs a good plan is to moisten them with milk and then squeeze out the greater part of the liquid. It is also important to feed these mixtures fresh. Do not mix up a large quantity and then allow it to stand around until it turns bad.

Johnnycake.—The following is a good recipe for johnnycake: One dozen eggs (wherever possible use clear eggs removed from the incubator), or one pound of sifted beef scrap, to ten pounds

(Courtesy Wisconsin Experiment Station)

Fig. 208.—Where flocks of different ages are likely to run together, the young broods should be given a fenced yard for the first few weeks so that they will not be molested by the older chickens.

of corn meal; add enough milk to make a dough, and one tablespoonful of baking soda; bake into cakes.

Milk.—Many breeders are having good results by feeding milk and no water the first week or ten days, and even longer. Sour milk should be fed just as it begins to thicken and before it separates, as the chicks like it better in that condition. Milk is more than a food; it contains lactic acid, which tends to prevent and correct white diarrhea and kindred bowel troubles. It should be placed before the chicks in shallow pans or founts, so designed that the birds cannot wade in it or contaminate it.

Scratch Grains.---Continue with the soft food for three or four days, then gradually substitute a scratch grain mixture consisting of cracked wheat, finely cracked corn and hulled oats in equal parts, to which may be added about five per cent of cracked peas or broken rice, and two per cent of millet or rape seed. At the same time a dry mash should be fed, which may consist of ten pounds of corn meal, ten pounds of wheat bran, two pounds of bone meal and one-half pound of granulated charcoal.

Only as much of the scratching feed should be given as the chicks will scratch out in ten minutes. If the chicks have range, the dry mash may be kept before them all the time, in shallow troughs or hoppers. If they are confined to the brooder, the mash should be left before them for about fifteen minutes at each feeding.

(*Courtesy Newtown Giant Incubator Company*)
Fig. 209.---Sectional view of brooder stove.

If it is impossible to supply the chicks with milk in any form, add a half pound of sifted beef scrap to the dry mash after the first week. Increase the proportion of scrap each week, until it amounts to about four pounds at the end of a month.

After the first week the number of feedings may be reduced to four a day, and after the second week to three a day. In the morning the feeds are light, so as to encourage exercise. Toward evening the heaviest feed is given, so that every chick may go under the hover with a full crop. The last meal should be given about an hour before sundown.

The best way to feed the soft mixtures is on boards or heavy sheets of cardboard. Tin pie plates are good, except that the chicks slip about on them. Never feed the soft mixtures in the litter, where they are likely to absorb filth from the droppings.

On the other hand, it is a bad plan to feed the scratch grains on a board, once the chicks have learned how to hunt for them, because they are likely to gorge themselves, which defeats the idea of the scratch grains. These grains should be sprinkled in the litter, where the chicks are obliged to dig and work for them.

Green Food.—Unless the chicks are given outdoor freedom where they have access to an abundance of tender green shoots, they must be given succulence in some other form, such as lettuce, sprouted oats, sliced onions and tops, or ground vegetables, such as mangels or turnips. Chicks are very fond of onions, which seem to act as a tonic. Slice a good-sized onion for each fifty chicks; cut it so as to form rings, and the chicks will tussle and fight for them as though they were worms.

Clover, alfalfa and rape are all excellent green foods for chicks, and wherever possible they should be given liberty where these crops are growing. This is the most convenient way to furnish green food, and the

(*Courtesy C. L. Opperman*)

Fig. 210.—Colony house fitted with gasoline brooder.

cheapest way. If a grass range is not available, substitutes must be found, because no brood will thrive without greens. Succulence supplies them with roughage for crop development, mineral elements to aid the digestion of concentrated foods, and with certain tonic properties which sustain the appetite. In short, if chicks are given unlimited freedom on a grassy range, they are almost certain to thrive even if the rest of the feeding system is seriously defective.

Avoid Moldy Feed.—Chicks are susceptible to bowel trouble if given moldy or musty grain or decomposed food, therefore it behooves the poultryman to investigate his feeds very carefully. Grain which has heated in the bin, or allowed to become damp and mildewed, is the equivalent to so much poison.   Bear this point in mind if you are offered cheap grain, because it is likely that the grain has suffered some such deterioration, hence the reduced price.   Bargain foods are usually inferior goods, and the most expensive in the long run.   It pays to get reliable stuff from a responsible dealer, and then to store it properly so that it will not spoil.

(*Courtesy Purdue Experiment Station*)

Fig. 211.—An orchard makes an ideal location for growing chicks.

Finely broken grit and oyster shells should be kept before the chicks at all times after they are a week old.   It is best not to put these articles before them in hoppers until this time, because some of the chicks are likely to mistake the grits for food and gorge themselves forthwith.   As previously mentioned, the early feedings of soft food should be sprinkled with the grit; sharp sand will answer the purpose.

If charcoal is not included in the mash, it is well to keep this before the chicks along with the grit and shells, also a box of granulated bone or bone meal. These articles are inexpensive and they will last a long time. The charcoal acts as an absorbent and aids digestion; it serves to keep the crop sweet. Granulated bone is rich in nitrogenous matter and makes bone and muscle; it keeps the chicks sturdy.

TABLE XXIV.—FEEDING RATIONS FOR CHICKS RECOMMENDED BY THE INDIANA EXPERIMENT STATION

SCRATCH GRAINS	Pounds	DRY MASH	Pounds
Cracked corn (sifted)	10	Wheat bran	10
Wheat (cracked)	10	Middlings	10
Oats	10	Corn meal	5
		Meat scrap	5
		Charcoal	2.5

Plenty of sour milk or buttermilk to drink. If milk is not available, the meat scrap may be increased to 15 pounds in the mash. Scratch grains are fed in litter from the first. The mash is supplied in a hopper when the chicks are five to seven days old.

TABLE XXV.—FEEDING RATIONS FOR CHICKS RECOMMENDED BY THE CALIFORNIA EXPERIMENT STATION

SCRATCH GRAINS	Pounds
Wheat	20
Oats (steel cut)	15
Millet	5
Rice	2
Cracked corn	6
Grit	10
Charcoal	5
Bone meal	5

The ingredients are crushed to the size of millet. The mixture is fed from the first. Toward the end of the second week the cracked wheat and oats are increased until by the sixth or seventh week these two grains have replaced all others, except the cracked corn. From the seventh week on the grains consist of cracked corn and wheat. A mash is given after the eighth day, consisting of either of the following mixtures:

MASH MIXTURE I	Pounds	MASH MIXTURE II	Pounds
Bran (wheat)	3	Bran (wheat)	4
Middlings	2	Alfalfa meal	2
Corn meal (coarse)	1	Corn meal (coarse)	1
Oatmeal	1	Meat meal	¼
Meat meal	¼	Bone meal	¼
Bone meal	¼	Charcoal	½
Charcoal	¼		

One teaspoonful of salt is added to each gallon of either mixture. The mixtures are intended for chicks from ten days old to six weeks. After this the meat meal and the bone meal should be increased gradually, until the meal equals one pound of the ration and the bone meal a half pound.

21

Fresh Water.—Needless to say, the fountains must be kept full of pure, clean water at all times. If they are permitted to go dry, though only for a short time, the chicks soon become famished for a drink, whereupon they fight for room around the refilled fountains and in so doing seriously wet themselves.

As soon as the flock can be weaned from artificial heat or from the care of the brooder, the chicks are old enough to be given a scratch-grain ration of whole wheat, cracked corn and other small

(Courtesy Purdue Experiment Station)

Fig. 212.—The hen and her brood must be given privacy for the first couple of weeks.

grains, such as milo maize, kafir corn and barley. The corn meal in the mash can be increased somewhat and other meals added, such as ground oats and wheat middlings.

Essentials to Health.—If called upon to suggest the proper care of baby chicks, and my expression were limited to ten words, it would be something in this fashion: *Keep them warm, dry, exercised, aired, watered, and somewhat hungry.* The question might then arise, how can one feed liberally and yet keep the chicks somewhat hungry? An answer is difficult without appearing to argue in a circle, which is really the case: If a chick is dry, warm, exercised, aired and watered, he is sure to be hungry, and when hungry he will search continually for food; and thereon hangs the secret of successful brooding. Keep all the conditions such that every chick is exercising for food, and always receiving a reward for its pains, from dawn to dark.

TABLE XXVI.—RATIONS AND METHODS OF FEEDING CHICKS RECOMMENDED BY POULTRY DEPARTMENT OF CORNELL UNIVERSITY

THE RATION	THE METHOD
Mixture No. 1	**One to Five Days**
8 pounds rolled oats. 8 pounds bread-crumbs or cracker waste. 2 pounds sifted meat scrap (best grade). 1 pound bone meal.	Mixture No. 1, moistened with sour skimmed milk, fed five times a day; Mixture No. 2 in shallow tray containing a little of No. 3 (dry) always before chicks. Shredded green food and fine grit and charcoal scattered over food.
Mixture No. 2	**Five Days to Two Weeks**
3 pounds wheat (cracked). 2 pounds cracked corn (fine). 1 pound pinhead oatmeal.	No. 2 in light litter twice a day; No. 3 moistened with sour skimmed milk, fed three times a day; No. 3 (dry) always available.
Mixture No. 3	**Two to Four Weeks**
3 pounds wheat bran. 3 pounds corn meal. 3 pounds wheat middlings. 3 pounds meat scrap (best grade). 1 pound bone meal.	As above, except that the moist mash is given twice a day.
Mixture No. 4	**Four to Six Weeks (until Chicks are on Range)**
3 pounds wheat (whole). 2 pounds cracked corn. 1 pound hulled oats.	Reduce meals of moist mash to one a day; Mixture No. 4 in litter twice a day; dry mash always available.
Mixture No. 5	**Six Weeks to Maturity**
3 pounds wheat. 3 pounds cracked corn.	No. 3 and No. 5 hopper fed. One meal a day of moist mash if it is desired to hasten development.

· Further instructions: Provide fine grit, charcoal, oyster shells and bone from the start. Give grass range or plenty of green food. Keep chickens active by allowing them to become hungry once daily.

The incubator chick is born with just as many instincts as the chick hatched under a hen, it is just as sturdy if the period of incubation has been properly conducted, hence there is no reason why it should not develop just as quickly and profitably. Like most of us, however, the chick is a creature of habits, some of which are pernicious, and generally inspired by simply *watching the other fellow do it*. For example, toe-pecking and feather-pulling are two of the most troublesome habits to combat in the brooder, and unless controlled at their inception, they will frequently lead to a heavy mortality. Therefore while it may be perfectly normal for a brood to develop toe-pecking, yet the habit must have been induced by an abnormal condition—lack of exercise, idleness due to exhaustion from overheating, over-feeding or improper feeding. Usually it is improper feeding, either as to quantity or an insufficient variety; not necessarily a lack of variety in the grains, but an improper balance of the nutritive elements—the greens, grains, grits and grubs.

Last, but not least, of the suggestions for brooding—be sure to get chicks on the soil at the earliest possible moment. No matter how well equipped the brooder, Mother Earth is the chick's natural habitat. The chick has an affinity for dirt,—and it won't be genuinely happy till it gets the dirt

CHAPTER XXIV

CARE OF YOUNG STOCK

The hatching season is admitted to be the critical period for the poultry keeper. It is the time when affairs are most pressing, and when the prospects for next season's flock are either secured or discouraged. As soon as this season is past, and the brooders and incubators have cooled off for the last time, say about the first of August, poultrymen are likely to relax their vigilance, which is natural enough, except that it must not be carried to extremes.

When chicks reach the age of four weeks, and are fairly well feathered, they are as good as grown, barring accidents, and providing they receive reasonable care. Most of them are independent of their mothers or brooders, and quite competent to shift for themselves; yet too much confidence must not be reposed in them. Young chickens are susceptible of indiscretions, just as are children. They need a watchful eye more or less at all times for best results.

The breeder who gets the most from his flock is the fellow with this watchful eye, to see that all are sufficiently fed, that they are not tortured by vermin, that they are not menaced by rats, hawks, crows and other pests, that none are killed accidentally and their bodies left to decay in some obscure place where the rest of the flock can eat this putrid matter, with its resultant ill effects. They must have an abundance of clean, cool water. They must have plenty of green food, and an opportunity to roam for insect life and mineral food.

Exercise.—They must have plenty of exercise, especially in the cool morning hours or in the late afternoons. Exercise sharpens the appetite and encourages eating large quantities of

325

food, so essential to rapid development, and it also wards off any tendency toward leg weakness which is apt to result from heavy eating without exercise. Exercise on a grassy range is the greatest boon of all. Give fowls range and clean living

(Courtesy Million Egg Farm)

Fig. 213.—Portable hover installed in a simply constructed coop.

quarters and their keeper can almost afford to throw away all the medicine bottles.

Over-crowding.—By all means avoid over-crowding at this season, which is saying a good deal. Where large flocks are kept it is not so easy as it sounds, even if sufficient buildings are available. Chicks have strange ways. They are gregarious;

they like to assemble in large numbers. Where one goes they all want to go, despite the fact that there may not be room for them.

However carefully the attendant may have been to distribute the young stock among brood coops and colony houses, if these buildings are on the one range, with no partition fences, the chicks are likely to desert some houses, and crowd into others. I never heard of a poultryman of any experience who was not bothered with this perversity. It is as sure to occur as two or three hens

Fig. 214.—A number of broods can be kept together if the mother hens are confined.

trying to crowd into one nest, though there may be a dozen or twenty other empty nests.

Keeping the houses fairly far apart tends to discourage this practice of over-crowding, but to do so is not always possible, nor practicable. Where large numbers of chickens are grown a great deal of ground is required. And when the houses are spread over a big acreage, it means considerable additional labor to distribute feed and water, and to perform the cleaning and other chores. Very large farms do this work with a team, which is the only practical, economical method.

Abandoned Houses.—Usually the houses that are farthest away from the central part of the farm are the ones most likely to be abandoned. The reasons for this are very apparent. Chickens soon learn the ways of an attendant and the hours when feed is distributed. Day after day they watch the feeder approach from the central part of the farm, and they go to meet him. The stock from the farthest points on the range comes in and joins the flocks close by; in a large herd they congregate, impatiently waiting for the dinner pail. It is natural enough, even if it is troublesome.

Change Feeding Ground.—Efforts should be made to avert the forming of these habits, though I confess, it is not always possible to do so. The first step is to keep the flocks guessing as to where the attendant is going to make his approach. In other words, if practicable alternate the routine as much as possible; approach the colony field from different points, so that no particular place exists as a feeding ground. On some farms this is easy to do, on others it is out of the question.

Another stunt is to avoid distributing feed near the houses that are nearest the central part of the farm, but to carry it to the farthest houses. In this way the flock will follow to the farthest points, and when the chicks have finished, especially if it is the evening meal, they are more likely to remain in the vicinity of the farthest houses, and to take shelter in them as night falls.

In our haste and efforts to reduce our steps it is natural to want to distribute the feed at the nearest point where the flocks can get it; but this is wrong, and will only pile up additional work in the long run. Short-sightedness is one of the worst characteristics a poultryman can have, yet it is strangely common in the matter of feeding. Avoid the spirit of doing a thing for the sake of *getting it done*. It is almost always fatal to success with poultry, just as it is a serious handicap in other lines of work. *Watch the flock and not the clock*, is a pretty good slogan for the chicken man.

Unlimited Feed.—Some of the most successful poultry raisers work on the assumption that the fowls know what is best for them in the matter of food, better than the man who does the feeding. Certainly this is true of some feeders, for I have seen some farm laborers who appeared to have as little interest in their tasks, and as little knowledge of the importance of their work, as they might have over a translation of Sanskrit. Where this condition obtains it is infinitely better to permit the stock to

(*Courtesy Purdue Experiment Station*)

Fig. 215.—A covered runway which can be moved from place to place is best for very young chicks until they are strong enough to battle with the older stock on free range.

exercise its own judgment, by keeping all feeds before them at all times, and giving them access to the feeds at will.

There is another virtue about this method: the birds do not establish any habits of waiting for the feeder two or three times a day, and then gorging themselves forthwith, only to go off in the shade somewhere, like a snake after it has swallowed a toad, and wait for digestion. Instead, if they have always been accustomed to feeding at will, they generally eat a little, run around, return and eat some more, run off again, and repeat the process all day long.   This is the best thing for them—a lot better than gorging at stated intervals.

When chickens have feed before them in hoppers situated at convenient places throughout the range, they have nothing to draw them in a herd in any one spot, consequently they are not so likely to find one location more attractive than another, and will remain pretty much as they are distributed over the range in the first place.   The "always filled hopper" principle goes a great way toward eliminating the bother of over-crowding due to the abandonment of certain houses.

Feed Hoppers.—There should be plenty of hoppers, of a non-wasteful type, and great care should be taken to make them water-tight and weather-proof.   Some of them can be placed indoors.   A good plan is to keep the mash hoppers inside the house, where the feed will not be wasted by high winds blowing the lighter meals away.   Scratch feed hoppers may be left outside, in sheltered spots, accessible to the fowls, but in such a way that the hoppers will have protection from sparrows and other thieves.

Watertight Covers.—Positively the hoppers must be made watertight, not only for the economy of the thing, but to avoid moldy, sour or spoiled food.   If the hoppers leak ever so little, it means musty grain, and musty grains mean bowel troubles, maybe serious ones, and heavy losses.   Take a little extra care in the making of the hoppers and provide tight lids or covers. And the covers must extend far enough over the sides of the hoppers to prevent driving rains from reaching the contents. The hoppers should have slatted sides through which the birds can reach the grain without difficulty, but not large enough for them to crawl through and perhaps soil the feed.

The pullet is the favorite—the "star boarder."   She is especially cherished on egg farms, and held in preference to the hen for fall and winter egg production—the periods of highest prices, which mean so much to the year's profits.

Every poultry raiser's experience will substantiate the belief that a fowl's greatest egg-producing capacity is in her first laying year—the pullet year, providing, of course, all conditions are equal and as they should be. Therefore, under favorable conditions the pullet is the most profitable bird on the farm; and as such she is deserving of special care and attention.

We are enjoined to get our hatches out early, in time to have pullets mature as autumn layers; but it is well to remember that age is not the only important consideration. While it is necessary to hatch chicks early to get mature pullets before cold weather sets in, especially among the heavier breeds, the feeding and general care of the growing stock have much to do with the flock's start as layers.

To be fitted for laying a pullet must be in full flesh, of normal size, with a fair amount of surplus fat, and these conditions are obtained only by an abundance of food of the right sort. See special chapters on feeding.

The pens from which the old stock has been removed should be carefully cleaned, sprayed or whitewashed before the pullets are turned into them; the yards should be plowed or spaded under, and if possible sown to green food to sweeten them. In short, everything must be made as fresh and comfortable as possible for the new tenants. Cleanliness and roominess tend to keep the growing pullets healthy and vigorous, which are essential to egg production.

The young stock should be kept growing steadily, yet it is a mistake to force pullets too rapidly. If their egg-producing organs are developed into a state of production in advance of their bodies having attained full growth, they will lay under-sized eggs, or they may lay a few eggs and then enter a molt, which will postpone further egg production until late in the winter. By no means force pullets by excessive feeding of highly concentrated animal protein foods.

Transfer Pullets Early.—Many poultrymen make the mistake of allowing their pullets to remain on the range, in colony houses, too far into the fall, sometimes until the birds are ready to lay. The error in this practice is this: Chickens are creatures of habit, and nothing disturbs them more than changing their accommodations, even though they may be moved to a more desirable. They will fuss and foolishly agitate themselves when placed in an unfamiliar building, which invariably results in a falling off in the egg yield. For proof of this, take a pen of fowls that are laying nicely and remove them to another building; then note the egg yield.

(*Courtesy Petaluma Chamber of Commerce*)

Fig. 216.—A husky brood; count them.

It should be a rule to get the pullets into their permanent winter quarters several weeks in advance of the time they are expected to commence laying, and to train them to accept their more regulated method of living and confinement with as much grace as possible.

Dangers of Poisons.—No experienced poultryman willfully feeds his flock on spoiled grain, because he knows that to do so is almost certain to result in sickness. No one with common sense would think of leaving poisons about, such as arsenate of lead or Paris green, where the fowls have access to them, or where children are likely to play with them. Common sense dictates that these poisonous things are in the class with high explosives, to be treated with the greatest caution and forethought, lest they result in some terrible fatality. In other words, we are impressed with the danger of explosives and poisons, therefore we handle them accordingly.

Decayed Animal Matter.—It is unfortunate that poultry raisers generally do not extend this caution in the matter of poisons, and make it cover all such risks, since others exist which are almost equally as potent as the arsenate of lead, despite the fact that little or no attention is paid to them. I am speaking of the decaying carcasses of fowls and rotten eggs which are so often carelessly left about the premises, thrown on rubbish piles, in manure pits, or in adjoining woods and fields.

Such carcasses, in fact, dead animal matter of any kind, really constitute just so much poison as soon as they start to putrefy. If death was caused by disease, the bodies are poisonous even before they start to putrefy, for reasons too obvious for further explanation. They are the carriers of contagion, which is the equivalent of poison.

Destroy the Dead.—Every authority who writes about poultry or gives advice on the subject is sure to say—"Destroy the bodies of dead fowls. Either burn them, put them in quicklime, or deeply bury them."

Perhaps poultry keepers have been told this so often that it has lost its power, for certain it is that the advice is not followed as a general practice, at least, not with the scrupulousness that is weighed against the skull and cross-bones poison label or the explosive. Yet it should be, every mite as carefully.

Chickens Are Scavengers.—It is not a very pleasant idea to contemplate chickens as scavengers, though in treating a subject of this kind we are obliged to deal with facts and not fancies. Chickens are scavengers. Almost all fowls are scavengers to a certain degree. Whether this is the result of intense domestication or a natural impulse, I am not prepared to say, but I do know that fowls will eat dead animal matter at the slightest opportunity, and, what is more, they eat it with apparent relish.

Fig. 217.—Growing coop for young stock constructed from piano cases covered with tar paper. Note runway in front of right-hand window.

In the early stages of decomposition, if the animal matter has not been infected with a malignant disease, little harm will result from eating it, unless eaten in large quantities, which will bring about bowel troubles. But as soon as an advanced state of putrefaction sets in, the carcass fairly swarms with bacteria—microbes of one kind or another, not to mention worms and the eggs of flies, which are highly poisonous. Taken into the bodies of the fowls these bacteria soon attack healthy tissue.

Since few fowls or animals meet with accidental deaths, or die without cause, it is well to consider all dead animal matter as being highly poisonous, therefore unfit for food.

All the disinfectants in the world are useless, and spraying, white-washing and cleaning go for naught if the carcasses of dead birds are left about the premises. We can treat disease till the end of time, but we can never hope to exterminate it so long as a single infected specimen remains on the plant. These are not the vaporings of a crank. They are plain truths. And the sooner we recognize them, the better it will be for our poultry and other folks' poultry as well.

If the evils resulting from the careless handling of dead animal matter were a little more tangible, no doubt we would be held accountable unto the law for spreading contagion. The fact that these evils are not tangible, and we are not held technically responsible, does not alter the moral obligation, however, consequently we owe it to the community, as well as to our own security, to provide every precaution.

Flies thrive and breed upon carrion. They are notorious germ carriers, traveling far and wide and doing untold damage. Your flock of fowls may be perfectly well, and the conditions under which the birds are kept may be the acme of sanitation, but if your neighbor's ways are negligent, it will be only good fortune if at some time or other your birds do not suddenly break down with illness of some kind, which, if it could be traced, would be chargeable to this neighbor.

The Easiest Way.—Dead bodies, especially those of little chicks, are improperly disposed of largely because it is deemed easier to get rid of them by the shortest route. This is a fallacy, if the bodies return in the form of disease, for nothing is more troublesome to combat than a flock of sick chickens. In back-yard flocks the bodies of dead chicks are frequently thrown into the garbage cans, many of which are without tops. With or without lids to the cans, this is a bad practice.

On farms where fowls are kept merely as a side line, the bodies are apt to be tossed into the manure pit. Sometimes an effort is made to cover them with the manure, but this does not remove the evil, if the bodies were infected with disease. At some time or other this manure is going to be spread about as fertilizer, and with it will go the diseased remains of the dead fowls. Most of the remains will be consumed, though not always the disease,

Fig. 218.—Inexpensive colony growing coop built on skids.

for some disease germs live for months in the soil, especially where there is heat, as in a manure pile, to further nurture them.

Hog Pen.—Frequently dead fowls are thrown into the hog pen. Even when they are consumed by the hogs, this is not fit food for hogs. Usually some parts of the remains lie about long enough for other fowls to find it, and run off with it. Maybe a dog will steal a body from the pen.

Pit or Well.—Some poultry keepers throw their dead into a

pit or abandoned well.   It may be that the hole is covered over
so that other fowls or animals cannot gain access to the carcasses,
but the chances are that flies will have no trouble in finding them.
Maggots, the larvæ of flies, worms and other "crawling things"
abide in putrid matter.   Later these insects may be eaten by the
chickens.

The most careless method is to throw the dead bodies under
buildings, into hedge rows, along fences, on rubbish heaps or
other seemingly out-of-the-way places.   Fowls and dogs, not to
mention rats, cats, skunks, crows and other flesh-eaters, soon

(*Courtesy Kansas Experiment Station*)

Fig. 219.—Choose a secluded spot   Fig.   220. — Open-front   colony
for the brood coop.   house with hinged front to exclude
                                       driving rains.

learn the whereabouts of such places, and thereafter they will
make a practice of haunting them, like scavengers.   If they
would consume all of the waste matter, it would not be so bad,
but they do not.   They eat portions, and distribute the balance
around the grounds.

Bad eggs, especially those removed from the incubator, are
often left lying about, together with the empty shells and a small
percentage of dead chicks gathered at the close of a hatch.   On
some of the largest hatcheries I have seen whole barrels of unhatched eggs, in various stages of decomposition, standing un-

22

covered outside the incubator cellars. In some cases the odors were so bad as to be sickening. Eggs are animal matter, and should be destroyed as thoroughly as dead bodies, especially during warm weather.

In the winter time, when snow is on the ground, it is not uncommon to find carcasses thrown into the snow right outside of the hen houses, on the assumption, I presume, that the cold will prevent their decay. No doubt it was the intention of these poultry raisers to gather the bodies before a thaw, but in most cases they were forgotten, or hidden by the snow, until decay had set in. Furthermore, cold does not kill all germs. Sometimes it simply suspends life, which will be renewed at the approach of warm weather.

There are but three really effectual methods of destroying dead animal matter: Incineration, quicklime and deep burial.

Placing the bodies in quicklime destroys them and all germ life utterly, but it is rather troublesome to do this every time a dead chick is found. Burial places the matter out of sight, and may or may not destroy it. In any event, the burial should be deep, so that other fowls or animals cannot dig it up.

Cremation is the best method, not only because burning puts an end to any possibility of infection, but because it is the easiest to perform. Contrive some kind of an incinerator out of an old garbage can or metal receptacle, raise it off the ground about eighteen inches, either by iron legs or a brick or stone foundation, so that a fire can be built underneath, and every time you have any waste paper or rubbish from the house, keep it handy for this purpose.

Aside from destroying the bodies of all dead fowls which are found in the brooders, laying houses and other buildings, do not forget to look around the range at regular intervals. Sometimes chickens are killed by strange causes, or they will go off in the brush and die as the result of sickness or exposure. If their bodies are not found by the attendant, they are sure to be found by the rest of the flock, which is likely to prove a serious menace, the importance of which cannot be over-estimated.

CHAPTER XXV

BREAKING UP BROODINESS

Hen's Business is to Lay Eggs.—On farms devoted to egg production it is the hen's exclusive business to produce this product in the greatest number, and it is her manager's duty to see that she is equipped with every facility toward this end, with no chance for even a temporary cessation of activities. Where poultry is raised on a large scale the hen is not held responsible for rearing next season's flock of pullets. It is far more economical to perform this work by artificial means—with the aid of incubators and brooders; in consequence the hen is denied any participation in the furtherance of her species, save the laying of the egg, and any inclination toward these maternal ambitions must be promptly discouraged.

Production is not Continuous.—Contrary, perhaps, to the opinion of the novice, egg production is not a sequence of certain quantities of correctly proportioned nutrients taken into the body daily, digested, assimilated and then converted into a regular supply of eggs—a continuous operation, as it were, uninterrupted so long as the hen's health and vigor are maintained, and her care is as it should be.

The egg cells, scarcely visible to the naked eye, of which there are many hundred in the well-bred normal fowl, and some authorities place the number of latent eggs at upwards of five thousand, are stimulated and developed in series or clusters, sometimes called "clutches" or "litters"; each series being ripened or held dormant in accordance with the fowl's general health and her capacity to consume sufficient quantities of nutrients essential to the stimulation of the egg-producing organs.

The number of cells in each series varies widely with different

339

breeds and with different specimens, and there seems to be no basis for an approximation. There may be a dozen cells in a litter, or five dozen, and in rare cases, such as the hens that have attained wonderful records at egg-laying contests, fowls will continue to lay almost without cessation, and continue to do so for a couple of years. Ordinarily, between clusters there is a period

(*Courtesy Cornell Experiment Station*)

Fig. 221.—Every laying house should be equipped with a broody hen coop.

of non-production, a sort of rest period, which varies in duration the same as the size of the litter. It may be a week or a month, or perhaps three months; and not infrequently a hen will lay but the one cluster of eggs and then stay off the nest for the remainder of the year. Such specimens are to be rated as drones, and dealt with accordingly. Birds that are impoverished and those that are not bred along the lines of heavy egg production are usually in this class, and for which the poultryman should keep a sharp lookout; they are not fitted for the highly organized egg plant.

Intervals Between Litters.—Generally speaking, hens that lay short litters take but a few days to the intervals between them, whereas those that lay from thirty to sixty eggs in almost daily succession will require a much longer period, which seems perfectly natural. Egg production is a severe tax on the hen's body; it is a secretory and a reproductive process combined, and as such it demands time in which to recuperate.

When a hen completes laying a litter, especially during the spring months, she is usually attended by a maternal instinct—a desire to hatch the eggs, all of which is very natural, indeed, but not in accordance with the poultryman's views on the subject. Producing eggs for table purposes does not concern Mistress

(Courtesy Missouri Experiment Station)

Fig. 222.—Outdoor coop for breaking up broody hens. Note the slatted bottom.

Biddy. She performs her labors in response to the highest ideal—that of reproducing her kind, and having completed the first step in the operation, the laying of the eggs, she cannot acquit herself of the responsibility until they are transformed into a fluffy flock of youngsters. It is a noble resolve, but, unfortunately for the hen, it has no place on the commercial egg farm.

Hens of the general purpose and meat varieties, such as Plymouth Rocks, Rhode Island Reds, Wyandottes and Brahmas are more addicted to this form of domesticity than are the lighter breeds; though the desire is pretty well founded in all classes of poultry, even to the so-called non-sitting breeds—Leghorns,

Campines and Minorcas. The non-sitting breeds, however, are not so habitually broody, nor so likely to be difficult to discourage.

In the fall of the year strict attention is paid to the requirements of the pullets to induce them to commence laying, and during the severe winter months that follow everything is done to sustain this yield. By March, which is the natural season for egg production, wherein almost anything that resembles a chicken is giving a good account of itself, the poultryman relaxes his vigilance over the egg basket and turns his energies toward other problems—mating, fertility, incubating and brooding. There is such an abundance of eggs at this time that this phase of the business seems to take care of itself.

Watch Out for Broodiness.—March, April and May are the months of heaviest production, after which the egg yield will fall off very rapidly if the poultryman is not watchful of his flock—on the lookout for broody hens. From the first of March and well into the summer the poultry keeper should make it a hard and fast rule to go over all the nests every evening, an hour after the last feeding time is best, and remove therefrom any fowls that show signs of broodiness. Very few hens lay after four o'clock in the afternoon, and inasmuch as they have no business on the nests after nightfall anyhow, it is a pretty safe practice to take up all birds found in the nests at that time, on the assumption that they are suspicious characters, and confine them in quarters specially built for their accommodation, which will be described in another paragraph.

Easily Broken at First.—Those who are inexperienced may reason that it seems unnecessary to make this a daily task, and that to go over the nests once a week or every few days will answer the same purpose; but such is not the case. A hen removed from the nest on the first day of her inclination to sit is very much easier to discourage than when she has been permitted to indulge her fancy for a week or more. She is usually rather indifferent about the matter at first and can be diverted with little effort, whereas at the end of a week the notion is a confirmed habit— a firm resolve, lodged crosswise in her mind and clinched on the inside, from which it is a tedious job to break her. Everyone who has raised chickens, no doubt, has had an opportunity to observe the tenacity and stubbornness of a sitting hen; her will-power is almost unconquerable.

Failure to discourage broodiness is probably the most potent cause for the low rate of egg production in the farm flock, and for which the farmer has nothing to blame but his own indifference or ignorance. The broody hen eats and drinks very little and takes practically no exercise, consequently she soon becomes thin and emaciated and we are apt to marvel how she sustains life at all. She could not survive if it were not for her ability to draw upon her internal store-house for sustenance. It is the depreciation of this store of energy that causes her egg-producing organs to become contracted and dormant, and in the same inactive, shrunken condition that we find in the immature pullet or the fowl that is going through the molt. Her entire attitude is that of sluggishness; the abdominal section that was once re-laxed and distended, is drawn well up into her body; the pelvic bones that were formerly pliable and spread far apart, are rigid and close together; and the comb and wattles that were pendu-lous and brightly colored, are now pale and shrivelled.

Fig. 223.—Egg-laying contest house for two pens of birds, Storrs, Connecticut.

Time Lost.—Briefly, when the hen becomes broody she reverts to the state of an undeveloped pullet, and the time involved is very short. Once she has been reduced to this condition, and with all things favorable, from four to six weeks are required to bring her back into laying. If conditions are not favorable; for example, if the weather is very hot, or she is not fed the proper ration, or if she was never but an indifferent layer at best, the chances are she will not resume laying until next season. Instead, it is highly probable that she will enter the molt.

Loss of Plumage.—When a hen is permitted to get in poor condition, especially in warm weather, either by illness, idleness on the nest, or by raising a brood of chicks, her plumage, like her body, dries up. It loses its sleek, glossy appearance; the oil in the quills is impoverished, and in consequence the fowl enters an early molt. The early molt is the longest, sometimes requiring four and five months, so that on farms devoted to egg production it is seldom considered profitable to carry the early "molters" over to the next season, and they are disposed of as meat.

Egg Production Is Secondary.—We might term egg production a supplementary function, for such it really is—secondary circulation, the result of over-stimulation. Strictly speaking, fowls eat to repair and restore the daily wear and tear to the body tissues—to maintain them in a healthy, normal state. If the amount of nourishment that they consume is merely equal to this task alone, there is none left for the work of stimulating the egg-producing organs to a point of activity. On the other hand, if there is an excess of nourishment, that which is assimilated over and above the daily requirements of the body, it goes into the development of the reproductive organs, which progress no faster than this nourishment is provided.

In the late winter and early spring the hen will instinctively labor to bring about this excess of nourishment for egg development. It is the natural season for her to commence laying. With the advent of milder weather she will forage for tidbits of greens, seeds, bugs, worms and other morsels in addition to her regular bill-of-fare, all of which is highly nutritious and places her in the pink of condition. When she has completed laying her first clutch of eggs, especially if it be a long one, the chances are she will manifest a desire to hatch them. Or, she may continue to lay another litter and then try to hatch it. Certain it is that she will try to hatch at least once during the spring months, and maybe three or four times.

If the poultryman is not alert, the hen will steal a march on him, for she is very persistent. The operator must thwart her plans immediately, and frustrate them in such a way that she is induced to continue to eat large quantities of food, and thus maintain her appetite, keep her body well nourished and prolong the life of her plumage, in which case she will continue laying throughout the summer.

Avoid Cruel Measures.—There are many ways of discouraging broodiness, but, remembering that the real reason for so doing is to induce further egg production, any practice or method that subjects the hen to cruelty or privation will only defeat the idea, therefore it should not be tolerated. The old-fashioned ideas of inflicting some form of punishment on the unoffending biddy because she responded to a natural impulse were wrong.

(*Courtesy Kansas Experiment Station*)

Fig. 224.—"A"-shaped colony house covered with tar paper.

Aside from humane reasons, to half starve or ill treat fowls, or to keep them from water, invites further loss in eggs, since these customs are sure to bring about the very condition that should be avoided—the reversion of the egg organs to a dormant state.

Small Flocks.—Where the birds are kept in small flocks, a good plan is to build a coop with a slatted bottom at the end of the roosting compartment, having it well ventilated and easy of access. See Fig. 221. Or an ordinary packing-case may be converted into a broody-hen coop: Remove the bottom and replace it with slats, mount the box on legs that will keep it about six inches off the floor, and then construct a simple wire-netting-covered-frame for a lid. As the *clucks* are taken from the nests they are placed in this coop, and the sensation of currents of air under them instead of eggs is disconcerting, to say the least.

The hens are unable to squat in a comfortable position, due to their legs protruding through the openings between the slats; they have no sense of privacy nor security, hence two or three days of this harmless pillory usually disgusts them with the idea of wanting to hatch a brood of chicks, and when released they are only too anxious to rejoin their companions in the laying house. It is understood, of course, that food and water are kept before them during their confinement, and that they are not to be treated like prisoners, but as hospital inmates.

In long laying houses of the continuous type, where the birds are kept in large units, a section of the roosting compartment may be given over to the broody-hen coop and fitted with a temporary slatted floor. It is better, however, to partition an end of the house in which there is the maximum amount of sunshine, and to remove all fixtures or nests that may offer secluded nooks. Green food in abundance should be placed in the pen along with fresh water and the regular grain rations, and if possible the inmates should be induced to exercise. The presence of a few lively cockerels in the pen will go a long way in breaking the obstinate *clucks*.

If the weather is mild, as it usually is when one has a large number of broody hens, another good plan is to place them in an open yard, giving them no access to a house even at night, except during a violent storm. Without refuge of any sort, and nothing to do but fuss with others of their kind, life holds very little enjoyment for them, so that even the most persistent members are readily converted.

However troublesome it may be to remove the broody hens every day, positively it must be done if eggs are to be secured in large numbers during the summer months. It is a part of the general scheme of intensive progressive poultry culture—equally as important as artificial incubation and brooding. Furthermore, it means greatly increased profits, for it should be remembered that August eggs bring about the same prices as January eggs.

CHAPTER XXVI

SURPLUS COCKERELS

Cockerels a Necessary Evil.—On poultry farms specializing in egg production the aim is to rear pullets. Cockerels are looked upon more or less as a nuisance. Yet, no matter how hard we try to mate our pens so that the hatches will run to females, as a general rule fifty per cent of the chickens are cockerels.

It is one of those natural laws over which we have no control. To the beginner this is sometimes discouraging, in view of the fact there is not a great deal of profit to be made from the average flock of surplus cockerels. In fact, some breeders complain that their cockerels actually become a liability, and sell for less than the cost of production. Where such is the case, there is something wrong with the management.

Poor Returns.—It is true, very few breeders derive any appreciable returns from their surplus cockerels, especially from males of the egg-laying varieties, such as the Leghorns. This is due in a large measure, however, to mismanagement. Either the birds are not properly fattened and prepared for market purposes, or they are not marketed at the right time. Many poultry raisers sell their cockerels at an early age, profit or no profit, regardless of the prevailing prices, in order to get them out of the way, retaining only the most promising specimens for future breeding purposes.

It is a mistake to force these birds on the market at a sacrifice, since they can be turned into easy money if one has sufficient space in which to segregate them, and then fatten and hold them for greater weight and better prices.

Separate at Early Age.—As soon as the cockerels are old enough for their sex to be determined they should be separated from the

347

pullets and kept by themselves.   If permitted to run with older fowls, they are constantly bullied and do not receive their proper share of food, which, of course, retards their development.   If they are allowed to mature with the pullets, the males not only get the *lion's share*, but they bother the pullets as well, which is not good for the pullets' growth.   If the pullets develop slowly, they are not fitted for early egg production.

Cockerels that are penned by themselves at the age of two

(*Courtesy Cornell Experiment Station*)

Fig. 225.—Wire-covered shed for housing cockerels in warm weather. Shed is built in the lee of a barn.

months live peacefully together, require very little attention, and if properly fed they can be made to put on flesh very rapidly. The quality of their flesh is greatly improved in this manner; instead of being dry and tough, it is tender and juicy, comparing favorably with the capon, and in place of angular bodies their carcasses will be plump and well-rounded.

The people of the United States are probably the greatest consumers of poultry and eggs in the world, and yet we are said to be satisfied with a very poor quality. The average quality of chicken seen in the retail store and on hotel tables in this country is far below that found abroad—in France, England, Belgium, Denmark and so on. This is largely due to the great consumption of broilers, which, however good they may be in some respects, lack the tenderness and abundance of flesh found on fowls that have been properly fattened before killing.

(*Courtesy Wisconsin Experiment Station*)

Fig. 226.—Home-made fattening crate located in the lee of a building. These crates can be utilized for broody hens as well.

In fact, as a general practice the fattening or finishing of poultry by special processes is virtually an unknown industry in this country.

The common plan has been to let the fowls eat all the corn they will consume for a couple of weeks before marketing, but this method does not produce prime table poultry in a strict sense of the term. Corn has a distinct tendency to put on weight—this is unquestionable; but this weight consists mainly of a heavy deposit of oily fat in layers under the skin, and in masses in the abdominal cavity, which is not particularly desirable inasmuch as it does not really constitute edible meat. In fowls that are properly fattened this excess weight is distributed in tiny globules of fat throughout all the body tissues, where it belongs; consequently in cooking this fat is not wasted, but renders the tissues soft and juicy.

The average American farmer is very careful to see that his steers and hogs are properly fattened before sending them to market, but to the poultry he pays little or no attention. It is difficult to find any excuse for this indifference, because a pound of grain can be converted into more poultry meat of greater value and in less time than through any *four-footed* medium.

Fattening Feeds.—Common sense dictates that if fowls are confined in small pens and kept quiet they will fatten much quicker than if allowed their liberty. The flesh of a chicken on unlimited range is tough and stringy, no matter how young the bird may be. Ground barley or oats, with one-third corn meal, thoroughly moistened with skim milk, makes a splendid growing and fattening food.

Cramming is practised by experts who wish to produce fowls of the highest quality table meat, and while it gives excellent results, the crate fattening method is almost as good, and will answer the purpose of the average poultry raiser. In the long run the crate method is probably the most profitable. See Fig. 227. It is practised very extensively in England, and in recent years many of the large American packing houses have specialized in it.

Rations.—The Ontario Agricultural College conducted a series of experiments in fattening poultry, and found that a mixture of two parts corn meal, two parts ground buckwheat and one part pearl oat dust, with an equal weight of skim milk, gave excellent results at a cost of three and a half cents per pound for the weight gained. Oats and skim milk made the gain cost slightly over five cents per pound.

Another good fattening ration is made up of 100 pounds of

corn meal, 100 pounds of wheat middlings, and 40 pounds of animal meal. The fowls should have access to plenty of sharp grit to aid digestion, and as a blood purifier, some poultrymen mix a little sulphur in the mash once every two weeks. At the close of the fattening period, which usually lasts about six weeks, a little tallow may be added to the feed.

It is unquestionable that the breeder of heavy fowls, such as Plymouth Rocks, Wyandottes and Rhode Island Reds, has the advantage over the Leghorn breeder when it comes to disposing of cockerels. If desired, the heavier breeds can be caponized, or they can be held for *roasters*, and made to weigh six or eight

(Courtesy U. S. Dep't Agriculture)

Fig. 227.—Pouring feed into the troughs of fattening crates.

pounds. Furthermore, breeders of heavy fowls usually hatch a month or two earlier than Leghorn breeders, consequently their young stock can be made to weigh four to five pounds to the pair at a time when broilers and fryers command top prices.

Squab Broilers.—March, April and May are the best months for hatching Leghorns, hence cockerels from these hatches are not large enough to be sold as broilers or fryers until June or July, at which time market prices have materially declined. In some localities there is a fair outlet for Leghorn cockerels as *squab broilers*, weighing about three-quarters of a pound each, during April and May. They are sometimes called *asparagus chickens*, and the price is about a dollar per pair.

Costs.—Leghorn breeders will do well to take advantage of this market whenever possible, since it means a very fair profit, and it also offers an opportunity to relieve any congestion in the colony houses. It is comparatively easy to bring Leghorns to weigh a pound in six weeks' time, and under normal circumstances the cost of production, including the original value of the egg and its incubation, together with the labor and expense of dressing and shipping, should not exceed twenty-five cents. This leaves a net profit of twenty-five cents per bird—nothing to brag about,

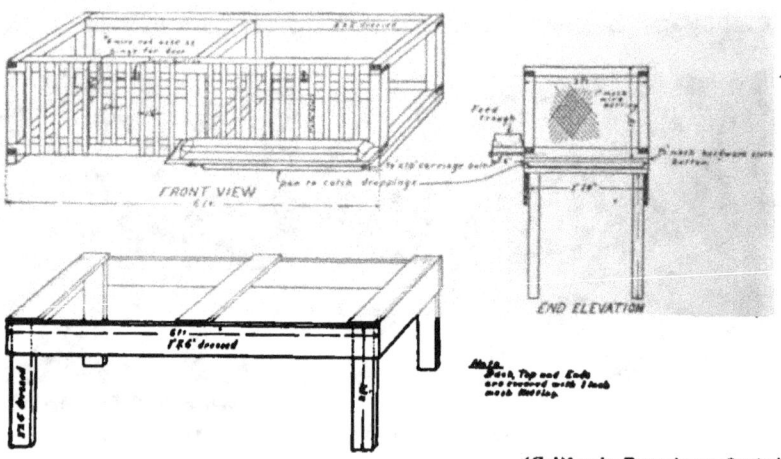

(*California Experiment Station*)

Fig. 228.—Plans for a 2-compartment fattening crate.

perhaps, yet all things considered it is a very fair return on the investment. This profit will defray the expense of maintaining a pullet for two months.

When Prices Are Low.—When prices on young chickens fall below twenty cents per pound live weight, as they do in July and August, there is very little profit to be had over the cost of production, especially for Leghorns weighing about two pounds each. The market is usually so glutted that buyers can afford to discriminate in favor of the heavier breeds. At such times, rather than sacrifice the shipments, it will pay the poultryman to fatten the cockerels for small roasters. True, the fact that they are Leghorns will always discount the highest prices, yet if they are properly fattened and neatly dressed their returns will be pretty nearly commensurate with their cost of production.

Quarters.—Some breeders argue against the idea of keeping surplus cockerels for the reason that they cannot provide suitable living quarters for them. Admitting that the poultryman's

(*Courtesy Million Egg Farm*)

Fig. 229.— Killing and dressing broilers—surplus cockerels.

space is limited, the cockerels come along at a time when it is usually advisable to dispose of some of the older fowls for meat, especially the breeding males, whereupon, by combining other pens, there will be left one or more empty pens for the accommodation of the cockerels. See Fig. 225. Before the breeding season opens again, or before these pens are required by the pullets, the cockerels will have been sold or used on the home table.

Quality Counts.—Many a consignment of poultry has brought

23

poor returns and bitter disappointment to the shipper simply because it was not uniformly graded, or because the fowls were carelessly dressed or improperly packed.   Nine times out of ten the poultryman has no one but himself to blame for poor prices. Specimens that are thin and emaciated, malformed in any way, those having crooked breast-bones, or those with bruised or mutilated skins, should not be included in a shipment intended to be sold at top prices.

It is far better to leave a few undesirable carcasses out of a shipment, to be used on the home table, than to include them, no matter how tempting it may be to add this additional weight.   Remember that the sales account is going to be returned on the basis of quality.   To keep the *undesirables* at home you may lose a few pounds, from which at least you derive a meal or two for the family; whereas to send them in the shipment may be the means

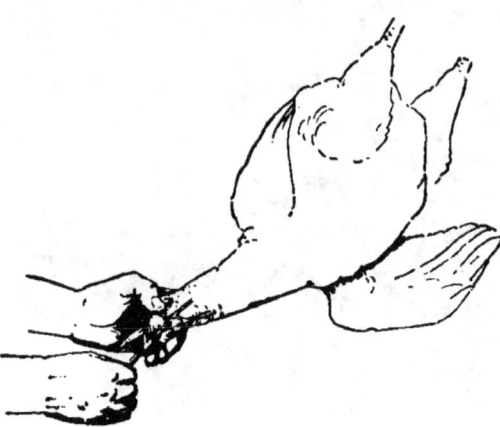

(*U. S. Dep't Agriculture*)

Fig. 230.—Correct way to grasp head of fowl for killing and bleeding.   Note position of knife in respect to veins.

of reducing the price a couple of cents per pound on the entire lot. Nothing appeals to a dealer so much as uniform quality.   Each grade should be uniform in size, shape, color of skin and shanks, age and degree of plumpness.

Scalding is the most rapid method of removing the feathers, and there is less loss in dressed weights than by dry picking, due to the absorption of a small amount of water by the body in the plumping process, but it is almost impossible to practise this method without destroying the natural appearance of the skin. If the water is too hot, or the fowl is immersed too long, the skin is partly cooked and the thin scarf skin peels off, which causes the flesh to become discolored. If the water is not hot enough, it is then difficult to remove the feathers without tearing the skin. The correct temperature is about 180 degrees. When the feathers are removed, singe the body with an alcohol flame to remove hairs, and then plunge it in cold water to remove the animal heat, and to plump the carcass.

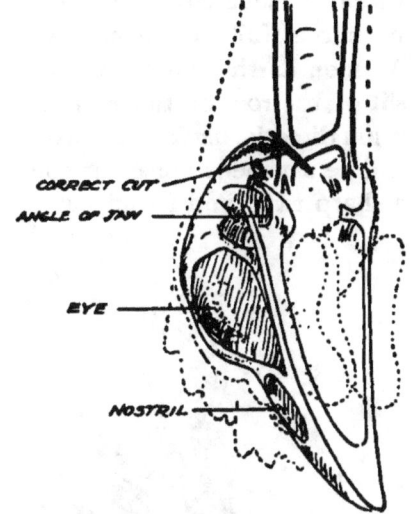

Dotted lines indicate comb and wattles.

Dry Picking.—It requires more time and skill to dry pick fowls, yet one is usually compensated for this additional trouble. Dry picked poultry is more attractive and brings higher prices, and in many markets no other sort will be tolerated, except to a cheaper class of trade.

The success of dry picking depends on getting the *right bleed* and the *correct stick.* Hang the fowl by a cord (see Fig. 229), or hold in the hands while sitting, seize the head in the left hand (see Fig. 230), and with the right hand run the blade of the sticking knife into the throat until the large artery in the left side of the

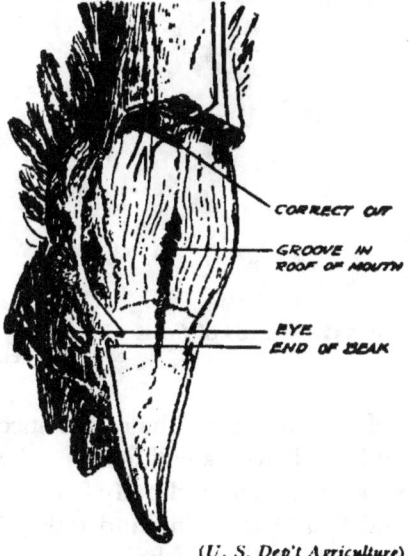

(*U. S. Dep't Agriculture*)

Fig. 231.—Anatomy of skull, showing position of veins and correct way to cut for killing.

throat is severed.   This operation is termed *bleeding*, and must
be successfully done before the fowl is stuck.   See Fig. 231.
As soon as the blood spurts freely insert the knife-blade in the
slit of the roof of the mouth and plunge it backward into the
brain directly back of the eye.   When the brain is reached, there
will be a violent muscular contraction, whereupon give the knife
a sharp twist and withdraw it.

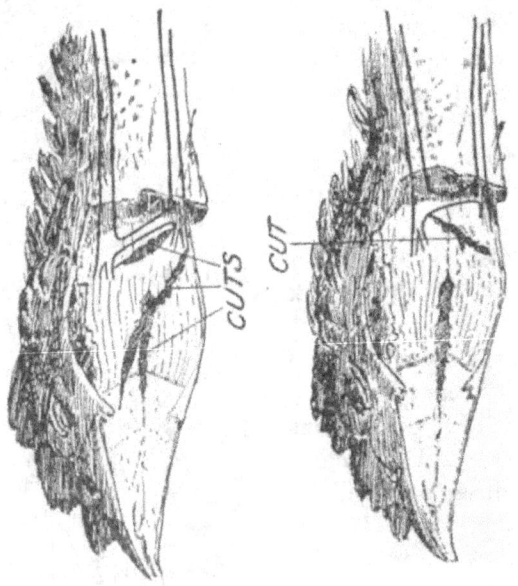

(*U. S. Dep't Agriculture*)

Fig. 232.—Heads of fowls with lower jaw removed, showing poor attempts at
severing veins in throat.

If the operation has been successful, paralysis will be induced,
and the feathers will loosen.   Begin picking immediately; rough
pick the breast and body feathers first, then wing and tail feathers
and finally the wing and tail quills.   Always remove the greater
bulk of all the feathers before attempting to clean off the small
feathers.   Avoid pulling too many feathers at one stroke to prevent tearing the skin, and remember that tearing is most likely

to occur on each side of the breast and on the neck. A neat job and rapid work are only acquired after considerable practice. To the uninitiated it looks very simple, indeed, until experience proves that there is quite a *knack* to be learned.

Give No Food.—All poultry intended for slaughter should be kept without food for twenty-four hours, and in many states this is required by law. In so doing the intestines are given time to become empty, which helps to prevent decomposition of food materials within the body, and adds greatly to the keeping quality of the carcass. Water should be supplied during this fasting period, for it aids in cleansing the intestines.

A few markets prefer poultry drawn, but that sent to New York, Chicago and cities in general is not drawn. Undrawn poultry keeps best.

Shipping.—Bearing in mind that the package frequently sells the product which it contains, the poultryman should prepare his shipments in the most attractive manner, so they will arrive in the best possible condition. Each box or barrel should be lined with paper, preferably parchment paper, which will help to prevent evaporation, or injury to the contents through rough handling. Clean rye or wheat straw may be used to advantage.

For icing poultry in barrels in warm weather, bore a hole in the bottom for drainage, then place a layer of ice, then alternate layers of poultry and ice until the package is full. Pack the poultry breast down and back up, with the legs straight towards the center of the barrel, making a ring of fowls side by side around the outside. The middle of the barrel may be filled with bodies or with cracked ice. Over the top layer of poultry place a layer of ice, then a piece of burlap, and finally a layer of ice on which the head rests. Poultry packed in this manner can be shipped long distances and should arrive at its destination in perfect condition. In cold weather it is seldom necessary to use ice.

Mark all shipments clearly, giving the name and address of the consignee, name and address of the shipper, and the contents of the package. A stencil is useful for this purpose; it is much neater than amateur printing. Besides, it is more business-like.

CHAPTER XXVII

CAPONIZING

Does Caponizing Pay?—There is a wide difference of opinion on the subject.

Without attempting to answer this question offhand, let us first consider the matter fully, and from different points of view.

Delicious Meat.—Capons are undoubtedly a more delicious meat than an uncaponized bird, which is especially true of fowls that are held past six months of age. The flesh is sweeter and of a superior flavor in the capon, consequently it brings much higher prices. The markets of the entire country are sparingly supplied with capon flesh, hence there is a constant demand for it at uniformly good prices. There is no definite capon season, apparently, but most breeders market their stock after the holidays, from January to March. Usually the highest prices prevail at this time. Even so it is almost impossible for the chance buyer to pick up any capons in the general markets, because the supply is seldom equal to the demand.

Advantages of the Capon.—The capon has many advantages over the rooster: In the first place he is very docile, his disposition is entirely changed, he seldom if ever fights, he declines any great amount of exercise, and will stand close confinement well. Life holds very little for him, except to eat and sit around and grow large and heavy. The meat of capons is more economically grown than that of cockerels, because more of the food consumed is stored up on the body as flesh and less is devoted to energy. What is most important, capons can be kept longer than cockerels, because they will continue to grow larger and heavier, without becoming coarse and staggy. They can be kept for a year or longer, and sold profitably when poultry is scarce and bringing

358

higher prices.   Not caponized, it is quite likely they could not be kept in prime condition for market longer than the fall of the year, at which time poultry is so plentiful that prices are usually low.   Capons command from twenty-five to forty cents a pound, depending upon locality and the season, while the ordinary rooster brings from twelve to twenty-five cents.   These are pre-war prices.

Surplus Cockerels.— On the average farm, especially those devoted to egg production, cockerels are taboo—unwelcomed guests.   Every year several hundred thousand male birds are sold at an actual loss to producers, because poultrymen believe they are a nuisance.   It is contended that they do not more than pay the expense of raising.   This is true if the young cockerels are marketed direct from the range without any special preparation.

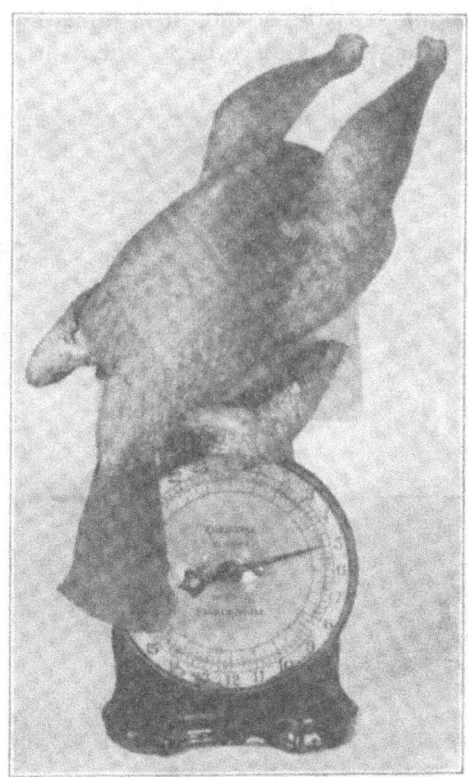

(Courtesy Wisconsin Experiment Station)

Fig. 233. — Well finished market fowl. Note the plump breast, well covered body and short thighs.

Cockerels sold off the range are too thin and muscular.   As prime table poultry they should be fattened for a couple of weeks in crates, as described in the previous chapter.

Increased Profits.—In our efforts to secure pullets for egg production, we cannot evade raising an equal number of cockerels, of which only a very small percentage are required for breeding purposes. Therefore, as a simple business principle, if we must raise surplus cockerels, we should strive to convert them into a profit, if only as a by-product. And if caponizing will bring this about, it is the strongest argument in favor of the practice.

Since the poultry department at Cornell University has been

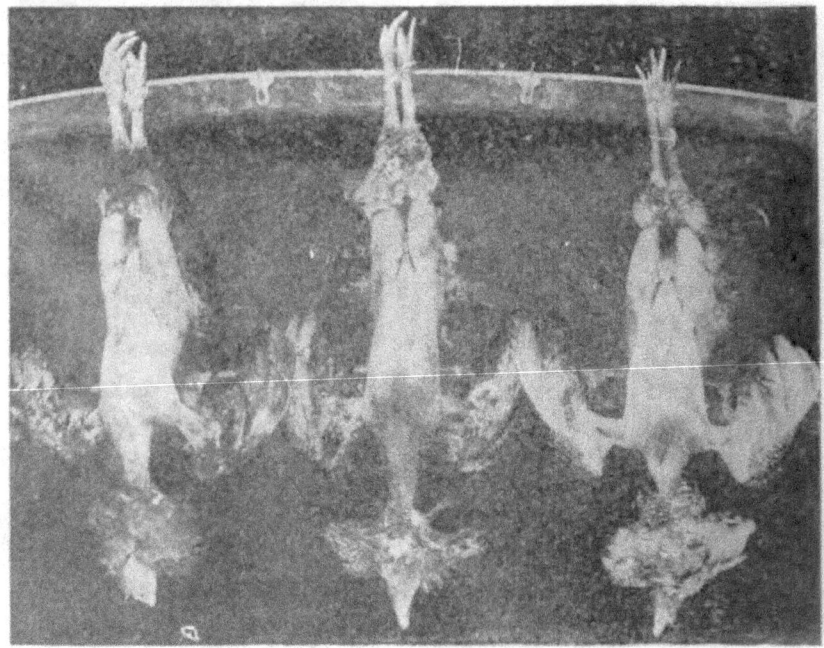

(*Courtesy U. S. Dep't Agriculture*)

Fig. 234.—Capons dressed for market. Conventional method of picking; the birds, however, are not in perfect condition of flesh.

caponizing their surplus males, it has added several hundred dollars a year to the net profits. Not only are better prices received for capons, but there is an increase in weight over the normal state for the same age and under precisely the same care and feeding. In the same length of time it is possible to raise capons that will weigh almost a half more than they would have weighed

as cockerels.   At six months of age cockerels have received their
most rapid growth.   The same birds, if caponized when about
twelve weeks old, can be made to continue their growth until
they are eight and ten months old.

Larger Breeds Are Best.—Cockerels of any breed can be made
to increase in weight by being caponized; yet the larger breeds

(Courtesy U. S. Dep't Agriculture)

Fig. 235.—Side view of capons dressed for market.

are by far the most desirable.   If a poultryman is going to specialize in capons, rather than develop surplus males as a side line, then
of course, the selection of the breed to be used is of primary importance.   The operator must keep before him the idea of well-
finished, well-rounded, solid meat, a bird that has the greatest
amount of flesh for the least amount of bone, and the shortest
shank.   See Fig. 233.   This ideal condition is shown by the use

of Asiatic breeds—Brahmas and Langshans, which are still
further improved by the admission of Plymouth Rock and English
Game blood.

Cross Breeds.—It is not uncommon for Light Brahma capons
to weigh from twelve to fifteen pounds each at eight or ten months
of age. Smaller capons, however, will more nearly supply the
needs of the average family, consequently the Plymouth Rocks
and Wyandottes will be found very satisfactory. White Plymouth Rocks, straight or crossed with White Cornish Game, give
excellent results, a long full breast, clean yellow legs, and no
dark pinfeathers.

Another practice is to cross the Barred Plymouth Rock with
the Light Brahma, which will give greater weight, but the dark
pinfeathers are an objection. A bird with handsome plumage is
more attractive as a capon, for in dressing this class of poultry
much of the plumage is left on the body. See Figs. 234 and
235.

The conventional way to dress capons is to leave the head and
hackle feathers, the feathers on the wings to the second joint, the
tail feathers, including those a little way up the back, and the
feathers on the legs halfway up the thighs. These feathers
serve to distinguish capons from other fowls in the market,
and for this reason partridge-colored birds are used to a great
extent. The undeveloped comb and wattles are other distinguishing features of the capon, also a long, rather pointed head.

Time to Caponize.—Cockerels may be caponized at any age,
but for the comfort of the bird and convenience of the operator,
it is not advisable to perform the operation when the birds are
too young or after they are more than six months old. In determining the proper time, the size of the bird should be the
governing factor, the most desirable time being when the birds
weigh from two and a half to three pounds, or when they are
about three months old. The operation will succeed on older
birds, but the percentage of deaths and *slips* will be greater. A
slip is a bird that is neither capon nor cockerel, and brings no
better price than a cockerel.

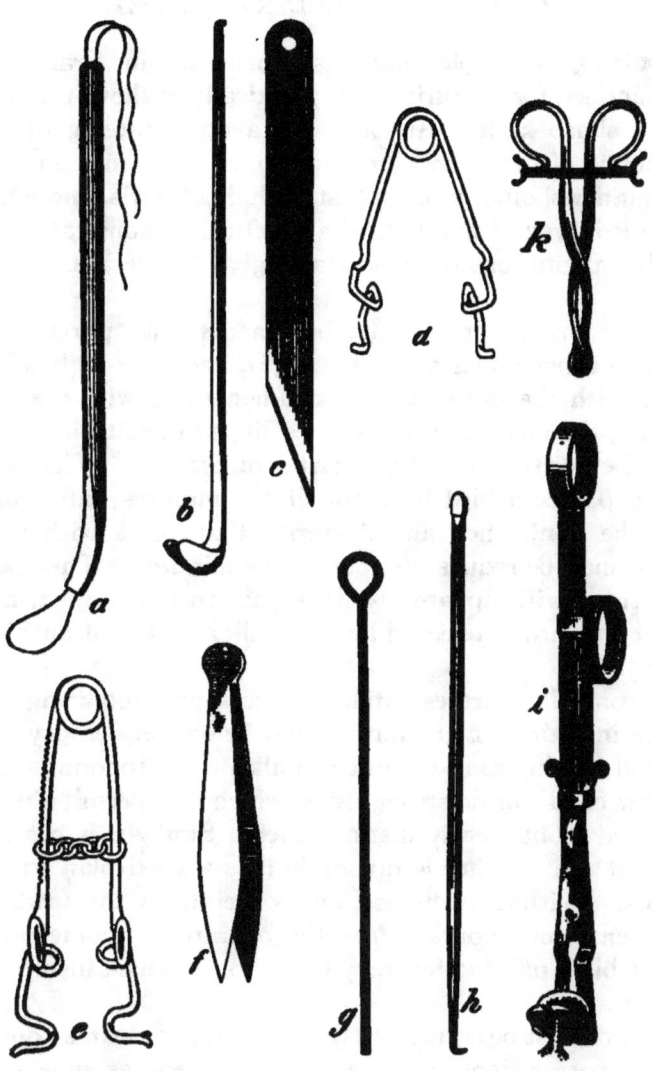

Fig. 236.—Caponizing instruments: *a*, hollow tube cannula; *b*, scoop cannula; *c*, knife; *d*, spring spreader; *e*, sliding spreader; *f*, tweezers; *g*, probe; *h*, sharp-pointed hook; *i*, sliding cannula; *k*, spoon forceps.

363

Caponizing is simple and easy to learn; in France it has been practised for centuries, and practically without instruments except a sharp knife. Anyone with average intelligence and a fair amount of dexterity can learn to caponize in a short time. The beginner should practise first with dead fowls, and wherever possible it is well to first attend a practical demonstration. The Agricultural Stations of some states give free lessons at certain seasons.

Reliable Instruments.—The beginner should purchase a reliable set of instruments (see Fig. 236), and once he has become familiar with the use of each instrument, and with the manner of making the incision, there will be little difficulty in doing the work. Dexterity is simply a matter of practice. The beginner should caponize a bird in about fifteen minutes; after one has gained the confidence and dexterity that come with practice, this time may be reduced to four or five minutes. The operation is performed with apparently little pain to the subject, and the moment the bird is released he will walk about as if nothing had occurred.

Two conditions are essential to success in caponizing: One is that the intestines of the bird should be entirely empty, so that there will be the least amount of bulk in the abdominal cavity; the other condition is strong light, which will permit the organs of the bird to be clearly distinguished. Sunlight is best, consequently if the weather is favorable it is a good plan to operate outdoors. Withhold all food and water from the fowls for at least twenty-four hours before the operation. Some operators keep the birds off food for forty-eight hours, which insures empty intestines.

The bird must be secured to the operating table in a convenient manner; pass a noose of cord about the legs, as shown in Fig. 237, and tie the wings in the same way. Attach weights to the ends of the cords, which will hold the bird in any desired position.

Have the instruments conveniently at hand, also a basin of water, to which have been added a few drops of carbolic acid, and

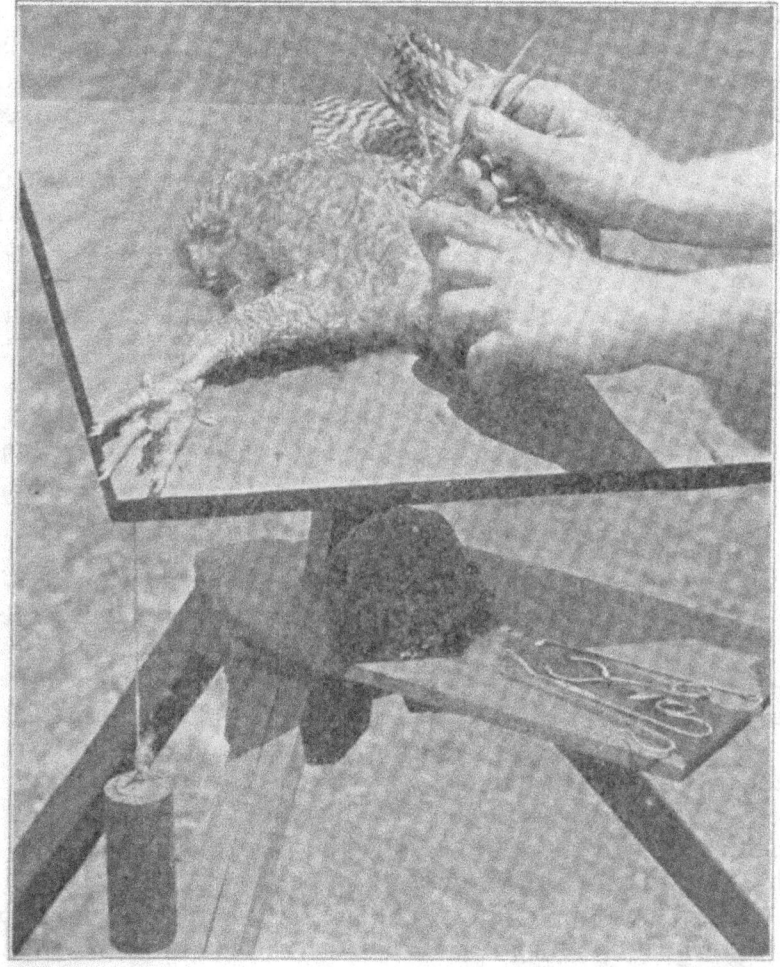

Fig. 237.—Layout for caponizing. First step in the operation is to locate the last two ribs and make the incision.

365

some absorbent cotton.   Once the operation is started, carry it through as quickly as possible.

Locate the area for the incision, between the last two ribs, as shown in Fig. 238, and then remove a few feathers where the cut is to be made.   Moisten the surrounding feathers to keep them out of the way.   Before making the incision, stretch the outer skin as far as possible toward the thigh; thus when the operation is completed the opening in the outer skin will not be over the cut between the ribs, inasmuch as it will have slipped back to its normal position.   Make the incision as neatly as possible, about an inch long.   There is little danger of cutting the intestines, providing the bird has been sufficiently starved.   Insert the spreader, being careful that it presses against the ribs, thus springing the ribs apart and exposing the intestines.   See Fig. 239.

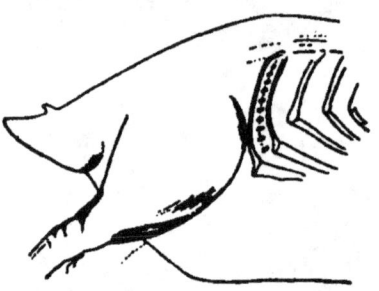

(U. S. Department of Agriculture)

Fig. 238.—Diagram of side of fowl, showing where incision should be made between the last two ribs.

The intestines are covered with a thin membrane called the *omentum*.   Tear this membrane apart with the point of a sharp hook (see Fig. 240); push the intestines aside with a probe, and up against the backbone the glands or *testicles* should be in plain sight.   See Fig. 241.   These glands are a creamy yellow and about the size and shape of a bean.   In very young birds the glands are little bigger than a grain of wheat.

Skilled operators remove both glands through one incision, in which case the lower gland should be removed first, so that any bleeding will not obscure the other gland.   Inexperienced operators remove only the upper or nearer gland, and then make a second incision on the opposite side of the body for the removal of the other gland.   This takes double the time, and is much harder on the bird.   It is not a good practice.

Back of the gland is a large blood-vessel, the spermatic artery,

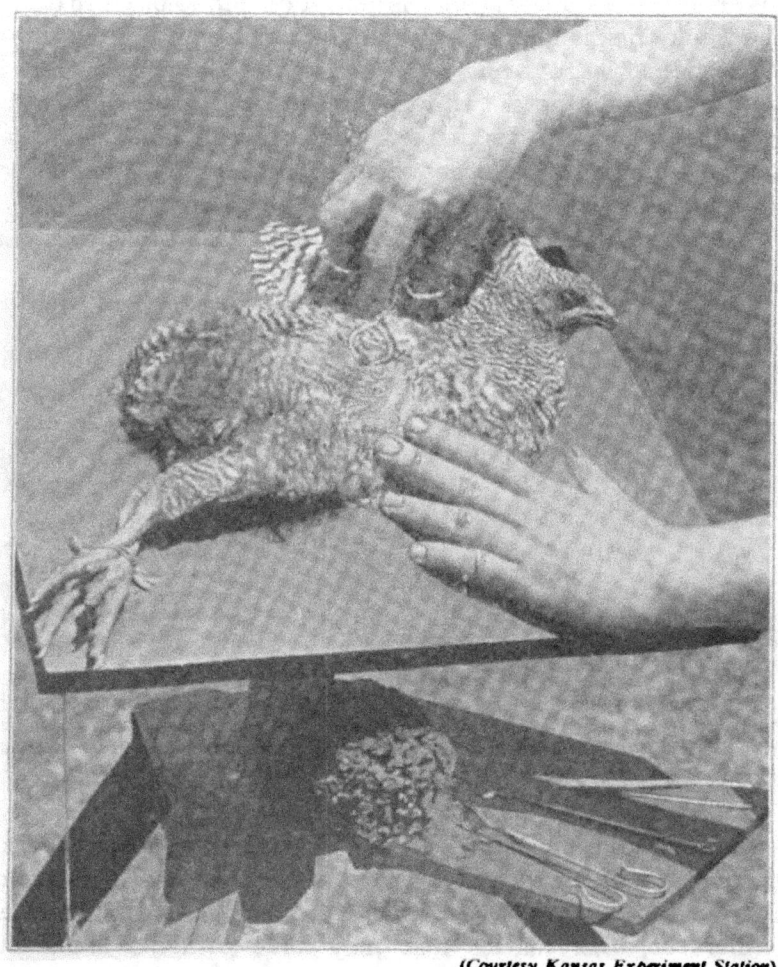

Fig. 239.—After the incision is made the spreaders are inserted to hold the cut open.

which constitutes the delicate part of the operation, because if this artery is ruptured the bird is almost certain to bleed to death. The whole trick is to grasp the gland without grasping the artery or the tissues surrounding it. The cannula or spoon forceps is used for this purpose. Having grasped the gland, twist the instrument around several times, then tear the gland from the body and remove it. Repeat the operation on the other gland. See Fig. 242.

After removing the glands, if the bleeding is at all profuse, it

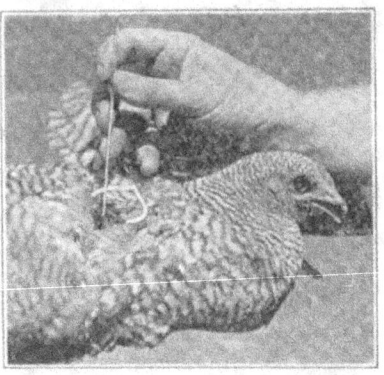 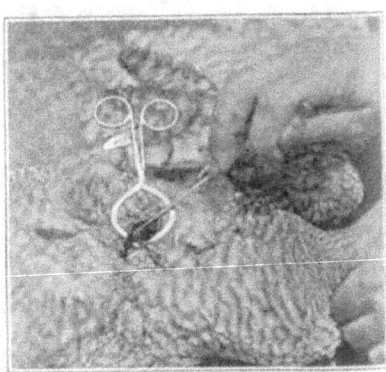

(Courtesy Kansas Experiment Station)

Fig. 240.—The hooked end of a sharp probe is then used to tear away the thin membranes covering the intestines.

Fig. 241.—When this is done the intestines are pushed aside, revealing the gland, which is about the size and shape of a small bean.

is advisable to remove a portion of the blood by means of small pieces of absorbent cotton, inserted in the wound with the aid of the tweezers or probe. Be sure to remove all blood-clots, feathers or other matter that may have gathered inside the wound, then take out the spreaders, thus allowing the skin to slip back over the cut. See Fig. 243.

Losses are likely to occur with the best operators. The mortality, however, should not exceed five per cent under average circumstances, and with a skilled person it will not be more than two per cent. If the birds are killed accidentally, they are perfectly good to eat, hence they are not wasted.

A Slip.—Sometimes the operation appears to be very successful, yet the bird develops much the same as a cockerel. This condition is due to the fact that a small portion of the gland has been left in the body. Such specimens are termed *slips*. They are neither cockerels nor capons.

Following the operation the birds should be placed in a clean pen by themselves. Give them all the water they want, and for

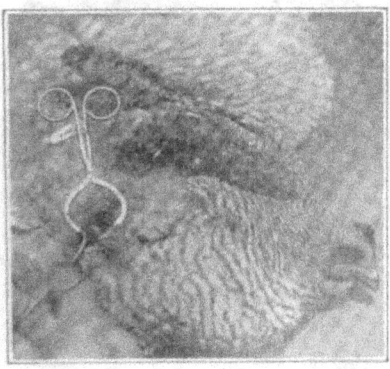

Fig. 242.—The gland is grasped by means of the spoon forceps or cannula, twisted round and removed.

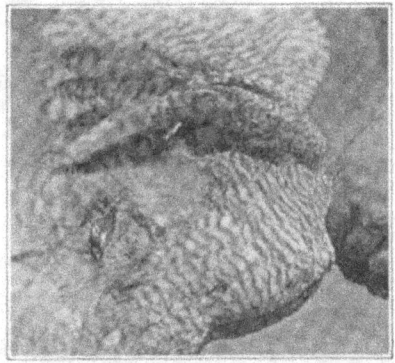

(Courtesy Kansas Experiment Station)

Fig. 243.—When the second gland is removed, or if both glands are removed at once, unfasten the spreaders and allow the skin to cover the opening between the ribs, as shown, and the operation is finished.

the next few days keep them confined on soft feed. A few of them will bloat during the first week, but this is not serious and can easily be remedied by pricking the skin with a needle. It is caused by air getting under the skin, raising a slight swelling or *wind puff*.

No stitching is required by the operation; the wound will heal in a few days, and in a couple of weeks it will be hard to find even a scar. Capons should be fed nourishing rather than fattening feed for the first few months, the object being to keep them

24

growing. About three weeks before marketing they should be fattened, either in small yards or crates. Corn meal and ground oats, equal parts by weight, or corn meal and middlings, moistened with water or milk, make a good mash for fattening capons or other poultry.

The one great mistake in raising capons is in marketing them too early and not having them fat enough. After caponizing they can be made to put on flesh in a surprising manner, and they

Fig. 244.—Class in killing and dressing at the Kansas Agricultural College.

should be kept growing as fast as possible until they attain full size, which will take from six to ten months.

After caponizing the comb and wattles cease to develop; the hackle, saddle feathers and tail feathers grow very long, and the plumage assumes a heavy, glossy appearance. Capons never crow, manifest no interest in the rest of the flock, and are generally despised for their peculiarities by other fowls. They are exceedingly gentle and tractable, and often show a fondness for little chicks. In fact, there are instances in which they have been utilized in rearing broods of chickens.

Their behavior with chicks, in the main, is very much like that

of a hen; such differences being of a minor character and difficult
to distinguish clearly from those of a broody hen. They will
cluck, though in an imperfect way, hunt food for the chicks, and
attack anyone who approaches the brood.

Darwin states that capons are said to incubate eggs as well
as care for chicks, but the writer has no proof of this statement.
In fact, tests of capons are not extensive enough to warrant definite conclusions, but it would appear entirely plausible that the
brooding instincts are after all not necessarily female characteristics. Male pigeons, for example, assist in brooding and rearing
the young, as do guinea cocks and many wild birds.

In housing capons very little space is required, providing the
quarters are kept clean. About two square feet per bird is
sufficient. They require very little care; one man can easily
take care of three to five thousand capons, which is a point in
favor of the practice.

The cost of feeding a capon to maturity, or for twelve months,
is from eighty cents to a dollar and a quarter, depending upon the
cost of feed, and whether a fair portion of the daily fare is picked
up on the range. Add to this the cost of the operation, let us
say, five cents, and the cost of the chick at hatching time, and
we have a total cost of about a dollar and a quarter. Assuming
an average weight of eight and a half pounds for a twelve months'
bird, which is very conservative, with a selling price of thirty
cents a pound, the market value at killing time is two dollars
and fifty-five cents. This leaves a net profit of more than a
dollar and a quarter—a profit that compares very favorably
with the results obtained from a heavy laying hen.

CHAPTER XXVIII

BREEDING PRINCIPLES

In the minds of many who raise chickens and other fowls there is but one idea and one kind of breeding—that of mating males and females, regardless of type, strain, variety, prolificness or relationship. Needless to add, that such matings sooner or later—*usually sooner*—prove of little value to their owner, and are finally completely dissipated. It is Nature's way of eliminating the unfit.

Definitions of Breeding Methods.—To be precise, there are the following methods: *In-breeding, line-breeding, out-breeding,* and *cross-breeding;* and—shall we say—no breeding at all, meaning rank mongrelism.

Line-Breeding.—Primarily, it is not advisable to make a practice of mating birds more closely related than first cousins, and the more distant this relationship can be drawn apart, the better the chances for success. In making a start with a flock of fowls, however, where one wishes to preserve the same strain of blood, or in creating a new breed, it is usually necessary to breed pretty close for a number of years, or until certain qualifications become intensified and fixed. If this breeding of related birds is done intelligently, with the view of fixing superiority in color, shape and so on, it is called *line-breeding.* If the breeding of related stock is done indiscriminately, and brothers and sisters are bred together for generations for no particular purpose, it is called *in-breeding.*

In other words, *line-breeding,* or .breeding in line, is keeping to the same ancestry—the same blood lines, without the disastrous effects of *in-breeding.* It is carefully selected, systematic in-breeding.

372

Out-breeding is a term applied to the practice of introducing new blood every year, but blood of the same breed. *Cross-breeding* is introducing entirely new blood of a distinctly different breed.

New Males.—Through fear of the flock degenerating many poultry raisers consider it absolutely necessary to bring in new males each year. Very often they make a practice of exchanging

(Courtesy Purdue Experiment Station)

Fig. 245.—Pullets should be placed in their permanent laying quarters as soon as they show signs of maturity.

males with nearby farms, which is the vogue among farmers, especially. This is inspired by the right idea, but it is likely to be accompanied by trouble. If it is desirable to introduce new blood, the rule should be to do so—not just because it is new blood, but because it is superior to your own in vigor and other ways.

Speculation.—It is very difficult to raise standard bred birds if new blood is added to the flock each year. You may buy a pure-bred male to mate with your pure-bred females, and later find that the two strains failed to *nick* properly. That is, the mating may throw offspring with defective combs, disproportion or poor color, which will take several generations of special breeding to eliminate. In short, the advent of new blood is a speculation.

A better way to introduce new blood is to take two years to do it, and experiment with individuals. Purchase a few hens of the desired strain and mate them to your best males, or secure a couple of outside males and mate them to your best females; then study their offspring for a year, and if satisfactory, mate the new blood to the balance of the flock.

There is no evidence to prove that *line-breeding* initiates degeneracy, providing reasonable care is exercised each year in selecting only vigorous breeders, and there is a large number of fowls from which to choose. The danger becomes even more remote if two divisions of the same blood are kept going year after year. This consists of keeping two distinct strains or matings on the same farm, both of which have a common ancestry, but which grow farther apart every year.

Every season the males of one line are mated to the females of the other line, and vice versa, these lines having been started by mating the best male to the best female, and continuing the second generation by mating the original male to his daughters, or the original hen to the son. Proceeding in a similar manner for the third generation, the original male is mated to granddaughters and the original hen to her grandson, which practically eliminates from each line its original respective sire or dam. It is difficult to explain this system of line-breeding in writing, but if you will make a chart of it and get down to actual figures, you will soon see that it is very simple.

Cross-Breeding.—Some time in the career of every poultryman there is the temptation to cross-breed with a view to improving certain qualities. In most instances the crossing of two pure breeds is a mistake. The appearance alone of a flock of cross-bred fowls when compared with the pure breeds whence

Fig. 246.—Three thousand White Leghorn pullets on colony range.

375

they originated should convince any one that this is a bad plan. The first cross is not so bad, as a rule, and occasionally it possesses some slight advantages in egg production or weight, but these hybrids should not be mated in any way, either among themselves or back to their parents. Therefore, to continue cross-breeding it is necessary to maintain two distinct pure breeds year after year, and to destroy the hybrids as soon as they cease to be profitable. This occasions many separate houses and yards, for the sexes of each pure breed, and for the crosses, a practice that is both expensive and troublesome.

Grading Up Mongrels.—Sometimes it is profitable to grade up

(*Courtesy Maryland Experiment Station*)

Fig. 247.—"Busy moment for the trap nests."

a flock of mongrels, such as are found on many general farms, by introducing pure bred males. Pure bred males of the same variety should be used year after year, however, and not the males from the offspring of the first cross. In the course of four or five generations, with careful selection, it is possible to grade up the original flock of mongrels to the level of the pure bred male in appearances, but scarcely in breeding qualities. There is always more or less chance of a reversion to type in breeding from mongrels, hence it is often cheaper in the long run to commence with pure bred stock.

Barnyard fowls are better than none, of course, but why keep mongrels when pure bred birds can be had for almost the same price. Those who appreciate the value of uniformity in body and eggs, and who realize the need of transmitting these qualities to the progeny, find no argument in favor of the *manure pile diggers*. There are more beauty and more dollars in the thorough-bred—be it hog, horse, cow or fowl.

Heavy Laying Strain.—The trap nest is the only positive index to the hen's performance as a layer. It has furnished the only means of establishing many facts leading to a more or less definite conception of just what characteristics belong to the heavy layer, also, the qualifications of the fowl possessing the faculty of trans-

(Courtesy Missouri Experiment Station)
Fig. 248.—Brood coop with slatted run for chicks.

mitting certain desirable qualities to its offspring. In fact, the trap nest has collected such a vast amount of data, that it is now possible to dispense with its service, if need be, and still profit from it. In other words, we are now able to verify certain external indications, actions and habits as belonging to this or that type of fowl.

Thus, the progressive poultryman who wishes to mate his birds along definite lines, but who is not in a position to *trap* them, can select his breeders so that their increase in efficiency compares favorably with flocks that are trap-nested.

Relation of Size and Shape.—No sensible person questions the importance of shape and size in the matter of breeding horses and cows.   One does not mate Percheron stock for speed in horses, nor Hereford cattle for a dairy farm.   The same idea holds true of poultry.   The general shape of the laying type of chicken is agreed upon as a V or wedge when viewed from the top, side and rear, the supposition being that in this conformation the egg-producing organs have the greatest opportunity for development.   This shape is sometimes·called by the term CAPACITY, which really amounts to abdominal power.

Capacity means the ability of the crop and digestive tract to receive, consume and assimilate large quantities of food.   Continuous egg production is an intensive, exhausting process.   It necessitates the consumption of vast quantities of food, otherwise it would be physically impossible for the hen to turn out an egg a day.   This is only common sense reasoning.   Beware of the small eater, or the hen that goes to roost on a crop half full of food.   She is either a defective, a drone or an invalid.

The depth of the abdomen, as well as the length, indicates CAPACITY.   Hence the good layer is described as having a long body, deep in the keel, which is another term for the breast bone. To ascertain or measure the abdominal capacity of a hen, the fowl is grasped by the legs in the left hand, and its head and wings are held under the arm in a horizontal position in what is admitted to be the correct method of holding or carrying a chicken. Then the fingers of the right hand are placed on the abdomen between the two pelvic bones on either side of the vent and the rear of the breast bone.   This distance will be found to vary quite considerably with different hens.   In some the width of one finger will be found to be sufficient to occupy the space between the pelvic bones and the breast bone, in others two fingers will be required, in others three, four, five and six fingers, and in rare cases seven fingers.

The pelvic bones are sometimes called the *lay bones or vent bones*.   They, too, are measured for the distance or spread between them; but do not mistake the abdominal measurement for the distance between the pelvic bones.   This latter test is for another purpose.   The tips of the fingers are used for this test, whereas for the abdominal measurement the widths of the fingers are used between the tips and the knuckles.

Further indications are found in a large comb and wattles of good color, a high tail rather than one carried low, medium-size head with a short, stocky beak, rather short legs well spread apart, and as much weight as the specimen should have to conform to the standard requirements for a particular variety.

Under-size specimens are seldom exceptional performers.

(Courtesy Petaluma Chamber of Commerce)

Fig. 249.—Leghorn pullets on a California poultry ranch.

The same is true of over-developed birds, though of the two conditions, small and large, the latter is the least objectionable. There are many exceptions to this rule, of course, just as there are many exceptions to every rule. The findings in this chapter are based on the general run of fowls, and must be considered as such, or averages.

Next to shape, color is probably the most reliable sign of a hen's ability as a layer. And the chief advantage of this test lies in the fact that it is discernible without having to handle the bird.

Color.—In virtually all of the yellow-skinned breeds of chickens the shanks, beak and flesh of the posterior parts of the birds are a rich yellow at the commencement of laying, and gradually undergo a fading out as the laying progresses, until these parts become a real pale color, sometimes pink, or white, as the laying season advances. This change in color is so consistent, in fact, and so quickly made, especially in the region of the flesh surrounding the vent, that it will be clearly apparent even to the layman.

The theory of this test is based on the fact that the same coloring matter that gives the shank, beak and skin its rich yellow look is also used in the color of the yolks of eggs. Heavy layers produce eggs faster than they can supply the coloring matter for the shanks, beak and so on, consequently the color becomes lighter and lighter, until it is frequently scarcely visible.

Furthermore, hens that are not in laying condition are prone to store up a certain amount, and sometimes a very large amount, of fat in the region surrounding the vent, and this fat, being of a rich yellow color, transmits its color to the flesh. When these hens start to lay and lay heavily, this fat is drawn upon to supply the body tissues with the necessary energy, until its supply is virtually depleted, in which case the skin loses its former rich yellow appearance.

In selecting hens by the color test, allowances must be made for the natural difference in color between different breeds and different individuals of the same breed. For example, it is manifestly unreasonable to compare the color of a Rhode Island Red or Barred Plymouth Rock with a White Orpington or White Leghorn. The color test should be made relatively and with the exercise of much common sense.

An examination should be made at the commencement of laying, and not during the molt or when the birds are immature, and the degree or shade of color carefully noted. Later, when the flock should be laying heavily, say about April first, the color

Rack for squatted and hanging dressed poultry.

(Courtesy U. S. Dep't Agriculture)

Stationary feeding battery used in Western feeding plants.

Fig. 250.—Packing house equipment.

is observed again, and the hens which have little or no color in the parts mentioned may be selected as the best layers.

To prove this, if you observe any birds which are noticeably yellower than the others, transfer them to a separate pen and note if they are not poor layers. If you have trap nests, these ideas are easily verified.

Conduct and Other Indications.—It is pretty generally established that pullets which begin to lay early in life, providing they are fully matured, are pretty sure to be the most prolific members of the flock. Ordinarily this means that the pullets which commence laying in the fall, before snow flies, are the most desirable. Fowls that fail to start production until after Christmas seldom attain high scores. By high scores is meant records of exceptional merit, say over 150 eggs a year. It usually follows, also, that pullets continuing to lay late into the fall, thereby postponing the molt until cold weather is at hand, are almost without exception heavy producers.

Late Molters.—Hens that have a nice new coat of feathers by July look well, but they are not often profitable. The late molters, those that look ragged and dirty when the others are sleek and clean, are almost always the best layers in the flock. Moreover, it will be noted that the late molters get through with this task very quickly, seemingly, which is an appreciable saving in time.

The appetite and general conduct of the hen are other indications of productiveness. The heavy layer is the first off the perch in the morning and the last to go to sleep at night. She is active, constantly searching for food, and when observed on the roost after dark, she will be found to have a very large crop tightly packed with food. It is also found that the hens which have the most confidence in their keeper, and are not foolishly disturbed or frightened, are the most consistent performers and the most reliable breeders.

Importance of the Male.—It is said that the male bird is half of the flock, meaning, of course, that his characteristics will be transmitted to the offspring in equal proportion to the hens,

(Courtesy Niagara Farm)

Fig. 251.—Breeders handled in large units on free range.

383

though the latter may outnumber the male fifteen to one. This is not an idle thought; it is true. In fact, in some cases it is highly probable that the influence of the male is even greater than a *half*—maybe three-quarters.

Prepotency of the Male.—This belief is certainly true of the male's ability to transmit type and color to his progeny. Therefore, if he can dominate certain important characteristics, is it not reasonable to suppose that he may dominate all of them? We assume, of course, that to so do, his vitality and constitu-

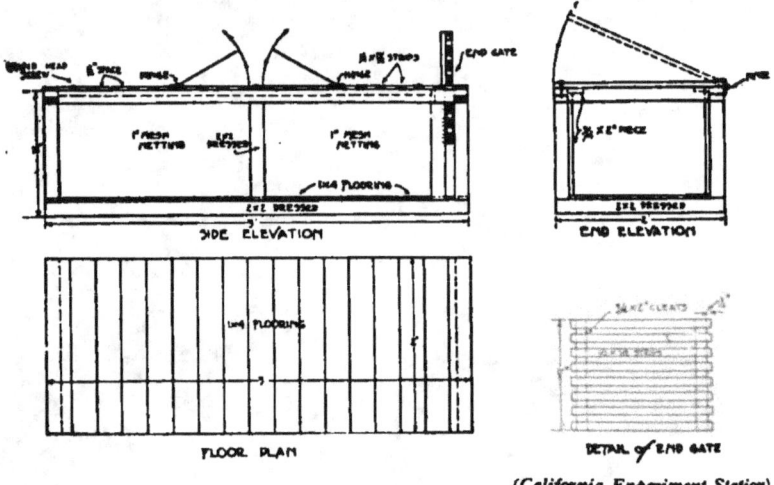

(*California Experiment Station*)

Fig. 252.—Plans for a catching coop.

tional vigor are equally as strong as the female's, if not stronger. More than that, we assume that he possesses the faculty or capacity of transmitting his qualities. This qualification is sometimes spoken of as *prepotency*.

Like Does Not Always Beget Like.—One of the greatest impediments to the successful breeder of poultry is the inability to select male birds of the required type. There is an old saying that "like begets like." In a general sense this is true, but not always so, because of which the idea becomes a subject worthy of study.   We often see children who resemble and act like one parent, let us say the father, while others resemble and act like the mother.   And yet some children will be like neither of the parents, nor like any of their immediate kin.   Breeders of horses and cattle and other branches of livestock are fully aware of the variations in offspring in this respect, and it is probably because of their persistent quest of knowledge along these lines that they have made so much more progress in scientific breeding than poultrymen.

Among those who have never given mating other than casual thought, and such persons are in the majority, there is the belief that heavy layers are bound to produce chicks which will develop into heavy layers.   In their minds, to produce a heavy laying strain all that is necessary is to trap-nest the flock, and breed only from such hens as have made a certain record.   This is a step in the right direction; the idea is a splendid one, and makes for careful selection.   But, unfortunately, it does not go far enough.   In the first place, we cannot trap-nest male birds; and pedigree, while useful and essential, is not alone a sufficient basis for the selection of that element which is to dominate future generations, and probably the success or failure of the poultry-man.

Maine Experiment.—Some years ago the Maine Experiment Station endeavored to establish a 200-egg strain of fowls by breeding only from those which laid the most eggs.   The experiments extended over a number of years, the plan was well conducted, it received the most favorable attention, and everything was done to further the idea.   The work was finally abandoned as a failure.   Those in charge of the work were trying to establish an extreme in egg production, by breeding from high-producing hens.   They failed to appreciate the need for that other faculty—*prepotency*—the capacity to transmit high-producing qualities.

Number of Males.—Let us first consider the ratio of males to females.   No fixed rule can be given, because the breed, size of the flock, living quarters, extent of range and the general health and vigor of the stock are all determining factors.   For example,

25

the Mediterranean breeds, such as the Leghorns, are usually mated one male to fifteen females, providing the flocks are single. In the American or general-purpose class, it is customary to mate one male to about ten females; and in the Asiatic or meat class, it is advisable to use one male to six or eight females.

Single Flocks.—As previously mentioned, these ratios apply to single flocks—pens in which but a single male is to be used. In larger flocks the ratio of males to females may be greatly re-

(*Courtesy U. S. Dep't Agriculture*)

Fig. 253.—Handy brood coop made from rough lumber, small sliding window serves as a door.

duced. The reason for this is easily understood. In a flock of fifteen Leghorns there may be two or three hens uncongenial to the male, or the male may be uncongenial to two or three females who will fight him away from them, in which case the fertility from such a pen will run about eighty-five per cent.

In a flock of 35 females to 2 males there is almost certain to be rivalry or jealousy, which tends to eliminate favoritism, and thereby increases the fertility. In a flock of 60 hens to 3 males there is still greater rivalry, while in a unit of 500 hens to about

20 males little, if any, discrimination is found, and the fertility of the eggs should run ninety-five per cent or better.

The same general ratio applies to the heavier breeds, but in no case can they be expected to equal the Mediterraneans in the matter of fertility. The heavier breeds have a more sluggish nature, and they are naturally less active fowls. From the writer's experience, sixty-five per cent fertility in the Asiatics is equal to seventy-five per cent in the Americans, or ninety-five per cent in the Mediterraneans.

In selecting males for breeding purposes the first qualities to be considered are those in plainest evidence, in other words, the general appearance of the birds. If a specimen has malformed feet, wry tail or serious squirrel tail, brassiness or other color defects, lopped comb or exceedingly ponderous or poorly shaped comb, under-developed ear lobes or wattles, or if a bird is abnormally large or small, noticeably disproportionate and ungainly, it goes without further argument that he should be discarded.

A fowl's actions is one of the best guides to its breeding ability. Males that are too greedy, or those that are so gallant that they will not eat until the hens have helped themselves, are likely to prove of little value in the breeding pen. The former are apt to become over-fat, due to over-feeding, and the latter under-fed and anemic. Crowing is an excellent indication of vigor and vitality, and should always be borne in mind in selecting breeding males. It characterizes physical strength and masculinity. The desirable breeding cockerel is always on the alert, strikingly erect in carriage, aristocratic and combative—a good fighter, and one who believes in crowing about it. Fear and physical weakness usually go together. A cowardly bird, or one that becomes unduly excited, which amounts to a *rattle-brained* nature, should never be placed in the breeding pen. He is too apt to be a degenerate.

Some Naked Truths.—Selecting breeders from the fancier's point of view—the show room—is more discriminating, perhaps, than for commercial purposes, yet it is also more superficial.

Color, shape, carriage, texture of plumage, comb, wattles and ear lobes are the essentials to success in the exhibition; but as they constitute a study in themselves, we will not attempt to cover them in this chapter, which is intended more for the commercial poultry raiser.

Having selected a group of birds of the desired appearance and most precocious habits, final judgment is passed upon their

Fig. 254.—Practical method of catching fowls. Crate is placed against small entrance door, through which the chickens are driven into the crate.

physical qualifications—literally speaking, their naked truths. There is a distinct correlation between the different parts of a fowl, make no mistake about this. The body of the vigorous fowl is broad, deep and blocky, as contrasted with the long, thin, slender type. And since the fowl's plumage is often very deceiving, they must be carefully handled. In a sense they must be measured.

In a foregoing paragraph on selecting pullets for heavy egg

production we emphasized the following requirements: Large crop and abdominal capacity; thin pelvic bones that are pliable and well spread apart; a fairly long back, depth in the keel and width between the legs.   In selecting males as breeders for heavy egg production, the same analysis should be applied to their anatomy—relatively, of course, for the male never has the spread of pelvic bones nor the abdominal capacity of a hen of the same size.

CHAPTER XXIX

DEVELOPMENT OF THE EGG

Embryology.—It is certain that the majority of poultry keepers do not know as much about the formation and development of the egg as they should. Yet this is a very important subject, a knowledge of which is essential before one can really exercise intelligent care and feeding. To do certain things blindly or on a guesswork principle is archaic; they may be correct, and the results therefrom may be entirely satisfactory up to a given point, when, without any warning, trouble may come. Then, if the foundation of one's knowledge is meager, or perhaps there is no foundation at all, which is frequently the case, the poultryman is at a loss for a solution or remedy. It is like trying to run an engine without some understanding of its construction; when trouble occurs, instead of being able to repair or adjust the defect, the situation becomes aggravated and serious.

Every phase of the poultryman's work should have a definite purpose, and in view of the fact that the egg is the first stage in the production of fowls, whether for meat, eggs or the show room, it behooves him to have at least a general idea of embryology.

Some hens are absolutely sterile; of this there is no doubt, but they are rare, and are to be compared with any other malformation. Others have the power to produce a few eggs in short litters, followed by long rest periods, whereas others have reproductive organs which are so strong and easily stimulated that they lay almost without cessation, and continue to do so for a couple of years. In fact, they seem almost to have a supernatural power.

Prolific Power of a Hen.—Some experts tell us there are more than 7,000 latent eggs in the normal hen, but whether or not this

390

is correct we need only concern ourselves with about 700.   The
number is not a fixed quantity, and those which will be developed
is still less certain.   The prolific power of a hen is largely an inherited tendency, the result of careful selection and breeding, made
potent by careful handling and feeding.   Both elements are absolutely essential, as we have shown in preceding chapters.

The ovary or egg cell cluster which contains the latent eggs is a
muscular tissue on the left side of the spine.   In it, in various
stages of development, from the full-sized yolk, ready to be de-
tached, to the cells which are so small as to be invisible without
the aid of a microscope, are the yolks or ova.   When a yolk is
fully matured and ripe, it bursts from the tough membrane of the
ovisac and enters the neck of the oviduct, a convoluted, muscular
tube some twenty inches long, wherein the albumen or white is
deposited, and later the shell is formed.   See Frontispiece.

The ovisac is lined with blood-vessels, yet provision is made in
the healthy, normal hen that when the yolk ruptures this membrane the blood-vessels are parted to one side and not broken.
It occasionally happens, however, either through an injury to the
fowl, fright or weakness due to a debilitated condition, that one
of the blood-vessels may become slightly ruptured, whereupon a
blood clot will escape with the yolk and later be incorporated with
the albumen.   This accounts for spots of blood found in strictly
fresh eggs, and which have led many consumers to believe they
have purchased partly incubated eggs.

Double Yolk Eggs.—It sometimes happens that two yolks
mature and burst through the ovisac at the same time; in this
event they are likely to become encased with albumen together,
and subsequently surrounded by the same shell, producing a
double-yolked egg.   Occasionally a mass of albumen will be deposited without yolk or shell, or it may be laid with a perfectly
formed shell but without a yolk; or a yolk will be laid without
albumen or shell, and in rare cases a perfectly formed egg has
been found within an outer egg shell.   These freak conditions
are brought about by improper care and feeding, but more
especially by fright, neglect or injury.

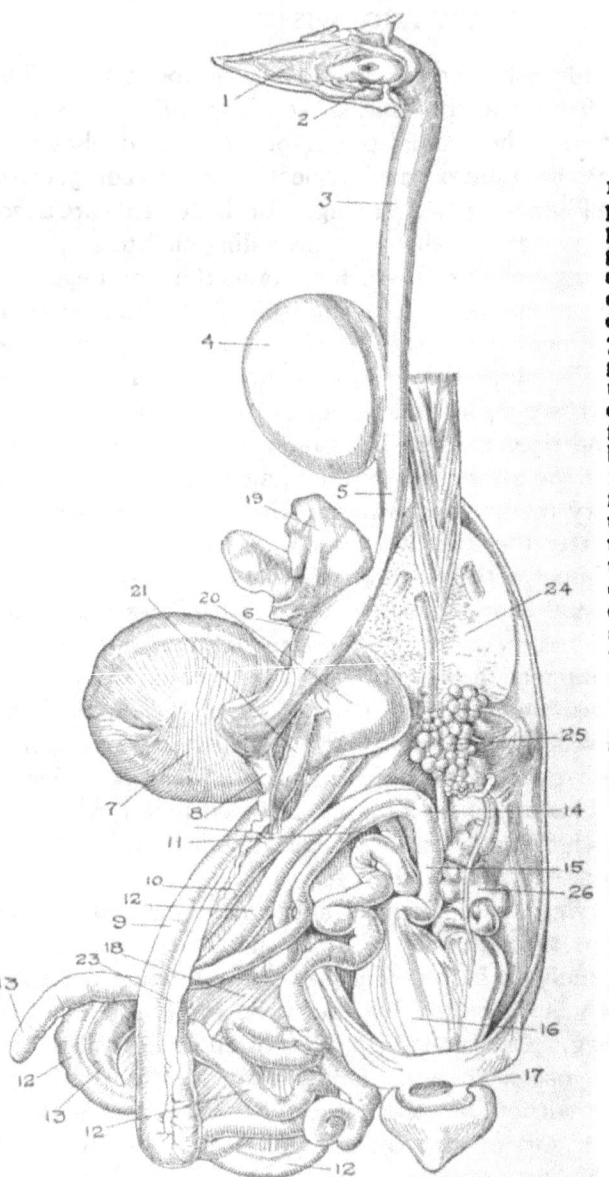

1, Tongue; 2, pharynx; 3, first portion of esophagus; 4, crop; 5, second portion of esophagus; 6, succentric ventricle; 7, gizzard; 8, origin of the duodenum; 9, first branch of the duodenal flexure; 10, second branch of same; 11, origin of the floating portion of the small intestine; 12, small intestine; 13, free extremities of the cæcums; 14, insertion of these two culs-de-sac into the intestinal tube; 15, rectum; 16, cloaca; 17, anus; 18, mesentery; 19, left lobe of liver; 20, right lobe of liver; 21, gall-bladder; 22, insertion of the pancreatic and biliary ducts; 23, pancreas; 24, diaphragmatic aspect of the lung; 25, ovary in a state of atrophy; 26, oviduct.

Fig. 255.—Diagram of digestive apparatus of a fowl.

392

Held Eggs Within the Body.—It is not positively established whether a hen can of her own will stop the development of the yolks prior to their entrance to the oviduct, and it is hardly likely that she can, but it is certain that she can control the egg after that period.   She can retain it for a considerable time after

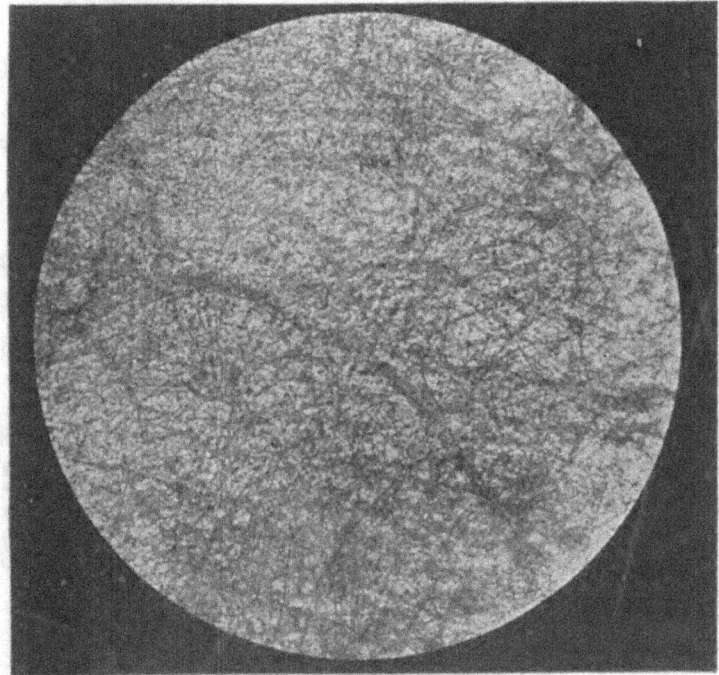

(*Courtesy U. S. Dep't Agriculture*)

Fig. 256.—Outer shell membrane of fresh egg.   Magnified 150 times.

it is completely formed, whereupon, instead of the life germ remaining dormant in the fertile egg after it is laid and until such time as it is subjected to the proper uniform temperature for incubation, it will commence to develop within the egg within the hen.   Obviously, although such an egg may be freshly laid, it is not a fresh egg.   On the contrary, it is a spoiled egg;

sometimes very badly spoiled, which is apt to lead to difficulties with the customer who has been unfortunate enough to receive it.

Egg-Bound.—Hens that delay their laying in this manner are usually egg-bound, a condition brought about through injudicious

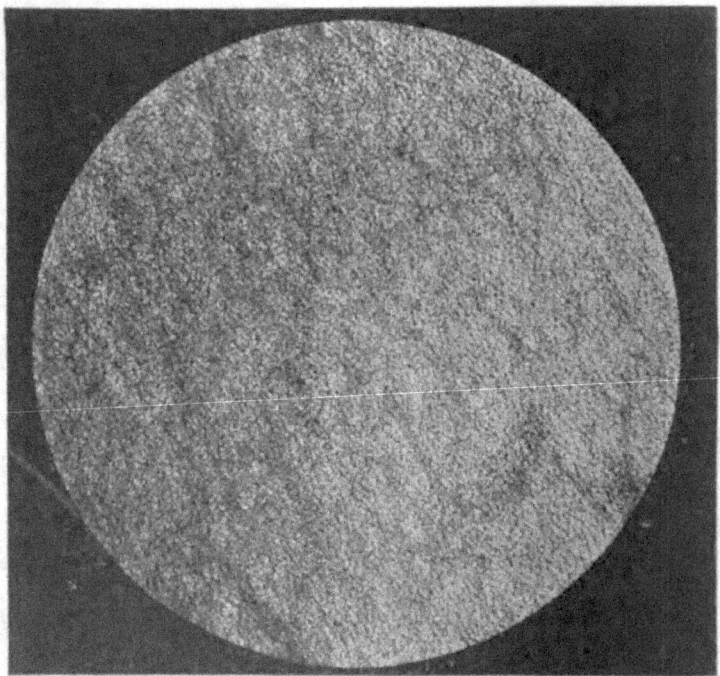

Fig. 257.—Inner shell membrane of fresh egg. Magnified 150 times.

feeding, injury and by the drying up of the secretions in the cloaca, where the egg rests before it is laid, thus failing to assist in the passage of the egg.   Over-fat hens and pullets producing their first eggs are apt to be troubled in this manner.

The most common symptoms are repeated trips to the nests, accompanied by prolonged squatting and straining, and in extreme cases a fowl so afflicted will crawl with her body upright and her tail dragging on the ground.

The shape of an egg is largely determined by the contours in the oviduct in which it is cast. During the passage of the yolk in the oviduct it is pushed forward by the muscles of this tissue, at the same time receiving a deposit of albumen. The oviduct being twisted and contracted, imparts a turning motion to the yolk as it advances, which causes the albumen to be formed in layers. These layers are sometimes visible in a raw egg, but are better seen in a hard-boiled egg.

Chalaza.—Two principal cords or fibers, technically known as the chalaza, support the yolk in about the center of the albuminous mass, and serve to protect the yolk from injury by undue jarring or rough handling. See Fig. 258. We have all noticed, perhaps, that whatever way an egg is turned the yolk quickly assumes its original position; this is due to the influence of the chalaza, and to the fact that the yolk, containing a large amount of fat, which is lighter than the albumen, has a tendency to float upward.

Shell Membrane.—When sufficient albumen has been secreted, at which time the entire mass has reached the lower part of the oviduct, the shell membrane is formed, after which it passes still further and the outer membrane is added. Here, glands which contain a secretion of carbonate of lime and other mineral substances, also the color pigment, deposit their liquid, which quickly hardens the outer membrane. This hardening process is very rapid, and frequently takes place while the hen is on the nest.

Bloom of the Egg.—The egg has now reached the lowest part of the oviduct, known as the cloaca, whence it is ready to be laid. While in this section it is covered with an oily secretion which, as previously mentioned, aids in the delivery of the egg. This secretion dries almost immediately the egg is laid, and gives it the *bloom* or fresh appearance found in a newly laid egg. When eggs are washed this *bloom* is destroyed, or partly so, which makes washed eggs rather easily detected by experienced handlers.

Texture of Shell.—It will be noted that the shell of an egg is exceedingly porous, which enables the embryo to take in oxygen through the shell, otherwise it could not breathe. See Fig. 256. During the early stages of incubation a network of blood-vessels surround the inner membrane of the egg, close to the shell. See Fig. 257. These blood-vessels absorb the oxygen and act as the respiratory apparatus for the embryo until about the nineteenth day of incubation, when the lungs are completed and brought into use. The oily secretion deposited on the shell in the cloaca tends to stop up the pores temporarily, so as to prevent undue evaporation of the contents of the egg, and to keep the pores clear. Obviously, a hatching egg should not be washed; and if very badly soiled it should not be used for hatching purposes at all.

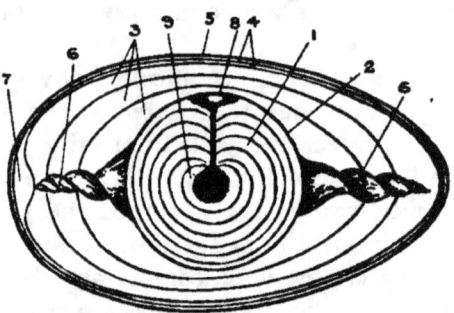

Fig. 258.—Sectional view of fowl's egg. 1, Yellow yolk composed of successive layers; 2, vitelline membrane; 3, layers of albumen (white); 4, two layers of the lining membrane of the shell; 5, calcareous shell; 6, chalaza; 7, air space between the two layers of the shell membranes; 8, cicatricula, with its nucleus, beneath which is seen the canal leading down to the white yolk cavity, or latebra, 9.

To return to the yolk, it must be very apparent that if production is to be successfully carried out, the yolk or ovum, which is the real beginning, must be carefully and normally developed, otherwise the succeeding processes are all thrown out of kelter. Egg making is a very exhaustive process, if we stop to consider that a profitable hen is expected to lay about 150 times a year, which is equivalent to almost five times her weight; hence the drain on her system is enormous. The activity of the ovary, then, depends upon the health of the bird.

The over-fat hen does not lay because over-fatness is an indication of improper or immoderate feeding, usually accompanied by lack of exercise. The poor, anemic, emaciated hen cannot

lay because there is not sufficient fat to develop the yolk.   Therefore, it will be found that the best layers are neither too thin nor
over-fat; rather those which are active and in good spirits from
dawn to dark.

Germ.—Although invisible to the naked eye, the yolk is

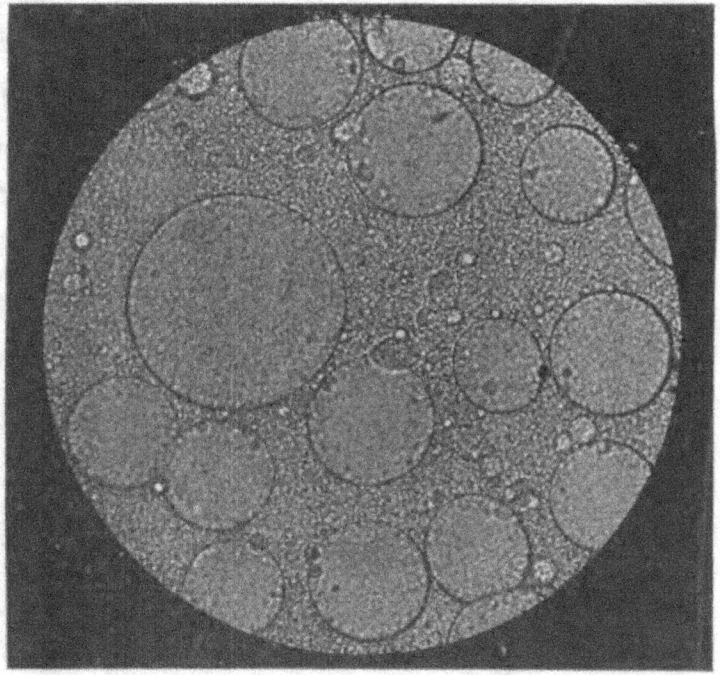

(Courtesy U. S. Dep't Agriculture)

Fig. 259.—Yellow yolk of fresh egg.   Magnified 250 times.

covered by a delicate membrane, called the vitelline membrane,
so named, perhaps, because clinging to this membrane is found
the life germ, the really vital part of the egg.   The contents of
the yolk is called the vitellus, upon which the life germ draws
for its sustenance.   When a hen's vitality becomes weakened it
is generally manifested in the composition of the vitelline mem-

brane, which is easily ruptured, causing the vitellus to escape and mix with the albumen. Naturally, this condition gives the egg an addled, unsavory appearance, undesirable as food, even though it may be strictly fresh. Nine times out of ten the housewife condemns it as a bad egg.

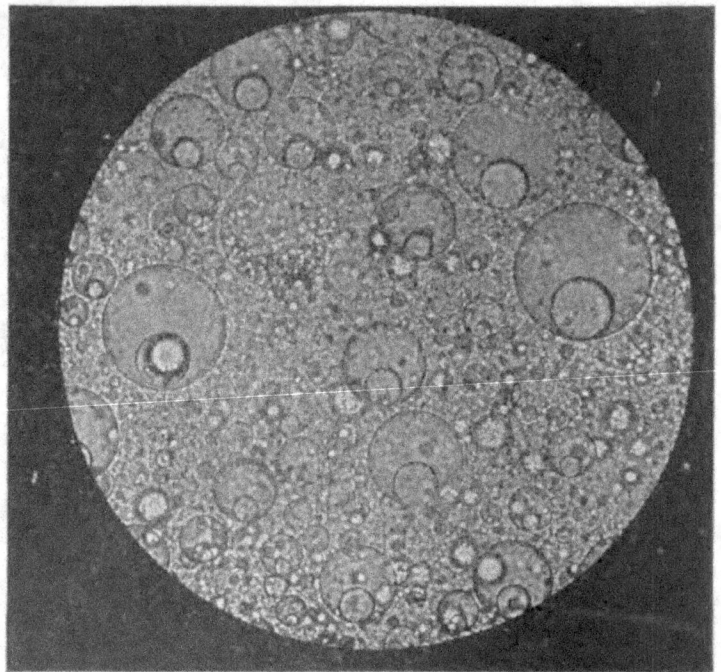

(Courtesy U. S. Dep't Agriculture)
Fig. 260.—White yolk of fresh egg.   Magnified 250 times.

Keeping eggs for any length of time weakens the vitelline membrane, also, consequently eggs intended for hatching purposes should be set as fresh as possible. An example of this deterioration is found in storage eggs, which, if kept for many months, frequently result in ruptured yolks as soon as they are opened. Moreover, in a stale egg the albumen loses its firm consistency and becomes watery, hence it fails to support the yolk, which then gravitates to the membranous lining of the shell and adheres to it.

Fertility.—All normal eggs contain a life germ, but all life germs are not fertile, and there is no way to determine the fertility of the embryo before incubation, except by breaking the shell and examining the contents very closely, and even this is not infallible. The fertile germ has a ring or film surrounding it, which is clear, and in the center may be seen little white dots—rudimentary cells. The sterile germ has a whitish appearance and does not possess the outer ring or the dots. After 24 hours' incubation life is perceptible if the egg is opened.

Shape and Color of Shells.—Notwithstanding many claims to the contrary, the shape of an egg has nothing whatever to do with its sex. We have shown that the shape of an egg is governed almost entirely by the shape of the oviduct, which is peculiar to every individual and practically constant. The same individuality applies to the color of an egg, which also remains more or less constant, except that as the laying season advances the color gradually fades to a lighter shade. The reason for this is plain: the glands which secrete the color pigment are more heavily drawn upon, consequently the supply is somewhat weakened.

Time for Development.—Just how much time is required for the development of an egg is not definitely known. The formation of the yolk is the longest period, and probably requires three weeks before it is ready to leave the ovisac. The second process, that of accumulating the albumen and forming the shell, is comparatively short, and requires about eighteen hours. It frequently happens that two eggs are under completion in the oviduct at the same time.

Like all secretory organs, these reproductive tissues, glands, and so on, are shrunken and very much contracted when not in use, and enlarge to many times their former size when stimulated to a point of productivity. It is this stimulation—the time required to overcome the inert condition—which is of vital importance to the poultryman's pocket-book.

CHAPTER XXX

MARKET EGGS

Quality in Eggs.—Housewives and consumers generally are seldom concerned with any but two kinds of eggs—good eggs and bad eggs. The term *good* in this sense usually means *fresh*, and has become synonymous with the idea of desirable quality. A *bad* egg—is a bad egg, commonly thought to be the result of old age, and as such it is condemned. The actual age of an egg, however, is only one of the factors that affect its quality. There are many other equally potent influences, a knowledge of which will be beneficial to those engaged in the production, handling or consumption of eggs. Strictly speaking, the term *fresh* should mean a definite quality rather than a definite age, for all newly laid eggs are not necessarily good eggs, in a sense that they are desirable as food. An explanation of the reasons for these peculiarities will be set forth in this chapter. In the succeeding paragraphs the term *fresh* is intended to express prime—superior —quality.

Strictly Fresh Normal Egg.—Eggs are one of the most difficult food products to grade, not only because each egg must be considered separately, but because an accurate knowledge of the contents cannot be ascertained without destroying the egg. They can be selected for size, shape, color, cleanliness and texture of shell, and freedom from cracks, from external appearances, which is the most common method of grading them. The best method of determining the interior quality is by the process of candling, which is used for commercial purposes. See Fig. 261.

Composition of the Egg.—The purpose of the normal egg in nature requires that it be of a fairly uniform composition; its contents must be so proportioned as to form the chick without

400

surplus matter, and naturally this demands a uniform chemical composition. When the egg is first laid it is completely filled, but as soon as it cools the contents contract and an air space or air cell is formed. This cell usually lies between the two shell membranes, and at the large end of the egg, where it is plainly visible with the aid of a candle. As the age of the egg increases evaporation takes place, which enlarges the air cell to considerable size, and therefore denotes, approximately, the degree of freshness.

The composition of hens' eggs is somewhat variable, with

Fig. 261.—Class in candling, grading and packing eggs at the Kansas Agricultural College.

breeds and with individuals, and also as the result of care and feeding. A general idea may be had from the following table:

	WHOLE EGG	YOLK	ALBUMEN OR WHITE
Water	70 to 76%	46 to 52%	80 to 88%
Fat	9 to 14%	30 to 35%	Traces
Protein	10 to 15%	14 to 16%	10 to 13%
Shell and its membranes	9 to 12%	....	....

The precise chemical analysis of a hen's egg is too technical for the subject of this book, and it is really unimportant so far as the average poultryman is concerned. The eggs of turkeys, geese, ducks, guineas and other birds vary slightly from the above table, and are, therefore, more desirable for certain purposes.

26

Size.—Although certain breeds are credited with laying larger eggs than others, as a general rule the size of an egg is controlled principally by selection of layers of large eggs and judicious breeding toward this end. In a number of tests conducted by Experiment Stations it has been found that care and feeding have slight influences in the sizes of eggs, but this does not establish anything beyond the fact that the condition and general health of the fowls are directly responsible. At the beginning of their laying period pullets lay a much smaller egg than those laid during the height of their laying season. Similarly, as a hen approaches the molt, her eggs become smaller. The difference in food value per pound is in favor of the large eggs, because they have a smaller percentage of shell.

Food Value.—It should be borne in mind, however, that there is considerable difference in the food value of eggs of different grades. Furthermore, the season of the year has something to do with the quality of eggs. Those produced in summer are of lower quality; the albumen is more watery than the eggs produced in the spring, hence they are not so desirable for storage purposes. In candling, the yolks of summer eggs float lower in the albumen, which is a sign of weakness, and the yolks appear slightly darker than in spring eggs. Packing houses always aim to store eggs produced during March, April and May for best results.

It is almost certain that some hens have an inherited tendency to produce eggs of poor quality, for the same reason that certain hens will almost invariably lay a malformed egg. If this is true, it is reasonable to suppose that this characteristic will descend to their progeny. Flocks should be culled for the quality of their eggs as well as for their productiveness. In no other way is it possible to develop a flock that will lay a uniformly high grade of eggs.

Abnormal Eggs.—To further illustrate the remark that a newly laid egg is not always a desirable egg, some of the most common abnormalities will be discussed. Double-yolk eggs result from the joining together of two yolk sacs during their development; their growth is identical, they have the same blood supply, and both drop into the oviduct at the same time.   The

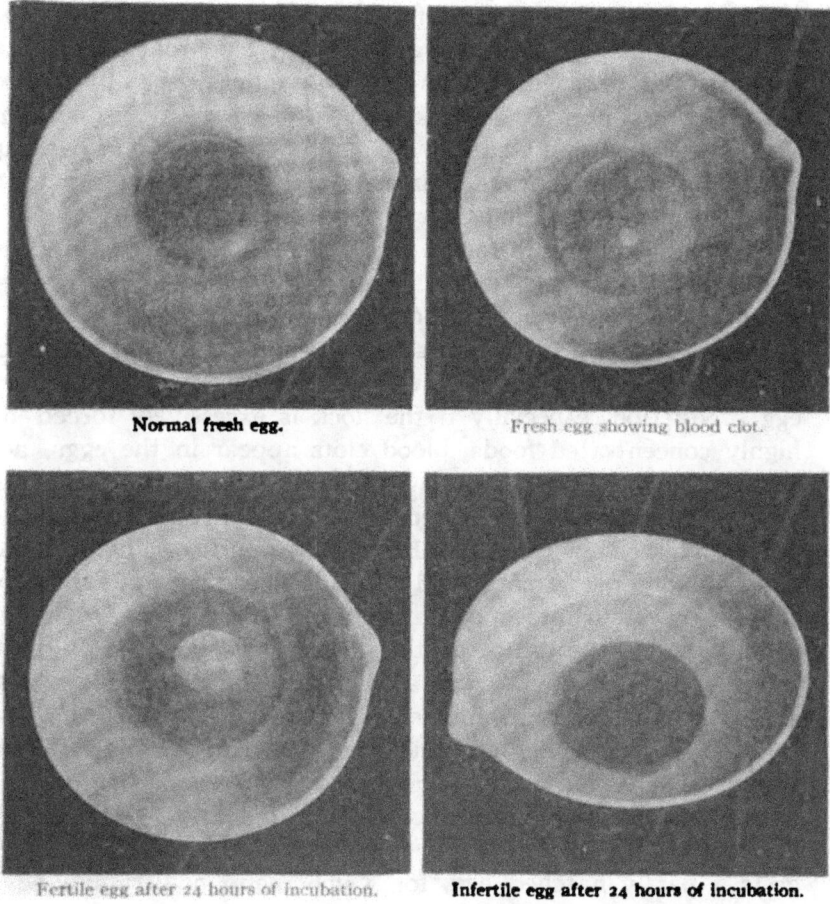

Normal fresh egg.   Fresh egg showing blood clot.

Fertile egg after 24 hours of incubation.   Infertile egg after 24 hours of incubation.

(Courtesy U. S. Dep't Agriculture)

Fig. 262.—Changes affecting the appearance of eggs.

formation of the albumen seems to be entirely automatic, hence the same mass surrounds both yolks, and later they are framed by the same shell.

Most double-yolk eggs have the same quality as normal eggs, but the poultryman should not try to pack them in ordinary cartons or fillers on account of their increased size. They will project above the level of the filler, or fit too snugly into the filler, and be broken by the eggs surrounding them. One broken egg in a shipment will damage perhaps four or five dozen eggs by reason of its leaking contents. A broken egg is a very messy, unpleasant looking article, consequently a moment of carelessness in packing is often responsible for a serious discount on the sales statement, or a disgruntled, indignant customer.

Yolks are sometimes forced into the oviduct before they are mature, and thus appear very small, and in some cases they are little more than specks, in the completed egg.

Blood Clots.—At certain seasons, usually during the first laying period of pullets and during the spring months of heavy egg production, especially if the flock is excessively forced by highly concentrated foods, blood clots appear in the eggs, adhering to the yolks. This is probably the commonest defect arising in the ovary, and often a very troublesome one for the poultryman, since he can not detect it without candling. Blood clots are different from bloody eggs. The former are usually caused by the rupture of a blood-vessel when the yolk sac splits to allow the escape of the mature yolk into the oviduct. The clot adheres to the yolk as it passes through the oviduct, and is encased by the albumen. It is easily detected by the candle. When the egg is opened the clot can be removed and the egg is suitable for food. For table purposes this is very objectionable, for the appearance of the blood is disagreeable. To those unfamiliar with the physiological reasons, it suggests a partly hatched egg, for which many poultrymen have been unjustly blamed. Obviously, when catering to a fancy retail trade, or when eggs are represented as being of the finest quality, they should be candled as a guard against this trouble.

Bloody Eggs.—An ordinary blood clot does not color the albumen. If the white of an egg is bloody, it is from a different cause, and such an egg is termed a bloody egg. When boiled, such eggs appear brownish, and they are distinctly unappetizing-

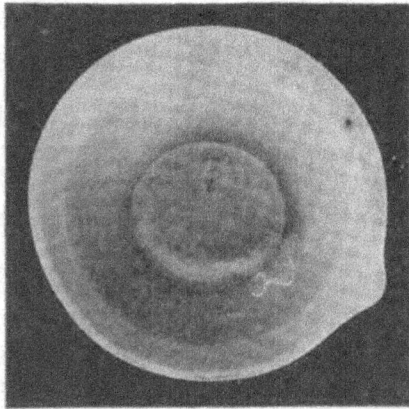 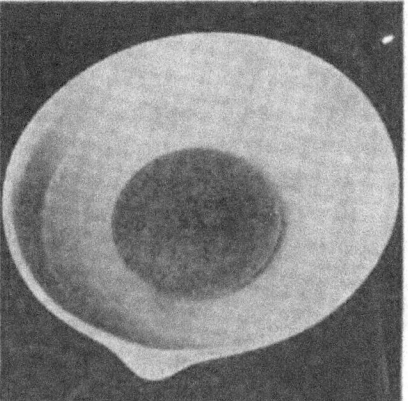

Fertile egg allowed to die after 36 hours of incubation.　　Infertile egg after 36 hours of incubation.

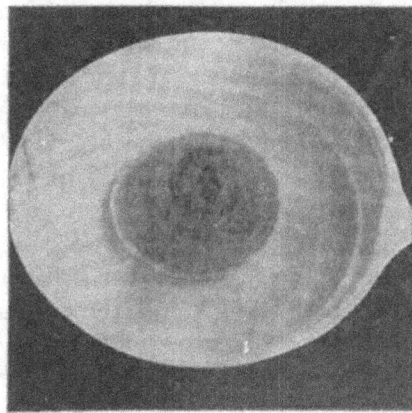 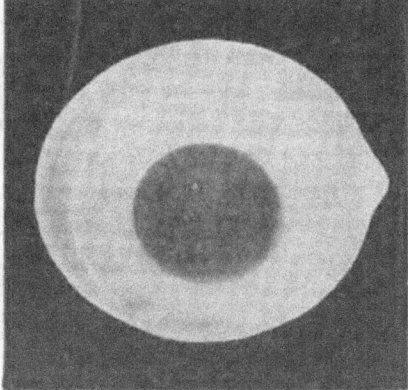

Fertile egg after 48 hours of incubation.　　Infertile egg after 48 hours of incubation.

(Courtesy U. S. Dep't Agriculture)

Fig. 263.—Changes affecting the appearance of eggs.

looking. It is virtually impossible to remove the affected part, as in the case of the blood clot, hence the egg is useless. Bloody eggs are caused by an irritation or injury to the wall of the oviduct, whereupon blood is exuded with the albumen, thereby making the white of the egg bloody-looking or streaked with blood. It is generally caused by over-exertion or constitutional weakness, and is not nearly so common as blood clots. Wherever possible the poultryman should strive to find out the hen that is responsible and to remove her from the flock.

Fresh Laid Stale Eggs.—It sounds incredible that a fowl should lay a stale egg, or a spoiled egg, but this frequently happens. The condition responsible for this peculiarity is sometimes chronic with certain hens, or merely occasional, and if the victims can be spotted they had better be killed for Sunday's dinner. If not, the poultryman will have to assume the burden of a tedious explanation to the customer who is unfortunate enough to receive such eggs.

After the yolk enters the oviduct it is forced through this tube by a circular movement or contraction of the muscles of the oviduct wall, and at the same time receiving layer after layer of albumen, and finally the shell, when it is ready for expulsion from the body. At some stage in this development, which under normal circumstances should only require about eighteen hours, from the time the yolk enters the oviduct until the finished egg is laid, this action ceases temporarily, maybe as the result of fright or due to an injury, and the egg is held in the body of the fowl for several days after it is completed.

It must be remembered that the life germ, or germinal disk, is complete when the yolk leaves the yolk sac, hence it is subject to heat for its development into the embryo chick. If the egg is held in the body, the life germ comes under the influence of the body temperature and incubation begins, providing, of course, the egg is fertile. If the egg is infertile, the results are not so disastrous; the egg has a stale flavor, or maybe the contents are of a peculiar color. In the fertile egg, when it is finally laid, the germ dies, and immediately it starts to decay. Such eggs are called body-heated eggs.

There are other minor abnormalities, those of such rare occurrence, such as an egg within an egg, foreign substances within eggs, intestinal worms within eggs, connected eggs, multigerminal disks, meat spots, soft-shell eggs and eggs with loose shell membranes, that space will not be devoted to their description.

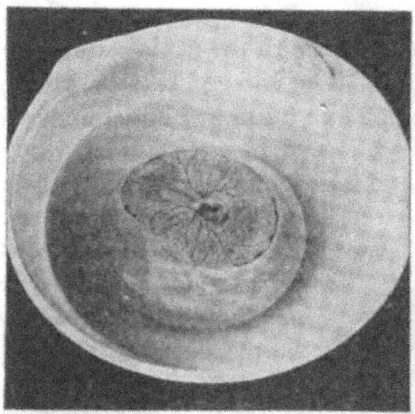
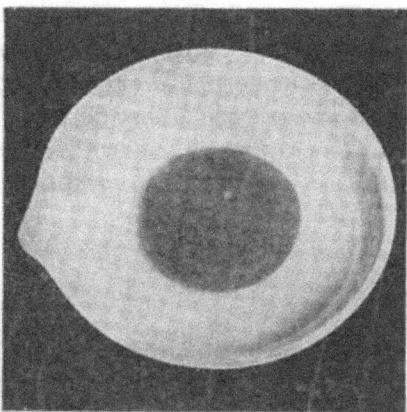

Fertile egg after 72 hours of incubation.   Infertile egg after 72 hours of incubation.

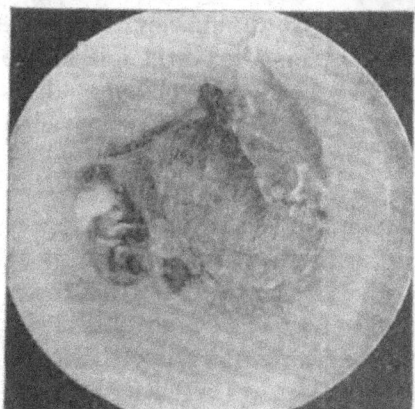
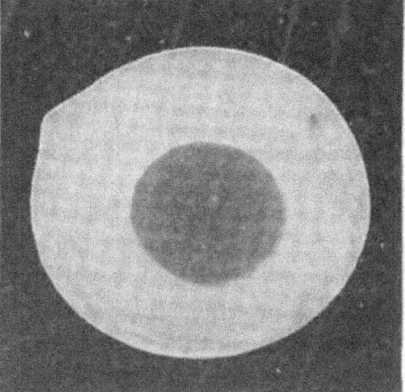

Fertile egg after 7 days of incubation.   Infertile egg after 7 days of incubation.

(Courtesy U. S. Dep't Agriculture)

Fig. 264.—Changes affecting the appearance of eggs.

External Influences.—In the foregoing paragraphs we have discussed only the internal factors affecting the quality of eggs.

There are as many, perhaps more, external influences. The most potent of these influences are evaporation, heating, mustiness, mold, bacterial contamination, moisture, bad odors and so on.

The egg shell is porous, so that the embryo chick may obtain air, and this porosity exposes the contents to the drying influence of the atmosphere surrounding it. The rate at which the interior moisture is lost depends upon the humidity and temperature of the air in which the eggs are stored, also the rate at which the air is moving over or among the eggs, and the texture of their shells and membranes. As evaporation continues, air replaces the moisture, which enlarges the air cell, and if the eggs are kept long enough, and at the same time protected against heating or contamination, all the moisture will evaporate and the yolk and albumen will become a dried mass.

Fig. 265.—A well-ventilated, clean vegetable cellar is a good place for the farmer to hold his eggs until ready for market.

Storage Place.—A cool, sweet cellar or refrigerator is the most desirable place for storing eggs, in which they may be kept for three or four weeks without serious evaporation. See Fig. 265. Never store eggs in the kitchen or shed, or where they are subjected to warm air or currents of air. The best receptacle is a pail or box, something that will prevent the free circulation of air through the eggs, and yet allow a moderate amount of ventilation, hung in a cool, moderately dry place where there are no odors to be absorbed by the eggs. Avoid excessive dampness, for this may cause mold or other changes to take place.

If eggs are allowed to remain in the one position for a long time, the yolks will rise and stick to the shell membranes, which is, of course, an objectionable feature. Moreover, if they are stored in a damp place, mold spots are very apt to form at the points where the yolks come in contact with the shell membranes, in which case they are unfit for food. It is sometimes possible to shake the yolk loose without rupturing the vitelline membrane of the yolk, but it more often breaks.

Water Test.—Some housewives attempt to ascertain the condition of their eggs by placing them in water, and they will argue that if the egg sinks it is fresh, and if it floats it is bad. See Fig. 266. This test is fairly accurate in determining the age of an egg, but only so far as the extent of its evaporation. It will not disclose the interior quality of the egg, nor determine the other peculiarities that we have discussed. A newly laid egg will lie nearly flat on the bottom of a dish filled with water. If the egg is slightly evaporated, the large end will tip slightly upward, and this tendency will increase with the degree of evaporation. If the egg rises to the surface of the water, it is pretty badly evaporated; sometimes they will float with half of their surface exposed.

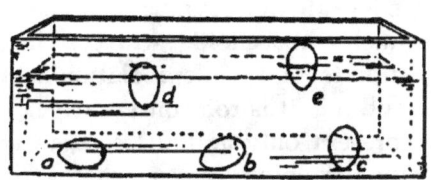

(*Cornell Experiment Station*)

Fig. 266.—Specific gravity test for determining the age and condition of eggs. *a*, Fresh egg; note smallness of air cell and the horizontal position of the egg when immersed in water; *b*, slight evaporation causes the egg to tip; *c*, increased evaporation causes egg to stand on end; *d*, badly evaporated egg which floats; *e*, an egg so badly evaporated that it is likely to be unfit for use.

The best way to inspect eggs is to candle them, a simple method of placing the egg before a bright light and looking through the egg toward the light. Anyone can become proficient in the use of this method in a very few trials.

Our Two Billion Egg Waste.—Immediately it is laid the product of the American hen is worth $50,000,000 more than when it is opened by the consumer. In other words, the value of the eggs produced in this country each year depreciates, *shrinks*, to the extent of $50,000,000 by reason of deterioration and damage due to improper care of the eggs on the farm, most of which may be attributed to gross negligence and inexcusable ignorance, and through careless packing and shipping, resulting in heavy breakage losses, and because of antiquated selling methods, such as the practice of exchanging eggs at the general store for merchandise.

Think of it, in a single commodity in a single country in a single year there is a loss of fifty millions of dollars. And it has been going on right under our noses year after year. Sounds incredible, no doubt.

The Department of Agriculture and other bureaus of statistics are authority for these figures. A journey through the egg trade—all the devious channels from the hen's nest to the consumer's household—will convince the most skeptical that this estimate of the depreciation in eggs is absolutely reliable. If anything, it is too conservative. For example, the above figures represent only actual losses, in addition to which there are certain intangible losses, which are none the less real because of their intangibility.

The intangible losses are those induced by curtailed consumption due to the presence of inferior eggs. Let me explain this point: All are agreed, I think, that nothing disturbs the appetite more than bad eggs. Perhaps this is because of the sensitiveness of our olfactory membranes. However, when we open a bad egg, even though it is only somewhat stale or of poor flavor, what is the result? Disgust—repulsion. We push it from us, or cause it to be removed from sight. At certain seasons of the year when fresh eggs are scarce this *dénouement* happens frequently. In fact, it is the rule rather than the exception, and what is the result? We regard all eggs with suspicion; they are *persona non grata*, and in consequence we turn to something else for our breakfast dishes. This cuts down the demand and lowers the price.

Statistics are sometimes wearisome, especially if they involve large numerals, yet they offer the only means of a convincing argument, outside of the actual conditions. Therefore, at the risk of seeming too statistical I am going to indulge in a few

Fig. 267.—Specially constructed freight cars for live poultry.

411

figures.   I want it to become vividly apparent and to *sink in*—just how appalling, how profligate, is this enormous waste.

Such a loss should be considered sinful, not alone because of the financial loss to the farmers, who bear the greater part of the burden, but because it is a community loss of so much valuable food, the most nourishing kind of food, of which thousands are in such urgent need.

A depreciation of fifty millions of dollars in the value of our eggs is equivalent to throwing away 2,000,000,000 eggs, or about 280,000,000 pounds of one of our finest food staples.   Reducing this loss to a *per capita* basis, it means that every man, woman and child in the country is deprived of about twenty eggs each year.   If we consider that very young infants do not eat eggs in any form, also that the majority of the poorer classes can seldom afford them except in the preparation of other foods, the loss *pro rata* to those who do eat table eggs would probably be in the neighborhood of four dozen each year.

Breakage.—Let us regard the waste in still another way: In New York City about 5,000,000 cases of eggs, or 150,000,000 dozens, are received each year.   Records show that in the spring and early summer months it is not unusual for 200,000 cases to be received in a single week.   The breakage on this egg supply, not the total damage, together with the depreciation resulting from such breakage, is about three and a half per cent, or 5,250,000 dozens annually.

Until a few years ago the railroads and other carriers were held accountable for the greater part of this breakage, and their claims in the New York district alone amounted to over a million dollars a year.   Their claims still amount to about a half million dollars annually.   This reduction in claims does not mean that the breakage has lessened, merely that the carriers have shifted a certain portion of the responsibility to the shoulders of the producers, shippers, packers and wholesale distributors, where it rightfully belongs, as I will explain later.

In a case heard before the Interstate Commerce Commission two years ago, between the New York Mercantile Exchange, a corporation of about five hundred merchants engaged in dealing in eggs and dairy products, which exchange corresponds to the Chicago Grain Exchange or the New York Stock Exchange, and six leading railway systems entering the metropolitan district, it was shown of record that one railroad's gross revenue on eggs and the total claims presented for loss and damage thereon amounted in the periods of 1912 below noted to the following amounts:

	Total Revenue	Total Claims	Per Cent
April	$34,014.54	$2,774.40	8.1
May	23,298.59	4,717.58	20.25
June	16,762.41	2,423.66	14.5
September, first week	5,589.58	3,190.71	57.25
September, second week	5,594.70	3,338.12	59.66
September, third week	4,578.70	2,661.28	58.1
September, fourth week	4,125.46	1,746.67	42.1

During the above periods there were no wrecks or derailments, and no unusual weather conditions or labor disturbances to account for these losses. At the same hearing it was also shown that in 1913 another railroad paid claims on eggs from its New York office alone amounting to $100,207.35. With this and similar losses going on constantly all over the country is it any wonder that the railroads find themselves financially embarrassed, or that they have had to forego dividends to their stockholders? In 1916 their total claim bill for loss and damage on all commodities amounted to thirty-five million dollars, which absorbed about two per cent of their total earnings.

For convenience in analyzing this subject we will divide the egg trade as a whole into four principal divisions or classes: First, the producers, second, the shippers, third, the carriers, and fourth, the distributors, both wholesale and retail.

All Are to Blame.—It cannot be said that any one class is responsible for our fifty-million-dollar egg-loss, and certainly no class is exempt from it. All are to blame, and all are equally culpable.

Losses in eggs occur at all stages of handling, on the farm, in the country store, with the local shipper, the egg-collecting center, the railroad, the packer, the jobber, the commission merchant, the teamster and the retailer. Strictly fresh eggs, generally known as *nearby hennery*, which tickle the palates of the well to do at the rate of eighty cents to a dollar a dozen, come largely from the environs of the large cities, where they are produced by poultrymen who make a specialty of fancy eggs. Such eggs constitute a very small percentage of the trade, however,—about ten per cent,—and their damage aside from breakage is not appreciable because most of the shipments are made by fast express. For the present we will not concern ourselves with this class of eggs.

(*Courtesy Million Egg Farm*)

Fig. 268.—"A full house."

The great bulk of the egg trade must travel a thousand or two thousand miles before it reaches its destination. These are the eggs wherein the serious losses occur. Not so much because of the distances transported, though this is an important factor, but because of the numerous hands through which they pass, and the personal equation in each instance, and because the character of the egg is such that any slight defect visited upon it at its place of origin rapidly accumulates further deterioration. A slightly heated egg on the farm, or an egg with a *blind check*, so termed because the crack is not visible without the aid of a strong candle, may reach the market in such an advanced state

Fig. 269.—Interior refrigerator car loaded with poultry in boxes and barrels.

of decomposition that, unless carted to a crematory or dump-heap, its only use is a manufacturing one, such as tanning leather, and for which eggs are worth about fifty cents a case.

All Bear the Burden.—While the burden of these losses falls upon all who handle the eggs, they are borne chiefly by the producers and the consumers. The producers' loss is represented in a decrease in price because of spoilage or poor quality; the consumers' loss is reflected in a decreased supply by reason of the spoilage, which compels the consumer to pay a higher price for the sound portions which finally reach him.

Investigations.—Exhaustive special inquiries have been conducted in different parts of the country for a number of years, both by the Department of Agriculture and other interested bodies, with the view to finding out the exact conditions of eggs in the numerous stages of handling, and to ascertain ways of remedying the defective practices.

A test was made in an egg-collecting center in the West made up of twenty prominent shippers, wherein it was found that the percentage of eggs that were so bad as to be an absolute loss amounted to 8.33 per cent. This was in November. In other sections during the warmer months as high as thirty per cent of the eggs were totally unfit for food.

An investigation among some country stores during October, which is considered a fairly favorable month for eggs, showed that only twenty-five per cent of the eggs collected from the adjacent farmers would rank as *firsts*, that sixty per cent were *seconds*, due to long holding on the farms, that five per cent were cracked, and that four per cent were rotten or stuck to the shell. In this experiment, as with many others, it was found that the majority of the farmers had held their eggs for four and six weeks before turning them into the village store.

Grades.—For the benefit of those who are not familiar with the different grades of eggs from a commercial standpoint,—and few are,—I want to explain the definitions of these grades as employed by the trade. The term *firsts* does not mean one hundred per cent strictly fresh eggs, or even good eggs, by any manner of means, though the term implies such a quality. The rules governing transactions in eggs on the New York Mercantile Exchange, which constitute the standard or basis by which all trading is carried on, and which are published in booklet form and distributed throughout the trade, are set forth in part as follows:

Rule I.—Classification and Grading:

1. Eggs shall be classified as "fresh gathered," "held," "refrigerator," and "limed."
2. There shall be grades of "extras," "extra firsts," "firsts," "seconds," "thirds," "No. 1 and 2 dirties," and "checks."

Rule II.—Qualities:

FRESH GATHERED EXTRAS shall be free from dirty eggs, of good uniform size, and shall contain reasonably fresh, reasonably full, strong-bodied, sweet eggs.

Quality A............................... 90 per cent
Quality B............................... 80 per cent
Quality C............................... 65 per cent

The balance, other than loss, may be slightly defective in strength or fullness, but must be sweet. The maximum total average loss per case permitted in "extras" shall vary with the requirement of reasonably full, strong-bodied eggs, as follows:

Quality A.................. 1 dozen maximum loss
Quality B.................. 1½ dozen maximum loss
Quality C.................. 2 dozen maximum loss

FRESH GATHERED FIRSTS shall be reasonably clean and of good average size, and shall contain reasonably fresh, reasonably full, strong-bodied, sweet eggs.

Quality A............................... 75 per cent
Quality B............................... 65 per cent
Quality C............................... 50 per cent
Quality D............................... 40 per cent

The balance, other than loss, may be defective in strength or fullness, but must be sweet. The maximum total loss per case permitted in "firsts" shall vary with the requirement of reasonably full, strong-bodied eggs, as follows:

Quality A.................. 1½ dozen maximum loss
Quality B.................. 2 dozen maximum loss
Quality C.................. 3 dozen maximum loss
Quality D.................. 4 dozen maximum loss

27

FRESH GATHERED SECONDS shall be reasonably clean and of fair average size, and shall contain reasonably fresh, reasonably full eggs:

Quality A.............................. 65 per cent
Quality B.............................. 50 per cent
Quality C.............................. 40 per cent
Quality D.............................. 30 per cent

The balance, other than loss, may be defective in strength and fullness, but must be merchantable stock. The maximum total average loss per case permitted in "seconds" shall vary with the proportion of reasonably full eggs required, as follows:

Quality A..................... 2 dozen maximum loss
Quality B..................... 3 dozen maximum loss
Quality C..................... 4 dozen maximum loss
Quality D..................... 5 dozen maximum loss

"Loss" as used in these rules, shall comprise all rotten, spotted, broken (leaking), broken yolked, hatched (blood veined) and sour eggs. Very small eggs, very dirty, cracked (not leaking), badly heated, badly shrunken and salt eggs shall be counted as half loss in all grades excepting "dirties" and "checks."

The foregoing rules cover but three grades of eggs. There are about twenty grades all told, with exceptions and modifications to each. Some of their names follow: Fresh gathered thirds, held firsts, held seconds, refrigerator extras, refrigerator firsts, refrigerator seconds, refrigerator thirds, limed extras, limed firsts, limed seconds, No. 1 dirties, No. 2 dirties, checks and so on.

To describe all of these different grades would take up too much space. The point that I wish to bring out is this, notwithstanding certain eggs are good enough to be rated as "fresh gathered firsts," they are still a long way from being perfect. And this *way* constitutes waste. As will be noted from the foregoing rules, the shrinkage in "fresh gathered firsts" runs from twenty-five to sixty per cent, with an actual loss of from one and a half to four dozen to the case.

A certain amount of shrinkage is to be expected in storage eggs, even when they are stored under the most favorable refrigera-

tion. It is a natural evaporation of the contents of the egg, resulting in an air cell which is familiar to all, and cannot be overcome.

Excessive shrinkage, badly shrunken eggs, is another matter.
Its responsibility usually commences on the farm, due to holding
the eggs too long, though the rural buyer or country store and
the egg-collecting center are in no wise innocent of the same poor
practice. Actual physical deterioration, or total loss, as de-

Fig. 270.—Candling eggs at the farmer's gate.

scribed in the foregoing rules of the Mercantile Exchange, is
still another matter, and includes such depreciation as heat affection, bacterial contamination, mustiness, mold and sour eggs.
The conditions which bring about these changes are almost
wholly due to negligence and antiquated methods. They are
unpardonable because in the main they are preventable.

Heat Losses.—As might be supposed, heat is the worst enemy
of the egg. The loss to the trade as the result of heated eggs is

greater than from any other source, especially if male birds are allowed to run with the hens.

Heat in connection with eggs does not necessarily mean excessive heat, which we know will injure the quality of eggs. The fertile egg is susceptible to even a moderate temperature. If allowed to remain in a temperature of 70 degrees F. for any length of time, it starts to incubate, slowly, of course, nevertheless development goes on, and every day it is exposed to this warmth it is hastened on its downward career. The first stage of this deterioration appears as blood on the yolk. Later a blood ring is formed, which indicates that the embryo is dead, and like all dead animal matter forthwith it starts to decay. Commercially, heated eggs are known as *floaters* and *blood rings*. *Light floats* correspond to about twenty-four hours' incubation at a temperature of 102 degrees,

Fig. 271.—Stolen nest in the hollow of a tree.

whereas *heavy floats* are equal to about forty-eight hours' incubation.

Infertile eggs, sometimes called sterile eggs, those laid by hens kept apart from male birds, withstand heat much better than fertile eggs, though it is an utter fallacy to assume that the former will not spoil. This mistaken notion is all too common. Without a vitalized life germ, there is no incubation in sterile eggs, hence there are no blood rings to develop; nevertheless they undergo certain chemical and physical changes which sooner or later impair the quality of their contents. They shrink as badly as the fertile eggs, and the yolks are likely to weaken and break or stick to the shells.

Farm Losses.—It is not uncommon for farm eggs to remain under a wheat shock or in some out-of-the-way corner for two or three days, maybe a week, before they are found, or before it is convenient to collect them, during which time probably they have been subjected to summer heat of perhaps 95 degrees. See Fig. 272. Frequently the eggs are stored in the kitchen or back porch, or in a closet in one of the outbuildings, where the thermometer hovers around 85 degrees at midday. Often the eggs are hauled to the village store or shipping point in an open wagon, maybe a wagon without springs, exposed to the direct rays of the sun and a temperature of 105 degrees, and then left on a truck at the railway station for several hours.

General Store.—If the eggs are delivered at the village store for credit, it is quite likely that the storekeeper, receiving the eggs over the counter, will let them remain in the store until the close of the day, and then carry them down to the cellar, where they will remain for perhaps a week, until a large enough quantity is gathered to ship to a local buyer or egg-collecting center. What if the cellar is warm or damp or musty or poorly ventilated, the storekeeper has no particular interest in the eggs, any more than as a means of keeping the farmer's trade in merchandise.

Local Buyer.—When the local buyer or egg-collecting depot receives the eggs, probably it is by way-freight, the eggs are again held for several days or a week until there is enough to make up a carlot, whereupon they are dispatched to a city jobber or to a packer. This last lap of the journey may be made in a refrigerator car, or it may not. In either event there is enough iniquity already stored up inside the eggs or their cases to account for a large portion of our fifty million dollar loss.

This system of marketing is not incidental. It is general. It is in vogue all over the country. Its evils are perfectly obvious. Exchanging eggs for merchandise or credit at the general store, as at present practised, is pernicious. It is the weakest spot in the egg trade.

A graphic idea of the loss due to heated stock can be obtained from the fact that in the South and Southwest the egg industry is practically dead from the first of June until cool weather in the fall. The losses are so great, and the net returns so small, that the majority of the farmers hardly consider it worth while to collect their eggs at all during the summer months. It is a common thing to find fully hatched, live chicks in cases of market eggs from these sections. In fact, the heat damage and other losses are so heavy that producers do well to receive a net price of five cents per dozen.

An average lot of summer southern eggs would candle about as follows: *Light floats*, 80 per cent; *heavy floats*, 15 per cent;

Weed nest.   Nest in straw stack.
                      (Courtesy U. S. Dep't Agriculture)

Fig. 272.—Stolen nests are responsible for a huge wastage of eggs.

blood rings and checks, 5 per cent. As bakers' stock the *light floats* might bring twelve cents a dozen, the *heavy floats* five cents a dozen, and the *blood rings* nothing. Deduct for breakage, commission, candling, freight and haulage, and the net returns are insignificant.

One in Four Lost.—As near as it can be estimated, the loss in southern and western eggs due to heat is one-fourth of the original value of the crop. For the entire country this element of waste is estimated at fully five per cent of the total valuation of the egg trade. Mind you, this percentage represents actual wastage, and nothing more. It does not include the losses resulting from decreased prices by reason of the inferiority of most farm eggs. The losses attributable to the inferior quality of the bulk of our egg trade are intangible, the same as the losses due to curtailed consumption are intangible; but they are real nevertheless. It is doubtful if one hundred million dollars a year would cover them.

The most significant testimony to this statement is found in the twenty or more grades of low quality eggs, selling at low prices, and their relation to a few grades of high class eggs, which sell at high prices. The farmer's eggs comprise the bulk of the low grade marks, which include storage eggs, and sell at prices from a third to two-thirds the value of strictly fresh, prime eggs. The following is a typical set of wholesale quotations in the New York market for December:

Nearby hennery whites, fine to fancy	76 to 82 cents
Nearby hennery browns	57 to 63 cents
Fresh gathered extras	54 to 55 cents
Extra firsts	52 to 53 cents
Firsts	48 to 50 cents
Seconds	42 to 46 cents
Refrigerator extras	38 to 39 cents
Refrigerator firsts	35 to 36 cents
Refrigerator seconds	32 to 33 cents
Refrigerator thirds	30 to 31 cents
Limed extras	28 to 29 cents
Limed firsts	26 to 27 cents
Limed seconds	24 to 25 cents
No. 1 dirties, refrigerator	23 to 24 cents
Checks	22 to 24 cents

It is not possible, of course, to bring the entire egg crop of the country up to the level of the "nearby hennery whites," but it is possible to create a vast improvement in the lower grades, which improvement will redound to the benefit of the farmers. Strange that they do not see it this way!

Breakage.—Next to heat, the greatest damage is occasioned by breakage. Indeed, it is nip and tuck which is the most influential factor in piling up our monument of worthless eggs— heat or breakage. A broken egg is such a malicious sort of breakage. It is not only a loss to itself, but it contaminates everything around it.

Trade Terms.—In the trade, breakage is designated by various names, some of which are descriptive, if not picturesque. Cracked eggs are termed *checks*. When the shells are pushed in without rupturing the inner membrane, they are called *dents*. If the eggs are partly open, or if they have lost a portion of their contents, they are known as *leakers*. When the eggs are completely broken—among the missing, as it were—they are spoken of as *mashers*.

Leakers and mashers not only produce smeared eggs, which

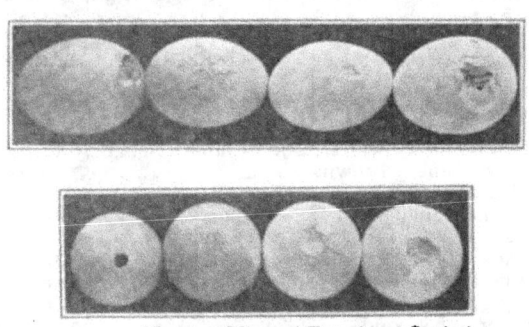

(*Courtesy Missouri Experiment Station*)

Fig. 273.—Typical forms of breakage which usually result from careless handling or packing.

are heavily discounted as dirties, but they account for a great many moldy eggs, also *rots* and *spots*. The *checks* and *dents* constitute a heavy loss because they do not keep in storage. It is estimated that thirteen million dollars' worth of eggs spoil in storage each year because of cracks, some of which are scarcely visible to the naked eye.

Mechanical Injury.—The responsibility for mechanical injury —*breakage*—is the bugbear of the egg trade. It has been the cause of so many claims, controversies and law suits between shippers, receivers and railroads that many of the latter have wanted to give up handling case eggs altogether. Even now

these relations are far from amicable. It is hardly likely that
they ever will be harmonious until we abandon some of our
obsolete practices.

If you listen to the shipper's or receiver's tale of woe in regard
to breakage, you gain the impression that the carriers are a set
of *beats*, who, for no reason at all, kick egg cases all over the
map, smash them into bits, and then refuse to pay claims for
damages.

True, the railroads, express companies and other carriers have
adopted rather drastic measures of late, one of which is the allowance of five per cent breakage on each case of eggs. Their tariff
files now state that they will not be held accountable for breakage
unless it exceeds five per cent, and that claims will be considered
only on the damage in excess of five per cent on each case or
crate.

Shippers and receivers regard this regulation as an unjust im-
position. They go so far as to call it confiscation. The carriers
have been put to it in self-defense. There was a time, and only
a few years back, when the greatest asset that many of the men
engaged in the egg trade had, was the volume of claims which they
collected from the carriers. Apparently, the carriers were regarded as legitimate prey. They were set upon and gouged in
a merciless manner by the most unscrupulous practices. The
carriers knew they were being gouged, and, like the proverbial
worm, they turned. They got the *goods* on various gentlemen,
and said gentlemen were indicted for fraudulent practices.

CHAPTER XXXI

SELLING EGGS ADVANTAGEOUSLY

Many poultrymen, especially those who raise but a few chickens, concern themselves chiefly with the problem of producing eggs, with insufficient attention to the best methods of selling them. They are accustomed to think that there is an unlimited demand, which is true—there is always a regular outlet for eggs, just as there is a regular market for corn, wheat and other staple products; but, to secure top prices, or to obtain the full value of the product, as much thought and care must be devoted to the selling end of the business as to the subject of production.

Eggs are probably the easiest commodity to sell.—In fact, the demand for strictly fresh goods of prime quality is always greater than the supply. If we so desire we do not have to stir from our doorstep to find a buyer. The buyer will come to us, pay spot cash if he is a dealer, or give us credit for merchandise at his store. Transactions of this sort are very common, and they are so simple and call for so little effort on the part of the producer that they have induced the poultry raiser to accept them as a regular practice—along the lines of the least resistance. In other words, it is so easy to sell eggs through the regular channels—the country store, commission merchant or wholesale dealer—that the practice has encouraged a form of laziness on the part of the producer. Instead of making him independent and resourceful, and stimulating a desire to seek the most profitable outlet for his wares, it has made him dependent and submissive —a sort of "take what I can get" spirit.

Middleman's Charges.—It is very convenient to dispose of one's output at the back door, but it is a great mistake to think that the producer is not paying for this convenience. He is

426

paying for it in the form of a reduction in price. Eggs for which the producer is paid twenty-five cents a dozen usually retail to the consumer for thirty-five and thirty-eight cents, or higher. Thus, the middlemen's profits and handlers' charges amount to over ten cents a dozen. These charges are legitimate, because

(Courtesy Million Egg Farm)

Fig. 274.—Grading and packing eggs.

these distributing agents have rendered service for the poultry-man. They have received, shipped, stored, graded, sold, de-livered and collected for his goods. On each operation the mid-dleman has been put to an expense, which is chargeable, and on each operation the middleman wants a profit above his expense, or there would be no incentive to do business.   The sum total of all expenses and profits is therefore quite considerable. Admitting that the poultryman is put to a similar expense when he markets his goods direct to the consumer, the profits on these expenses, at least, are diverted to his own bank account.

In talking to poultrymen on this subject I have heard remarks as follows: "It doesn't pay me to sell eggs retail. I haven't the time to bother with small shipments. I'd rather take a lower price and ship to a commission merchant—let him peddle the eggs. A private trade is all right in a way, but the pay is too

(*Courtesy U. S. Dep't Agriculture*)

Fig. 275.—Well-equipped feeding plant. Note the complete lighting and ventilating system.

slow. I've got feed bills to meet, and I've got to get my check for a shipment promptly. Besides, when you ship to customers direct, and there is breakage or a delay of some kind, there are too many complaints. Let the other fellow do the worrying."

Retailing Troubles.—There is a great deal of truth in the fore-going remarks; it is troublesome to sell eggs retail; yet it is the only way to get the greatest profits from the business. The details of any enterprise require close attention, and it behooves the man in charge to evolve a system or scheme whereby they are simplified as much as possible. Certain it is that if the middlemen find it profitable to distribute retail or direct to the consumer, the producer should find it equally profitable, providing he follows practical methods.

In addition to the loss in price due to the charges and profits previously mentioned, it should also be borne in mind that the more eggs are handled the more their quality depreciates, consequently their value is discounted, as we have shown in a preceding chapter.

Number of Handlings.—Though transportation facilities are relatively simple, it is surprising to note the number of times a shipment of eggs is handled even when the eggs are produced within a hundred miles of their ultimate market. Eggs produced on a Pennsylvania or New Jersey poultry farm and consigned to a New York commission merchant are packed in 30-dozen crates. The crate is delivered to the express agent by the poultryman, the agent puts it aboard the train, and while it is in the care of the express company it may be handled four or five times before it reaches New York. At the railroad terminal the crate is removed from the car to the warehouse, and later loaded into a truck and delivered to the commission house to which it is consigned.

Commission Merchants' Work.—Most commission merchants find it necessary to unpack, candle and grade the shipment, after which the eggs are sold to a retail store or hotel, or frequently the wholesale house will sell the shipment to a jobber, who will distribute to the retail stores, restaurants and other consumers in one or two case lots. The retail store distributes the eggs to the ultimate consumer in dozen lots. The eggs have probably been handled twenty times, which does not improve their quality, particularly if they have been kept in warm temperatures or where the atmosphere is not as fresh as it might be. Such is the devious route by which a local shipment of eggs is received, and if they are of prime quality, they are known to the trade as "Nearby Hennery Fancy."

Eggs of this grade, however, are more or less limited, and if the public had to depend upon them alone, there would be an egg famine. The greater number of eggs consumed in the large

Eastern cities come from the Central States, over a route about as follows:

The farmer collects his eggs whenever he is of the mind to do so—maybe it is every day or twice a week, and from nests scattered about the farm buildings, under wheat shocks and in the brush. If they are fairly plentiful, he will take them to the general store (just as he has found them), in a basket, on an average of once a week. Or, perhaps the farmer lives in a section visited by an agent who drives from farm to farm gathering eggs in small

Fig. 276.—Interior of a Western feeding station for handling live poultry.

quantities. See Fig. 270. When the local merchant has collected a number of cases he disposes of them to a wholesale buyer, who is usually the representative of a large commission house in the East. The local merchant probably makes his shipment by freight to save expense. The wholesale buyer collects in case lots until he has enough for a carload, which is generally sent East in a refrigerator car. He may have repacked the eggs in new cases after candling them, or sold them *case count*. When the commission house receives the carload, the eggs are sorted into a number of grades, which are sold to jobbers, and thereafter the same route is followed as for the local eggs, only to a different class of trade. These eggs seldom come in competition with local eggs, for they are usually of inferior quality.

Selling Direct.—If eggs are worth producing, every effort should be made to get their full value, and to do this they must be marketed as quickly as possible, so as to avoid any deterioration, and sold wherever possible direct to the consumer, all other arguments to the contrary notwithstanding. Hotels, restaurants of the better class, clubs, steamships, railroads and soda fountains are among the highest bidders for strictly first-grade eggs, and many are supplied directly from large poultry farms. The prices at which these eggs are sold are usually at a given premium over top market quotation, or at a fixed rate per dozen on a year's contract for a given number of crates per week. The premium method is probably the best for all concerned, and may be made from three to ten cents above the market.

Business of this character is generally done on a standing order, and the shipments are seldom more troublesome than dealing with commission merchants. Collections may be slower, of course, but the progressive poultryman must endeavor to take a position where he can extend a certain amount of credit; it is one of the principles of modern business.

Another class of trade may be found among the first-class grocers who make a specialty in eggs of known quality, and to dealers who carry on a strictly high grade butter and egg delivery. Special terms can be made with a trade of this kind, similar to the hotel patronage—a premium over market quotations, and if those dealers who cater to a discriminating class can rely upon the poultryman for an absolutely uniform product, they will take special pains to create a high price market.

Unscrupulous poultrymen have abused the practice of receiving premiums, hence those who have suffered by the abuse are apt to be a little skeptical, and justly so. The weakness of the plan is this: The premium offered sometimes tempts the poultryman into buying eggs from his neighbors and other outside sources,

Fig. 277.—Metal chilling racks for "squatted" poultry.

432

all in good faith, no doubt, but unless the eggs are produced on his own farm, the poultryman has no real warranty as to their quality. He must depend upon the word of others, and frequently such assurances are not dependable. The quality becomes uncertain, complaints are made, and confidence is shattered. Altogether, the practice is a risky one, and in a measure it is a misrepresentation—unfair dealing.

Lost Confidence.—It is strange, perhaps, how quickly a customer will lose faith in the source of an egg supply, especially if the dealer is making attractive claims for the product. You may serve a household with eggs of the best quality for weeks and months, with never a complaint, when suddenly there is trouble. Maybe two or three eggs in a consignment are slightly stale, or they have a peculiar flavor, or they contain blood clots; immediately the consumer loses confidence, the loss of which is no easy task to recover.

Private Trade.—In the outlets just described the business is more or less of a wholesale nature, yet many of the middle profits, notably those of the commission merchant and jobber, have been eliminated. There still remains, however, the profit taken by the retail dealer, which usually amounts to about five cents a dozen. Surely this profit is worth striving for, and may be had if the poultryman will make an effort to reach a private trade direct.

If the farm is situated within easy driving distance of a large community, the problem is a simple one: the poultryman can make regular deliveries about twice a week. If this is not practicable, shipments can be made by parcel post, or by express, or the poultryman can establish a delivery in the city and have his eggs shipped to the city in large quantities, whence they are distributed.

Parcel Post.—It is practicable to ship eggs by parcel post, so far as the security of the packages is concerned. If the containers are returned and used over and over, the charges are greatly reduced. There are many types of carriers on the market, generally made of corrugated paper, which are strong and durable and yet

28

very light.   The prices range from ten to thirty cents, depending upon the style and size, and from the writer's experience each carrier will make on an average of eight trips before it becomes too badly worn for further use.   The customer is requested to hold the *empties* until four or five are collected, when they are then tied together and returned for the one charge.   Thus, a carrier

(*Courtesy U. S. Dep't Agriculture*)

Fig. 278.—Bench killing and picking.

holding four dozen eggs and costing twenty cents can be made to transport the eggs for about one cent a dozen, which includes the return charges on the carriers.   The parcel post charges will amount to about three cents a dozen, bringing the total cost to four cents a dozen.

Charges on Case Eggs.—At a glance this seems very high and to defeat the idea of shipping by parcel post; but if we consider the cost of shipping eggs wholesale in case lots there is not a great deal of difference.   Consignments made to commission merchants are usually made in gift crates and the express charges must be borne by the shipper, the cost of which comes to about two cents a dozen.   The difference of two cents a dozen can be made up in the price charged for the eggs when catering to a select trade, over and above the retail profit of five cents a dozen.

The club plan is probably the best method of selling eggs retail, and the easiest for the poultryman, although he should not expect to receive quite as high a price for his product.   The plan is this—two or three families residing close together place a standing order for a case or a half case of eggs once a week; the shipment is made to one address and the bill collected from there; the work of distributing the eggs being left to the consignee, who must also collect for them and remit to the poultryman.

The additional profits accruing to the poultryman who sells retail are probably equal to the regular profits to be made on wholesale shipments.   The practice requires more work and closer supervision, of course, but if one is engaged in the business for what can be taken out of it, why not try to take all the profits possible?   If one thousand layers can be made to show a profit of two dollars a year per bird by selling retail, why raise and maintain two thousand layers at a profit of a dollar per bird per year selling wholesale?

CHAPTER XXXII

PRESERVING EGGS

No article of diet of animal origin is more commonly eaten by all classes in all countries and in a greater variety of ways than eggs. They form one of our principal elements of food, and as such they are always in great demand. Unfortunately, however, the daily supply is not in proportion to the daily demand. We want just as many eggs in the fall and winter as we do in the spring and summer, but unless we store them, they are not available. Dealers realize the importance of this—witness the stocks they place in cold storage each year. Why should the individual not exercise the same forethought? It is entirely practicable, and the investment required is small, outside of the cost of the eggs themselves.

Virtues of Cold Storage.—It is strange and interesting to note how some of our conventions and prejudices have evolved, and how far removed from the truth are some of the popular beliefs. Many people regard cold storage eggs, and all those connected with the business, including the producer, middleman and retailer, as being traffickers in a more or less illegitimate product. Among some the very name *cold storage* is as odious as adulteration; and they rail against the practice as though it were a form of knavery. Such a belief is absolutely false.

The cold-storage industry is a development of the past thirty years, and as a whole it has been of enormous benefit to both the producer and the consumer. Of this there can be no doubt if one will only take the trouble to look into the subject a little. Placing eggs in storage has tended to make the prices of this commodity more uniform throughout the year, and to increase fall and winter consumption.

"How is that?" someone will ask. "I remember when you

could buy eggs in the spring for ten cents a dozen." True; but at that time it was difficult to buy eggs in the fall and winter for any price. They were not to be had in any quantity. Hens lay very few eggs in the autumn, especially on general farms; which farm flocks produce about 80 per cent of the country's supply. It is a perfectly natural condition. The old hens are in the molt, and the pullets have not reached maturity; and not until scientific methods were perfected, such as artificial incubating and brooding, and improved housing and feeding, was it possible to obtain any sort of a yield except in the spring and early summer.

(Courtesy Petaluma Chamber of Commerce)

Fig. 279.—California type of laying house. Note that there are no dropping boards.

In earlier times eggs were marketed at the time of and near the place where they were produced. During the spring and summer months there was an over-production; farmers found it difficult to find a market for their eggs, and at times were lucky to get five cents a dozen. In the South and Southwest it has only been in very recent years that the farmers ever bothered to collect their eggs during these seasons, they were that unprofitable. To-day conditions are very different. Since the advent of the cold storage system and improved methods of transportation, we are able to take advantage of this over-production, and to preserve it for our needs when there is no production.

The farmer is therefore paid a fair price for what he previously wasted—usually from fifteen to twenty-five cents per dozen.

Eggs are preserved in a number of ways. For convenience they may be divided into two general classes: The use of low temperature, from 31 to 32 degrees F.; and by excluding the air by coating, covering or immersing in a solution or dry substance. The first classification is the only way they can be preserved on a commercial scale, i. e., cold storage. Two methods are followed: storing the eggs in crates in a fairly dry atmosphere, and removing them from their shells and freezing them in bulk in cans containing about fifty pounds each.

Under proper conditions, when fresh-laid eggs are placed in storage, very little change takes place in their quality, except evaporation. But they must not be allowed to remain long out of storage before they are used. It is failure on the part of the consumer and retailer to observe this point that results in most of the difficulties with storage eggs.

As for the other methods, their aim being to exclude air conveying micro-organisms to the interior of the egg, and for suppressing the growth of those already present, the results obtained are by no means uniform, which is largely due to the condition of the eggs at the time they are placed in storage.

One of the old-fashioned domestic methods was to pack the eggs in bran, or in salt, or by covering them with limewater. Sometimes the eggs remained in good flavor, other times they spoiled. Their degree of preservation was commonly referred to as *luck;* whereas it was chiefly due to ignorance. Only eggs of known freshness and quality, and preferably non-fertile eggs, should be preserved by these methods.

Twenty methods of preserving eggs were tested, with the following results, according to the Department of Agriculture: Those preserved in salt water, brine, were all bad, not rotten, but unpalatable, the salt having penetrated the eggs. Of the eggs preserved by wrapping in paper, 80 per cent were bad; and the same proportion of those preserved in a solution of salicylic acid and glycerin were unfit for use. Seventy per cent of the

Fig. 280.—Elevated laying houses. Sometimes called two-story houses, inasmuch as they provide shade and shelter beneath.

439

eggs rubbed with salt were bad, and the same proportion of those preserved by packing in bran, or covered with paraffin or varnished with a solution of glycerin and salicylic acid. Of the eggs sterilized by placing in boiling water for 12 to 15 seconds, 50 per cent were bad. One-half of those treated with a solution of alum or put in a solution of salicylic acid were also bad. Forty per cent of the eggs varnished with water glass, collodion or shellac were spoiled. Twenty per cent of the eggs packed in peat dust were unfit for use, the same percentage of those preserved in wood ashes, or treated with a solution of boracic acid and water

Fig. 281.—Preparing water glass for preserving eggs.

glass, or with a solution of permanganate of potash, were also bad. Some of the eggs were varnished with vaseline; these were all good, as were those preserved in limewater or in a solution of water glass.

Of the three methods that were entirely successful, the water glass treatment is to be recommended. Covering eggs with vaseline requires too much time, and the idea is not a particularly pleasant one. The limewater treatment sometimes communicates an odor to the eggs.

Water glass, or soluble glass, is the popular term for sodium silicate or potassium silicate, the commercial article often being a mixture of the two. The commercial product is generally used for preserving eggs, if it is of a good grade, inasmuch as the chemically pure article, which is used for medical and other purposes, is very much more expensive. Inferior grades are likely to be alkaline, which should not be used, as the alkali will impart a bad flavor to the eggs. Moreover, they do not keep well in it.

Water glass is offered for sale in two forms—a thick liquid having the consistency of molasses, and in a powder. See Fig. 281. The former is perhaps the most commonly used, and may be purchased at any drug store for about forty cents a quart. In larger quantities it may be bought of wholesale druggists very much cheaper. In buying it, it is well to state the purpose for which you intend to use it, and to receive some assurance as to its quality.

The North Dakota Experiment Station conducted a series of experiments with water glass which were very successful, and their reports recommend a solution of 1 part water glass in liquid form to 9 parts water. If the powder is used, a smaller quantity of the water glass is required for the same amount of water. Only pure water should be used in making the solution, hence it is well to

(*Courtesy C. L. Opperman*)

Fig. 282.—Earthenware crocks make the best receptacles for preserving eggs in water glass.

boil the water for about twenty minutes and then allow it to cool before mixing it with the water glass. One gallon of the water glass should make sufficient solution for covering 50 dozen eggs, if they are economically packed; hence, at a cost of $1.25 for the water glass, the cost of storing eggs, including the cost of the container, should not exceed 3 cents a dozen.

Earthenware crocks make the best containers (see Fig. 282), though good results have been obtained with wooden kegs and barrels. In any event, the container must be thoroughly cleaned, scalded, scrubbed and then rinsed. The receptacles should be stored in a cool, clean place, preferably in a well-ventilated cellar, one that would be suitable for the storage of preserves.   If they are placed where it is too warm, the eggs will not keep well, the solution will evaporate rapidly, and the silicate is likely to form a deposit on the egg shells.

Only clean eggs should be preserved, and by that is meant, only eggs which have always been clean, and not washed.   Washing the eggs removes the natural mucilaginous coating on the shell, which was intended by Nature to make the egg more or less impervious to foreign substances.   The fresher the eggs the better, naturally, for there is less likelihood of their having been contaminated in any way.   Eggs known to be older than a week should not be used as a general practice;   and wherever possible use non-fertile eggs — those from flocks having no male birds.   Sterile eggs do not contain an active life germ or embryo, consequently they are safe from any state of animal growth, if at any time, no matter how short the period may be, they were subjected to a temperature that would start incubation.

(Courtesy Cornell Experiment Station)

Fig. 283.—Examining eggs by means of an electric candler.

As a further safeguard, it is well to candle all eggs before they are stored, which will determine their freshness and detect any eggs containing blood clots.   See Fig. 283.   The importance of this is not over-estimated, if we consider that one or two spoiled eggs may ruin the entire container.

Packing.—The eggs should be packed with the small end down, which will help to keep the yolks from gravitating and adhering to the shells.   The solution is then poured over the eggs, covering them to a depth of about two inches.   Later, if much of the liquid has evaporated, it may be necessary to add more of the solution.   If one is unable to fill a receptacle with eggs at one operation, which is hardly likely except on a large egg farm, the eggs may be packed a layer at a time and covered with the liquid, adding more eggs and more of the solution until the container

(Courtesy Kansas Experiment Station)

Fig. 284.—Houses and runs should be arranged to render the greatest facility in caring for the flocks.

is filled. If earthenware crocks are used and they have lids, place the lids on, for this will reduce the amount of evaporation. Otherwise, the containers should be protected in some way, and a good plan is to cover them with paper, glued fast, and then shellacked.

Influence on Eggs.—The eggs should be removed as they are desired, and not kept out of the water glass for any considerable time. If the eggs and all other conditions are right, they may be kept for six to nine months and be perfectly edible.   That is, they may be used for any purposes, but they will not have the consistency of fresh eggs.   The white or albumen will be more watery than the strictly fresh egg, and the yolk will not have its former firmness. These are the conditions found in the storage egg, and they are to be expected. The membranes of the yolk and the entire structure of the egg are weakened by its age, yet their value as food is in no way impaired.

Preserved eggs will not stand the handling that fresh eggs can receive, consequently the housewife should not be disappointed to find some of the yolks broken. When boiling preserved eggs a tiny hole should be pricked with a needle in the air cell end, to prevent the shell from cracking. This is easily done by inserting the point of the needle once. It is sometimes difficult to poach preserved eggs, because the yolks are apt to run into the whites, and the same difficulties may be met in frying them; but for cooking purposes generally these shortcomings are unimportant. When we consider the saving made over eggs purchased from the store at certain seasons, which, by the way, are not always as represented, the deficiencies of the preserved eggs are amply compensated.

This subject should not be construed to mean that poultrymen should adopt this means for other than home use; for they should not. It is a mistake to think that eggs can be preserved in this way and then sold for fresh eggs. They cannot be made to deceive anyone. If sold, they should be offered for just what they are,—*preserved eggs*,—and no attempt made to misrepresent them. If sent to an egg dealer in one of the large cities, they will be candled and their contents will be noted. As previously mentioned, since they do not withstand handling well, their contents are very apt to be addled, and a price paid accordingly.

Preserved eggs are intended for home use, or for sale as such, and the results obtained for a number of years in every way warrant a more widespread practice. The subject is now being taken up by many of the leading Women's Clubs throughout the country, and the State Experiment Stations.

CHAPTER XXXIII

BY-PRODUCTS OF POULTRY

The progressive poultryman is interested in any device that will add to his profits. He will spare no expense to make his fowls comfortable; he will take great pains with the feeding and watering, and he will devise every conceivable plan to increase egg production, if that is his specialty. Not for a moment would he tolerate conditions that might endanger the welfare of his establishment. No doubt he prides himself that he is on the lookout for opportunities to buy and sell to the best advantage, and perhaps he also flatters himself that he can detect a leak in any of the farm's operations.

At the same time it is likely this very same poultryman may be neglecting one of the most important, or at least one of the most stable, sources of income—the revenue to be derived from by-products—those things which are usually held as a nuisance.

Many thousands of dollars are wasted each year in the careless handling or neglect of poultry droppings, while still other thousands are wasted because no attention is paid to saving feathers.

The manure from fowls is rich in nitrogen; it heads the list of farm manures, being worth four or five times the value of stable manure.

Quantity Produced.—The Maine Experiment Station conducted a series of tests on the subject of hen manure, and in one of its bulletins it states that the weight of night droppings from a fowl will average thirty pounds a year, and that this manure contains 0.8 pound of organic nitrogen, 0.5 pound of phosphoric acid, and 0.25 pound of potash. Ordinarily, the value of these elements would amount to a trifle over twenty cents. The way prices have soared on fertilizers, these elements are now worth considerably more.

445

There are no reliable data to show the total quantity of matter voided by the fowl, yet because the chicken is more than half its time off the roost, it is safe to estimate that the weight of its manure while off the roost will be at least thirty pounds a year, probably nearer forty pounds, of which a fair percentage can be conserved in the litter of the poultry buildings.

The writer has visited large poultry establishments where no effort was made to preserve this by-product. In fact, I have

(*Courtesy Purdue Experiment Station*)

Fig. 285.—Indiana poultry house erected on concrete walls which extend two feet above the ground level. Walls are built of novelty siding, making a very neat exterior.

been on plants where the droppings, as they were gathered from beneath the perches in the laying houses, were thrown outdoors, actually cast on an open pile, exposed to rains and winds which quickly rob such matter of its chemical value.

On one farm the disposal of the droppings was flagrantly in error. I might add that it was *fragrantly* wrong, as well. The houses were of the continuous type, located on a hillside, and erected on piling several feet from the ground. In the front of each building, in the yards proper, there was a huge mound of manure and discarded litter, probably the accumulation of several years. Needless to say, these mounds furnished anything but sanitary scratching quarters for the fowls. They were objectionable enough in dry weather. In wet weather they were abominable—a slimy, unsanitary, steaming, stinking mass. The proprietor of this place was not only wasting several hundred dollars annually in the loss of this by-product, but he was menacing the success and health of the flocks by imposing such unsanitary conditions.

Yet many persons wonder at the number of failures in the chicken business.

Loss of Nitrogen.—Some poultrymen profess to take care of the droppings, and really do go to considerable trouble; but their care consists merely in storing the manure under cover. Keeping it from the weather, of course, will preserve it to a certain extent, because there is no leaching; but it will not prevent the escape of a large percentage of the nitrogen, which is its most valuable product, and the most expensive element.

Physical Condition.—Undoubtedly, the chief reason for the great waste of poultry manure is due to its physical condition, which is such that it requires special treatment to conserve it. Its greatest value, as mentioned before, is in its nitrogen content, which is subject to what are termed putrefactive processes, which convert it into ammonia compounds. These compounds are highly volatile, and unless they are conserved in some way, a third or more of the nitrogen will pass off as ammonia gas. You can smell this gas in almost every poultry house, and it is particularly noticeable in damp, humid weather, especially in the winter months.

By itself, hen manure, like all other natural manures, is not a well-balanced fertilizer. It contains too much nitrogen in proportion to the amount of potash and phosphoric acid. Used alone it is wasteful, because of this excess of nitrogen. To properly balance this manure, and thus afford an economical distribution of the nitrogen, suitable amounts of potash salts

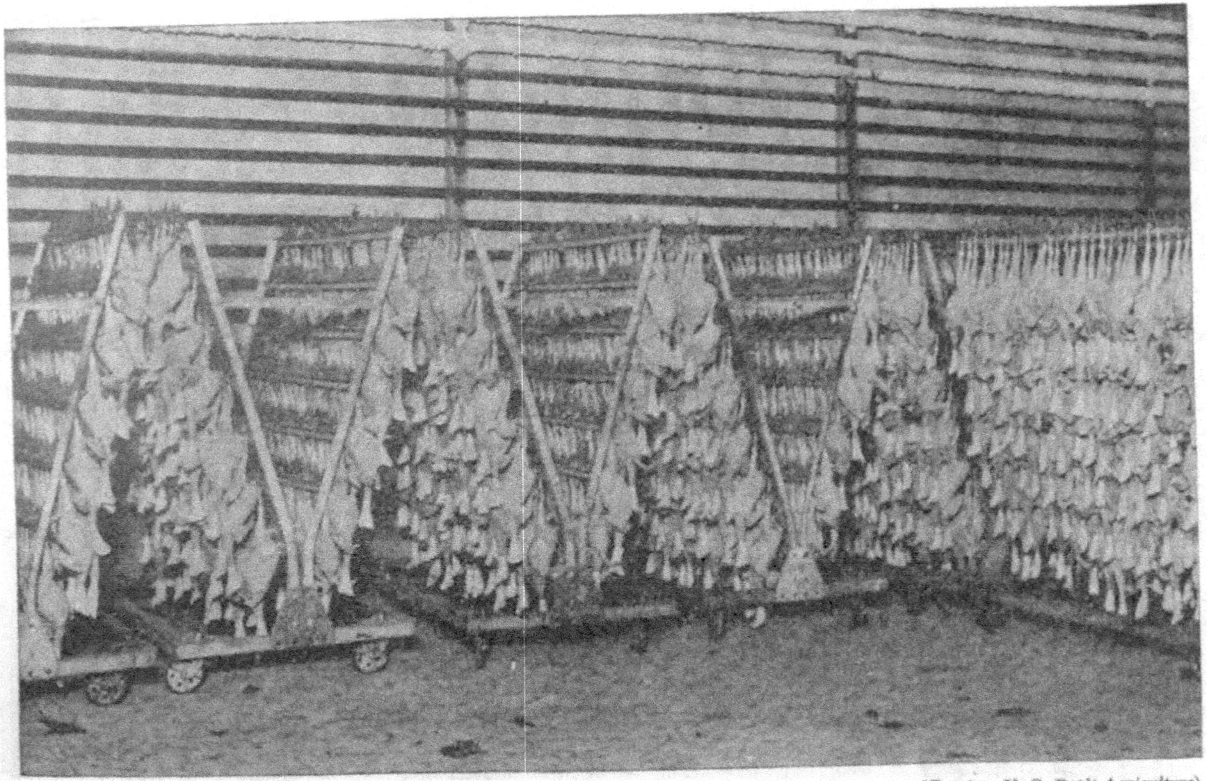

(Courtesy U. S. Dep't Agriculture)

Fig. 286.—Metal chilling racks for hanging poultry, standing in a mechanically cooled chill room.

and phosphoric acid should be added.   This will make it a more efficient fertilizer generally.

To Preserve Manure.—From an agricultural standpoint, the successful treatment of poultry manure resolves itself into three problems: First, to prevent the loss of the nitrogen; second, to add sufficient phosphorus and potassium in forms available for plant food to make a balanced fertilizer; and third, to improve the physical condition so that it can be applied to the land in an economical manner, either in a fertilizer drill or with a manure spreader.

From the poultryman's point of view, these same questions are of interest, and the idea is to work out a scheme that will serve every purpose.   The poultryman is anxious to prevent the loss of nitrogen, because in so doing offensive odors are kept down; and it is desirable to improve the condition of the droppings by drying them out, because they are handled easier at cleaning time, and the condition of the roosting compartments is vastly more sanitary.

Acid phosphate and kainit both prevent the loss of nitrogen, and if these are added to the manure in connection with sawdust, land plaster or some other absorbent (good dry loam or peat will answer nicely), there will result a well-balanced fertilizer.   For example, a mixture of 30 pounds of hen manure, 10 pounds of sawdust, 16 pounds of acid phosphate, and 8 pounds of kainit will test about 0.25 per cent nitrogen, 4.5 per cent phosphoric acid, and 2.0 per cent potash.

Need for Absorbent.—The addition of kainit or acid phosphate by itself makes the manure quite moist and sticky, hence the necessity for a drier.   Any absorbent may be used that can be obtained at low cost, for the amount of plant food added by the drier is of small consequence.   Because of its slight acidity, peat has some advantage, since it will help to preserve the nitrogen.   As a general rule, however, the farmer need only concern himself with the selection of a material that will keep down odors and absorb the moisture, since the addition of kainit and acid phosphate will prevent the loss of the nitrogen. Do not use wood ashes, for they tend to liberate the ammonia.

29

Roosts and perches should have tight platforms under them, popularly known as dropping-boards, which should be cleaned daily, or as often as it is necessary to maintain the quarters in a sanitary condition. If the houses are not crowded, and the weather is mild and dry, once a week will probably suffice for cleaning.

The absorbent should be kept conveniently at hand, and each time the dropping-boards are cleaned, if they are cleaned daily, the platforms should be sprinkled with the drier. If the boards are cleaned weekly, then each morning the droppings should be

Fig. 287.—Small fattening station with feed room in the rear.

sprinkled with the absorbent material. This is quickly done, and will work wonders with the general improvement of the house. When cleaning time comes the waste matter is easily removed with a hoe or scraper, leaving the boards comparatively clean and dry. It is a big help in the winter months, for it will prevent the droppings freezing to the boards, which condition makes cleaning exceedingly laborious.

Each time the droppings are collected they should be treated with the kainit and acid phosphate, and then carefully stored in a sheltered bin or shed. At first it will be necessary to weigh the ingredients, to insure the correct proportion, after which it will be possible to make a fairly accurate guess at the desired amounts.

Any form of shelter can be used, though on a poultry farm of large size it will pay to erect or remodel a small building for this special purpose. One having a watertight floor, which will prevent the entrance of moisture from without, and the escape of any liquids from within, is the ideal shelter, and will soon pay for itself many times over in the increased valuation of the manure.

Fertilizer Formulas.—To aid those who wish to compound their own fertilizer mixtures, the following analysis of hen manure is given, which is in accordance with the investigations of the Massachusetts Experiment Station:

COMPOSITION OF POULTRY MANURE

	PER CENT
Water	65.00
Nitrogen	1.56
Potash	.44
Phosphoric acid	1.09
Calcium oxide (lime)	1.62

FORMULA FOR LAWNS

	POUNDS
Hen manure	1800
Muriate of potash	75
Acid phosphate	125
	2000

Approximate analysis: Nitrogen 1.4 per cent, phosphoric acid 1.9 per cent, potash 2.2 per cent. Apply from one to one and a half tons to the acre.

FORMULA FOR CORN

	POUNDS
Hen manure	1510
Acid phosphate	340
Muriate of potash	150
	2000

Approximate analysis: Nitrogen 1.1 per cent, phosphoric acid 3.2 per cent, potash 4.0 per cent.

FORMULA FOR FRUIT TREES

	POUNDS
Hen manure	1500
High-grade sulphate potash	170
16 per cent acid phosphate	330
	2000

Approximate analysis: Nitrogen 1.1 per cent, phosphoric acid 3.3 per cent, potash 4.5 per cent.

FORMULA FOR BEANS AND PEAS

	Pounds
Hen manure	550
Ammonium sulphate	100
High-grade sulphate potash	350
Acid phosphate	1000
	2000

Approximate analysis: Nitrogen 1.4 per cent, phosphoric acid 8.0 per cent, potash 8.8 per cent.

FORMULA FOR STRAWBERRIES

	Pounds
Hen manure	1000
Nitrate of soda	100
Ammonium sulphate	100
High-grade sulphate potash	200
Acid phosphate	600
	2000

Approximate analysis: Nitrogen 2.5 per cent, phosphoric acid 5.0 per cent, potash 5.2 per cent.

The disregard of the value of feathers is another source of waste to many poultry raisers. There is a uniformly steady demand for feathers in all sections of the country, some dealers send out buyers for this purpose, yet thousands of dollars are lost annually because farms pay no attention to this product. With some system of saving, sorting and curing the feathers, they can be made to defray the cost of dressing and marketing the poultry, which is an item worthy of consideration. As proof of the demand for feathers, government reports show that in recent years nearly $2,000,000 worth are imported annually, exclusive of ostrich and similar ornamental varieties. Evidently, the foreigner finds that it pays to save feathers. Why should the American farmer not follow this example?

There are many kinds and grades of feathers, and prices vary accordingly. Geese feathers are the most highly prized, and bring about sixty cents a pound. When we consider that geese may be plucked twice a year, maybe three times, it is easy to see that the feathers are a source of profit, similar to the wool of sheep. All white feathers sell for more than colored ones or mixed feathers, and dry-picked feathers are preferred to scalded ones. Usually duck feathers are rated next to goose feathers, though very often carefully selected white turkey feathers are sold as high as the best grades of goose feathers. The body feathers from white chickens come next to duck feathers, and some grades compare favorably with goose feathers.

The down and very finest feathers from geese often sell for a dollar a pound, and are used for quilts. The finer body feathers of ducks and chickens are used for pillows, beds, cushions and so on. Tail and wing feathers, those with quills, are used mostly in making dusters and screens; feather boas are made from

(*Courtesy U. S. Dep't Agriculture*)

Fig. 288.—Removing small feathers while the birds hang by the feet.

hackle feathers, or from feathers curled with a hot iron; thousands of pounds of feathers are glued together in the forms of wings and breasts in imitation of birds of paradise, which are no longer permitted to be imported, and sold to milliners; and large quantities are used in the manufacture of feather flowers, fans, muffs and toys. For decorative purposes the feathers of peacocks, large turkey feathers and the tail plumage from chickens are in great demand. If you stop to think a moment, numerous other uses for feathers will present themselves. Featherbone, for example, is made from the shafts or quills of large flight feathers.

Prime feathers are those that are dry-picked, clean and fully cured, and properly sorted. By all means keep the different grades separate. Tail feathers should not be mixed with body feathers, not even with wing feathers, unless they resemble them. Make bundles of the quills, and either tie them together or pack

(*Courtesy U. S. Dep't Agriculture*)
Fig. 289.—"String" killing and picking.

them in boxes. One-sided quills, usually the flight feathers from the wing tips, are not worth as much as the full feathers, hence they should be kept separate.

Remember, that an inferior article packed with a superior one will invariably reduce the value of all to the level of the inferior article. The call for feathers is constantly changing, consequently it is well to consult the dealers as to just what is wanted, and for any details in packing and shipping. Commission men who handle poultry can usually dispose of feathers to an advantage, or they will be glad to put you in touch with firms who make a specialty of this business.

Feathers should be thoroughly dried before they are packed, or they mat together, turn musty and maybe spoil. Only a few precautions are necessary to save feathers properly. In the first place, the poultry must be dry-picked, but then, dry-picked poultry always brings a better price than scalded poultry, consequently it is to the grower's advantage to follow this method regardless of the feathers.

Provide barrels or boxes, have them arranged beside the picker, and as the pluck-ing is done drop the body feathers in one receptacle and tail feathers in another, and so on. This also helps to keep the feathers clean, for if they are allowed to fall at the picker's feet, it is quite likely they will be saturated or spattered with blood. Before the feathers are stored away, spread the different kinds in trays or on the floor of a dry,

(*Courtesy U. S. Dep't Agriculture*)

Fig. 290.—Holding birds on the lap to remove small feathers.

well-ventilated room, to a depth of four to six inches, and every day for a week, or until they appear to be thoroughly dry, stir them with a stick. They are then ready to be packed or shipped. If they are packed green, the animal heat will make them damp and moldy, and greatly reduce their value.

Cleaning.—Most of the feathers shipped to the dealers are just as they come from the fowls, and quite naturally they must be cleaned. Clean feathers, of course, bring higher prices than soiled ones, but whether it will pay the poultryman to go to this extra work is a question that each person must find out himself. I am inclined to think that it pays to clean them at home, since they bring almost twice the price, and the work is not difficult.

Manufacturers and feather houses have facilities for washing the feathers by machinery, treating them with live steam, and then drying them with wringers and subjecting them to strong drafts of air from fans which lay all animal odors and leave the feathers in a fine, fluffy condition.

Shipping Methods.—Feathers are shipped in burlap sacks, tightly compressed, and the quills in boxes. Or the quills may be tied securely in bundles and packed in sacks. Do not pack the quills loosely in bags, all jumbled together, for they will not bring a good price. And I repeat, keep the white feathers separate; they are the most valuable. White chicken feathers are worth about eighteen cents a pound; colored chicken feathers about six cents a pound.

Feathers which are too badly soiled for use as such can be utilized as fertilizer. They are valuable for this purpose, though they decay slowly and are therefore a long time in the ground before they become available for plant food.

CHAPTER XXXIV

PREPARING BIRDS FOR THE SHOW

Educational Value.—Next to the poultry press, the *show room* has done more for the poultry industry than anything else. It is probably the most potent educational factor, and one of the greatest advertising mediums. Without these annual displays interest in poultry affairs would be dwarfed. The strongest proof of their popularity lies in the fact that the number of shows increases every year.

Every one who raises good poultry should take an active interest in poultry shows, especially in the local shows. This includes the utility breeder as well as the fancier. The man who discredits the value of the show room simply because he raises chickens for eggs and meat is short-sighted. The progressive utility breeder is one who opposes mongrelism. For any purpose he appreciates that pure breeds are superior. In the show room not alone fine feathers and correct markings are displayed, but the qualifications that go to make the egg or meat type of bird are also shown.

Exhibit whenever possible, but whether you are an exhibitor or not you will find it to your advantage to patronize the shows, to be in attendance, and to contribute any assistance at your command. One may not carry off blue ribbons, and yet win many things of even greater value. You can obtain a closer friendship with your fellow breeders, a broader view of the conditions that make for success, poultry wisdom, some new points on salesmanship or advertising, a better knowledge of how to mate next season's contestants, an exhibition of the latest improved apparatus, and last, but not least, a good time.

It is not always the poultry association with the largest membership that is the most successful. One of the best exhibitions in Pennsylvania has only a dozen members, but these men are

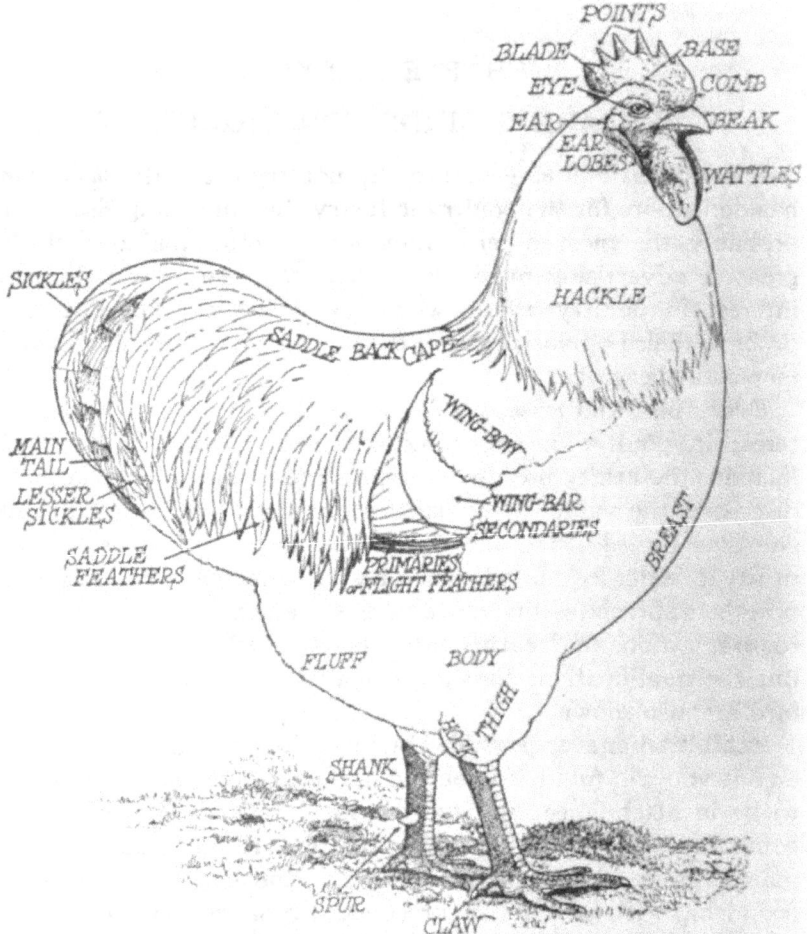

Fig. 291.—Glossary chart giving the names of the various sections of a male fowl. In the female the cushion takes the place of the saddle of the male, and the sickle feathers are absent.

up-to-date, hard workers and they work with a co-operative spirit. Every year they set aside the week of the show for that particular purpose. They don't stand round with their hands in their pockets, giving advice, and paying some one else to do the work. Except for the judges' fees, prizes, hall rent, feed, light and similar expenses, there are no other charges; in consequence the show is a financial success.

Selecting Specimens.—There was a time when a breeder could

(Courtesy Missouri Experiment Station)

Fig. 292.—Catching coop for fowls. It is placed in front of a small door in the main house, through which the birds are driven into the coop.

look over his birds, select the most promising specimens, and without any further ado pack them off to the show and win. That day is past. To-day, though a specimen ranks high in size, shape, color and most of the other points that contribute to the ideal, if it is not shown in perfect condition the chances for a ribbon are limited. In fact, it is the art of conditioning specimens, as well as breeding them, that brings success in this generation.

Some breeders are opposed to conditioning, and condemn it as faking. There is a vast difference between the two, however, even though it may be difficult to establish in some cases. The mere fact that you have grown a fowl that conforms to certain requirements, such as weight, shape, the angle at which the tail is carried, length of shank and so on, is by no means as far as it is possible to exert your skill. The other fellow, let us say, has done the same thing, and a little more. He improved on the fowl's ability to keep itself sleek and clean. Maybe he influenced the bird's conduct, by training it to pose and strut about, to exhibit its virtues to the best advantage, and not object to handling. A wild, unruly bird is an abomination, and will try the patience of the most skillful judge.

Show Conduct.—To properly demonstrate his or her points of superiority, a bird must be docile, accustomed to pose when the judge's stick is placed on its back, and in all ways assume an aristocratic air. These characteristics are seldom born in a specimen; they are the result of careful training on the part of the exhibitor, and as such they are worthy of consideration. Sometimes it requires weeks of careful handling and training before a bird is in perfect show condition. Many a superior bird has failed to be placed because it was impossible for the judge to keep it quiet long enough to form any idea of its correct type.

In the selection for the show room every section of the bird must receive exacting scrutiny. In some breeds color must be more carefully considered, in others, as, for example, the Mediterranean varieties, the head points are of great importance. In the Hamburg breeds, if tail be not full and flowing, its long sickles well curved and carried at the right angle with the back, or if it approaches a squirrel position, the whole Hamburg character is lost or seriously marred. If the sickles are short and the tail pinched, or carried in a trailing low carriage, they fail to attract the necessary attention.

Like the tail, unless a fowl's shanks are of the right length, and the angle formed by thigh and shank shows apparent strength and perfect poise, the specimen appears at a disadvantage. The shanks in the Cochin classes, because of their excessively heavy plumage, will look short, but they must not look dwarfed. In the American classes we describe the length of the shanks as medium, meaning that they free the specimen from any stilted appearance; at the same time they must not look short for the size of the specimen.

Thus, different breeds require different judgment, and the only way to gain complete information about a particular breed is to study the American Standard of Perfection, which is the authority by which show specimens are judged.

Birds intended for the show room should be selected a month or two before the show dates, and the males separated from the females to avoid broken feathers. They should be given more or less isolated quarters, where they cannot fight, or in trying to do so, injure their combs in any way. The specimens should be carefully dusted with an insect powder to free them of lice, and if they are badly soiled they may require a preliminary washing.

Washing.—It is seldom necessary to wash the dark-colored breeds, such as the Rhode Island Reds, unless they have been reared in an atmosphere of coal smoke, in bare yards, or have become very dirty, in which case a thorough washing will mean a great improvement. Washing is sometimes used to improve the shape of certain breeds, such as Cochins and Orpingtons, which should have loose, fluffy plumage. By drying before a fire, one that is not too hot, for this will make the feathers curl, and fanning the feathers while they are drying, the plumage will remain loose and fluffy. This gives the birds a fuller, rounder appearance, and adds much to their beauty.

White birds are the hardest to condition. To insure a good job they should be washed twice, the first time about two weeks before the show, and again a day or two before shipping. The idea of washing chickens sounds like a difficult task, but after a little experience it becomes a pleasure. It is surprising how placidly most fowls submit to the work. If the specimens have

Fig. 293.—Well arranged poultry exhibition." The show room has done much toward the advancement of the industry.

462

received proper training, which is but another term for sufficient petting and handling, they will allow their master to perform these ablutions in a very orderly fashion. A warm room, plenty of soft water, good soap, a couple of tubs, some towels or cloths, a sponge, patience and common sense are the requirements.

Quarters.—On many of the large farms, those that make a specialty of exhibiting, special quarters are provided for conditioning the fowls. In fact, some of the conditioning rooms resemble exhibitions themselves, being fitted with show cages in much the same manner. The small breeder may obtain equally good results without going to this expense. An outside kitchen or sheltered porch, some place where members of the household frequently pass, is a good location to erect temporary cages, and as often as one can spare the time, handle the birds and make them accustomed to persons coming close to the cage.

Always handle fowls with quiet movements yet with a firm grip, being careful to keep their wings closely folded against their bodies when removing them from the cages, so as to avoid any damage to the feathers. Sometimes it is best to commence the training at night, for light seems to fascinate them, and they are less wild. Teach the specimens to assume certain poses, and by gently stroking them between the wattles they can be made to retain a pose indefinitely. A few minutes spent in training every morning and night for a couple of weeks will usually conquer the wildest of birds. If you find a specimen that no amount of training will tame, better discard it; the chances are it will try to pull down the cage in the judge's presence.

Preparations for Washing.—Everything should be in readiness before washing the birds, and a start made in the morning, so that by night the fowls are dried off. Fill two tubs with warm water of a temperature that is comfortable to the hands; use the first for washing and the second for rinsing and sponging. Provide a third tub of cold water for the final rinsing, to which a small amount of laundry bluing may be added if desired. This is a very particular part of the operation. The tub should contain enough water to entirely immerse the bird, except the head, because if one portion of the plumage is submerged longer than another, the bird will not be evenly blued. Use about as much bluing as would be proper for laundering clothes.

Lather Each Section.—When all is ready, carefully immerse the specimen, and then starting with the head, thoroughly lather each section until every particle of dirt has been freed. Cover each feather with lather clear to the skin and then rub the feathers well between the hands. Don't be afraid of damaging the feathers; once wet they are very pliable and may be rubbed much the same as shampooing one's hair. After washing the upper part of the bird place a clean board across the sides of the tub and stand the bird on this while you wash the breast and body. A nail-brush or discarded tooth-brush should be used for washing the comb and face, legs and toes, especially the claws.

Rinsing.—When you are assured of a good job of cleaning, squeeze off the greater part of the lather into the first tub, remove the bird to the second tub, and with the aid of a sponge or dipper thoroughly rinse every trace of soap-suds and dirty water. If any soap is left in the plumage it will stick together. This rinsing operation is the secret of a satisfactory result. If any traces of soapy water remain the plumage will dry blotchy and streaked, and if bluing is used in the third tub, any presence of soap tends to prevent the feathers from taking the bluing evenly.

Third Tub.—When the specimen is thoroughly sponged and rinsed, plunge it into the third tub of cold water, agitate and ruffle the feathers so that the clean water, especially if bluing is used, comes in contact with every section; then drain the bulk of the water from the plumage by squeezing it; take the specimen in your lap and wrap it in an absorbent towel or cloth.

When it has ceased to drip, return the bird to the conditioning coop, which should be previously replenished with clean shavings, straw or other material. A good plan is to cover the top, sides and back of the cage with muslin to prevent drafts, and leave only the front open. If convenient arrange the cages around a stove while the birds are drying, or in a room where the temperature is about 80 degrees. As previously mentioned,

Fig. 294.—Prize-winning black Langshan cock.

if too hot the feathers are apt to crimp and curl, which is undesirable.

Black and red varieties may be improved in lustre and brilliancy of plumage by rubbing the feathers with a little sweet oil. Use very little—a couple of drops on the palm of the hand are sufficient.

Shanks.—Many exhibitors are careless in not cleaning the shanks. If the legs are soaked in warm water and well washed with a stiff brush, and a wooden tooth-pick is inserted under the scales that lodge dirt, every bit of discoloration can be removed as readily as one can clean his finger-nails. A little oil applied to the shank, or carbolated vaseline, rubbed with a woolen cloth, will work wonders. This brings out the true color nicely, and gives them a fresh, immaculate appearance. It is not generally known, perhaps, but fowls molt the scales on their shanks and toes about the same time they molt their feathers. Look carefully to see that any dead scales are removed.

The comb, wattles, face and earlobes should be rubbed with vaseline, using very little, but rubbing it in well. Those are the finishing touches and should be given attention at the last minute.

Under Weight.—If show aspirants are a little under weight, careful feeding for a few weeks will usually bring them up to specifications. Vary the birds' rations so their appetites are not cloyed, using a mixture of some of the following articles: boiled potatoes, cornmeal, boiled rice, buckwheat meal, barley meal, middlings, ground oats, wheat, skimmed milk, a little beef tallow, linseed meal or cottonseed meal. At the same time give them a little sweetened water to drink, and a good tonic or regulator to offset any ill effects from the forcing. If the birds have lost weight in shipment to the show room, as they are likely to do, feed them liberally on the regularly cooked and seasoned bologna sausage.

In conditioning old hens that are over-fat and inactive, no soft food should be given except green vegetables. The whole grain should be largely oats scattered in deep litter.

Many beginners make the mistake of cooping the birds too

closely and too long before the show, which makes them dull and sluggish. A pullet is at her best just before she lays her first egg; after that she loses her bloom and goes off shape.

Buff color is the hardest to maintain at an even shade. The secret of getting even, rich, golden buff color is never to allow the sun or rain to touch the surface of a showbird. Some of the most successful breeders of buff plumaged birds do not give their birds freedom in the open from the time their final plumage begins to show, but keep them in shady runs or under sheds.

Shipping Crates.—Above all things, do not try to save space or a few cents on expressage by crowding show specimens into small crates. There is no economy in it. Use standard exhibition crates, which may be purchased at a reasonable price. If made at home, build them high enough for the birds to stand upright, and wide enough for them to turn round without injuring their tails. Label or tag the crates neatly, and in strict accordance with the instructions of the show secretary. See that the crates are bedded with clean straw or some material that is free from dust. We have seen instances where birds that had been carefully washed and groomed were hopelessly marred by a deposit of dust on face, shanks and plumage, caused by scratching in dusty litter.

To insure identification every specimen should be leg-banded, and the number or distinguishing feature of the band should be marked on the entry blank and shipping crate.

Sportsmanship.—It has been said, any one can win blue ribbons, but it takes a real fancier to lose. If there is one thing in a show room that is detestable, it is the exhibitor who has not sufficient sportsmanship to abide by the decision of the judge. If you can't understand an award, have it out with the judge in a gentlemanly way, but if he can't convince you that his opinion is correct, take your medicine like a man. A *good loser* is always respected and admired.

Chickens are great imitators, and their imitations frequently lead to habits which are very troublesome and difficult to combat, among which are egg-eating, feather-pulling, cannibalism

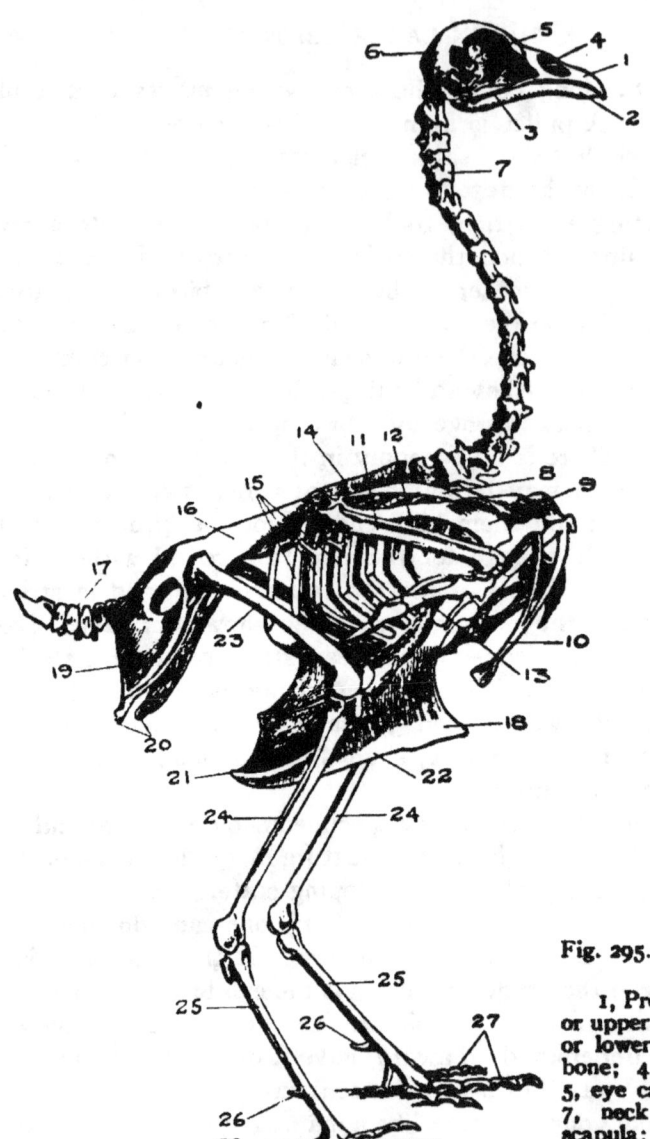

Fig. 295.—Skeleton of a
fowl.

1, Premaxillary bone
or upper jaw; 2, maxilla
or lower jaw; 3, jugal
bone; 4, nasal cavity;
5, eye cavity; 6, skull;
7, neck vertebræ; 8,
scapula; 9, humerus; 10,
clavicles (wishbone); 11,
ulna; 12, radius; 13,
bones of forewing; 14, backbone; 15, ribs; 16, ilium; 17, pygostyle or tail
bones; 18, breastbone; 19, ischium; 20, pubis; 21, sternum; 22, keel; 23, femur;
24, tibia; 25, tarso-metatarsus; 26, spur; 27, digits or toes; 28, rear toe.

and high-flying. Nothing is more distressing to the poultry keeper than these four habits once they have become confirmed practices, for then they amount to vices. They usually start through accidents, or from the example set by a chief offender— a ringleader, which should be removed as soon as the trouble is discovered. Then, if the offender cannot be reformed, rather than return it to the flock, it should be made to pay the death penalty. Usually, these habits can be broken up if taken in time and dealt with accordingly.

Flying over the fences is likely to become one of the troubles among the Mediterranean breeds, which is not serious in itself, except that where two or more varieties are kept it is almost certain to result in cross-breeding. And even if one breed of chickens is kept, in all probability the poultryman has mated his pens with a definite purpose, hence he cannot tolerate promiscuous changes. For one thing, it may lead to inbreeding; and then again it may lead to unpleasant difficulties with one's neighbors—their gardens or flower-beds.

(Courtesy U. S. Dep't Agriculture)

Fig. 296.—Commercial feeding station, 300 feet long, accommodating 30,000 birds.

Clipping Wings.—It is natural that we should hesitate to clip a fowl's wings, as is customarily done to prevent high-flying, for the usual method is sure to disfigure the bird. Clipping off the flight feathers completely is not necessary, however, and if a little care is taken in cutting, the wings can be deprived of their power in such a manner that the mutilated feathers will not be detected unless the fowl is caught and closely examined.

Flight Feathers.—The *primaries* or flight feathers are the long quill feathers that grow on the first joint of the fowl's wing, and are hidden, or nearly so, when the wing is folded against the body. The *secondaries* are the quill feathers that grow on the second joint of the fowl's wing, which are visible when the wing is closed, and which form the section known as the wing-bay. Together with the *primaries* they constitute the main feathers of the wing or the flying feathers.

The ordinary way of clipping a fowl's wings is to clip off both *primaries* and *secondaries* within a few inches of the fleshy portion of the wings, and which always leaves a ragged, badly disfigured appearance, and seriously detracts from the bird's sleekness. The practice is unnecessary and should be condemned.

Clipping Without Disfiguration.—The following method is equally effective in restraining high-fliers, and while it takes a little more time to perform the operation, the results seem to warrant the additional trouble: Take the bird under your arm, or better still, sit down to the work and hold the fowl between your knees; then spread the feathers of a wing wide open, and with a pair of scissors clip the *web* or *plumed* portion of each *primary* close to the *shaft;* but do not cut off the *shaft* and do not strip the *webs* of the *secondaries*. Repeat the operation on the other wing.

When the fowl resumes its natural poise and the wings are folded against its body, the clipped portions of the *primaries* will be hidden from view by the *secondaries*, and it will take a very acute observer to discover that the wings have been tampered with at all.

CHAPTER XXXV

AILMENTS AND DISEASES

When to Doctor.—It has been said that the best cure for ailing fowls is a sharp hatchet.

The writer will not take issue with this treatment as an effective remedy for some ailments, yet as a hard and fast rule to be recommended for general practice, the *hatchet cure* is a little too stringent.

There is no question but that a sickly flock of fowls are a constant source of vexation and financial loss to their owner, and while it is very often unprofitable to expend much time and trouble doctoring them, in my experience it will pay to administer first-aid treatment in the early stages of a disorder; and if the *patient* responds within a reasonable time, continue the treatment; otherwise, call in the assistance of the hatchet.

Health Is Everything.—It does not matter how valuable a strain of blood there may be in a flock of chickens, how long a pedigree, or how many blue ribbons are back of them; how splendidly equipped are their buildings; nor yet how attractive and convenient may be the location and environment of a farm, the foundation for success with poultry is built on perfect health, —a strong, vigorous vitality,—and to this all else is subordinate.

Sickness in some form, though it may be of a trivial nature, visits every flock at some time or other, and whether the trouble finds a permanent abiding place, a home, so to speak, in which to thrive, or whether it is met with an aggressive inhospitality and promptly driven off and exterminated, depends almost wholly upon the caliber and energy of the attendant in charge of the flock.

Unfortunately, the beginner's farm usually endures the greatest

471

hardships on this score,—not because he neglects to give his fowls the best of care, rather because, lacking the experience and practical knowledge of the more seasoned poultryman, the novice frequently fails to detect the first symptoms of a disease. Then, too, when the novice does discover trouble, very often he has not the courage to sacrifice a few birds as a precautionary measure. Yet drastic measures are sometimes necessary for the safety of the rest of the flock, just as in a serious conflagration dynamite

(Courtesy Kansas Experiment Station)

Fig. 297.—Open-front house with irregular gable roof designed to afford sufficient head room inside where it is required.

is used to raze whole blocks of houses, and thus check the spread of the fire.

Trivial or Contagious.—Poultry diseases may be divided into two general classes: Those of a more or less trivial nature, which will respond to flock treatment through the drinking water or feed, and those of a highly contagious, virulent action, which require individual treatment—if treated at all, for in badly infected birds it is usually advisable to kill them at once and destroy the carcass, thereby preventing the spread of contagion.

Burying the carcass, unless it is buried very deep, is not always a safe method; for at some future time it may be dug up by a dog, or accidentally plowed up, and the infection again spread about the premises. Instead, such a carcass should be burned, or put in quicklime.

Indications of Illness.—The competent poultryman makes it his business to note the physical condition of every bird, every day—which is a much simpler task than it sounds. In the early morning and at feeding times, it is an easy matter to detect a bird that is feeling out of sorts. If a fowl remains on the perches, with ruffled feathers, head drawn close to its body, and is otherwise sluggish and disinterested, it should be taken out of the pen and examined. Or if a fowl is seen to rub its eye on its wing frequently, or if a soiled spot appears on its wing; if it sneezes often, gasps for breath, mopes in the corners and has a purplish comb—these conditions are all abnormal, and the fowl should be removed for a closer examination and diagnosis.

Detention Coop.—On a farm of any size there should be a small house or coop isolated from the rest of the buildings, which should be equipped with a few cages, or crates, and designated as a hospital, or detention ward. When a fowl is discovered out of condition, if only with a cold, it should be removed to this hospital, carefully examined for the nature of its ailment, and treated accordingly. A small closet should be convenient, fitted for the storage of bottles, clean rags, a sponge, basin, and so on. With the knowledge of a few simple drugs and remedies many of these hospital inmates can be promptly and permanently cured.

Determine the Cause.—When an abnormal condition is first discovered, the poultryman should analyze every symptom, not only with the view to ascertaining the nature of the ailment, but to determine the cause of the disorder. There is sure to be a cause, and of course it should be immediately removed or corrected, to prevent further trouble. It is then up to the attendant to decide if the ailment is curable, and what the chances are for an early recovery. If the chances are poor, he had better stifle all sympathetic feelings and sentence the victim to be executed—the hatchet.

Simple Remedies.—The following drugs are simple remedies that should be found in the poultryman's dispensary, and their actions are no doubt familiar to everyone. They are all inexpensive, and a supply should be kept on hand for instant use. Like the "stitch in time," first-aid treatment in the early stages of a fowl's ailment is the secret of a successful cure.

Castor oil is one of the most commonly used remedies for disorders of the crop and other digestive organs. It not only removes the irritant, but also helps to soothe and heal any inflamed tissues. In fact, it is a pretty good rule to commence the treatment

(Courtesy Atlantic Farm)

Fig. 298.—Pen of Pekin breeders. Duck houses are simple affairs.

of every ailment with a dose of castor oil. Give about two tablespoonfuls to a grown bird, administering it with the aid of a dropper or glass syringe.

Epsom salts and Rochelle salts are both splendid laxatives, and will correct liver troubles and relieve diarrhea. Half a teaspoonful is the correct dose. In treating a large flock, the salts may be mixed with the mash,—but the fowls should first be allowed to become quite hungry, to insure their eating a sufficient quantity of the mash.

Bi-carbonate of soda, or Baking Soda, will relieve a sour or distended crop, which is equivalent to indigestion. An amount sufficient to cover a dime is the average dose for a single bird. For flock treatment, dissolve a half-teaspoonful in a quart of water, and keep it before the birds for several days.

Tincture of aconite is a well-known drug for the relief of colds, catarrh, and in allaying fever. About five drops is the customary dose for a bird.

Sulphate of magnesia, ten drops to a pint of drinking water, will relieve costiveness.

Spirits of camphor is another good remedy for slight cases of diarrhea. Add a few drops to the drinking water.

Quinine will work wonders with colds and chicken pox; and Iron, Quinine and Strychnine tablets make an excellent tonic for birds whose vitality has been lowered through loss of blood, excessive breeding or illness.

Bismuth nitrate will frequently cure the more serious intestinal disorders,—bloody diarrhea or enteritis.

Tincture of nux vomica, about ten drops to a quart of drinking water, is recommended for cases of leg weakness, and it also stimulates the digestive organs.

Calomel is another excellent corrective for liver troubles, and a strong laxative. Use a quarter of a grain to a grain.

Carbolated vaseline, or some one of the petroleum products, should be on hand to use in anointing wounds and sores, and for chicken pox, scaly legs, and frozen combs.

Gas tar, or one-third carbolic acid mixed with two-thirds glycerine, are two other remedies for scaly leg. Cover the affected shanks with the gas tar, and allow it to remain until it wears off. The scales will come off with the tar.

Liver Pills.—When a fowl is somewhat mopish, the ordinary family *liver pill* will very often correct this sluggishness, much the same as it does with the human being.

Spirits of turpentine and sweet oil will usually relieve any bronchial affection, such as a *rattling* or *bubbling* sound in the throat. One part turpentine to five parts oil is the proper mixture: administer about ten drops daily to the individual bird.

Fowls seem more susceptible to roup, canker, and other respiratory derangements in the fall of the year than at other seasons, and as either preventative or balm for these ills, the Missouri State Experimental Station reports excellent results from the use of the following mixture:

Magnesium Sulphate	10 ounces
Sulphur	3 "
Magnesium Oxide	1 ounce
Sulphate of Iron	2 ounces
Ground Ginger	2 "

A tablespoonful is fed in moist mash for 12 birds, for three days. The *Magnesium Sulphate* acts on the intestines, as pre-

Fig. 299.—Good type of feeder. Note the short, thick head.

viously mentioned; the *Sulphur* is a general antiseptic; the *Magnesium Oxide* acts on the kidneys; *Sulphate of Iron* stimulates the blood; and *Ground Ginger* is beneficial to all organs.

A strong disinfectant, preferably one of the coaltar products, should be included in the poultryman's dispensary, and it should be used freely as a spray whenever an infectious disease breaks out.

Permanganate of potash is frequently used in the drinking water for antiseptic and disinfecting purposes, and will prevent the spread of colds, coughs, bronchitis and similar troubles. Use enough of the crystals to turn the water a deep purple color. About as much as would cover a dime is sufficient for a gallon of water.

It would seem as though chickens were heir to as many ills as man yet since most of them are but very occasional, we need only concern ourselves with the more common troubles hereinafter mentioned, together with their principal symptoms and causes.

Asthenia, or "Going Light."—A term applied to fowls that persistently lose weight, become emaciated, anemic, weak and unproductive. It is not a form of consumption, as some suppose, but may be due to several causes. Sometimes it is the result of lice or mites; at other times it is due to worms—this is the most common cause; or the birds may be afflicted with bacteria harbored in the small intestine, which subsist on the food consumed by the fowls. On investigation, if no defect is found in the rations, one of the birds should be killed and examined for worms. If many are found, the emaciation is probably due to this cause, and the flock should be treated accordingly. See chapter on Worms.

Fig. 300.—Indifferent type of feeder. Note crow-like shape of head.

If no worms are found, and the fowls are not bothered with lice and mites, and their rations and living quarters are O. K., the cause of the trouble is probably infection by microbes mentioned above. In this case a slight inflammation may be noted. The treatment should be, first, the removal of the bacteria by purgation, using Epsom salts, castor oil or calomel, following which the flock should be given a tonic to build up the system, and an abundance of easily digested foods. At the same time look carefully into the housing conditions; clean and disinfect everything.

The following tonic is recommended: 30 grains each of powdered fennel, anise, coriander seed, cinchona; 1 dram each of powdered gentian and ginger; and 15 grains of powdered sulphate of iron. Mix these ingredients thoroughly. Dosage: from two to four grains of the mixture added to the food twice a day for each fowl.

Bagging Down.—The posterior parts of a fowl hang down and even drag on the ground. It is not a disease, but the result of improper feeding and lack of exercise, an over-fat condition, and very hard to correct. Better kill such specimens for the table; they are not sick birds, remember.

Blackhead (*Entero-Hepatitis*).—This is a disease that attacks young turkeys, and is seldom found in other fowls. In the course of the disease the head generally becomes dark colored or nearly black, hence the name. It is an infection of the liver, similar in its nature to human dysentery. It is highly contagious and very difficult to cure. If the birds are in an advanced stage of the disease, they had better be killed, and the bodies carefully disposed of, because thus far the treatment for blackhead has not given satisfactory results. If treated at all, the afflicted birds must be isolated.

The symptoms are loss of appetite and condition, diarrhea, and finally prostration. The liver is found to be more or less enlarged, and spotted with yellowish or greenish-yellow nodules.

The remedies used are sulphur 5 grains, sulphate of iron 1 grain; or benzonaphthol 1 grain, salicylate of bismuth 1 grain; or sulphate of iron 1 grain, salicylate of soda 1 grain. These remedies are followed by a purgative. Fifteen grains of catechu to the gallon of drinking water is found to be beneficial.

Do not be afraid to use the axe in this disease, because sacrificing a few birds may be the means of saving the balance of the flock. Dissemination of this contagion has made some localities almost impossible to raise turkeys. Every effort should be made to check the disease.

Bronchitis.—Usually caused by exposure to sudden changes in temperature, dampness or irritating particles of dust, like lime. The symptoms are dullness, loss of appetite, coughing, which is accompanied by a whistling or bubbling sound in the throat. Though simply an inflammation of the breathing tubes, which may be treated successfully in large flocks, if only a few birds are affected, it is well to isolate them and administer the turpentine remedy.

Baldness.—See Favus.

Blood Spots in Eggs.—This trouble is due to the escape of blood from ruptured blood-vessels, which generally occurs at the time the yolk is freed from the ovary and enters the oviduct, where it receives the coating of albumen. These hemorrhages are thought to be the result of great functional activity and congestion induced by the excessive use of stimulants or highly concentrated foods; or they may be caused by the general breaking down of a hen's vitality.

Fig. 301.—Poor feeder. A cripple, or bird off feed.

Relieving any causes which might lead to congestion or inflammation is the logical way to attempt a remedy. Reduce the amount of grain or animal food, and increase the green food. Give a little Epsom salts in the drinking water. Some hens habitually lay eggs with blood spots or streaks, and these should be killed for the table, to escape this nuisance.

Bumble Foot.—One of the minor ailments. A callus or corn that forms on the bottom of the foot and later becomes a painful swelling attended by ulcerations. It is caused by the birds' jumping from perches that are too high, bruises and irritations from splinters. A fowl so afflicted will limp and stand on one foot. Painting with tincture of iodine will dissipate the callus if taken in the early stages; but if it has ulcerated, open it, remove any pus, cleanse the wound with an antiseptic, and then bind the foot in an application of carbolated vaseline.

Canker.—This is probably the most disgusting ailment, and is usually caused by contaminated food, also chronic cases of roup. It is very contagious, hence the sick birds should be isolated, and if they are badly afflicted, it is advisable to kill them and destroy their carcasses. A yellow, cheesy, foul-smelling matter forms in the corners of the mouth and in the windpipe, which when removed reveals the raw flesh. Remove this foreign matter and apply carbolated vaseline, or sulpho-carbolate of zinc on the sores. See Roup.

Catarrh.—See Roup.

Chicken Pox (*Bird Pox*).—While this is contagious and will spread rapidly throughout an entire flock, it is not necessarily a serious trouble. It is usually caused by dampness or filth. It may be introduced by new birds, or exhibition birds which return from show rooms infected. Or it may be carried into the flock by pigeons, sparrows and other birds, or by the attendant. Scabby, yellow nodules or pimples appear on the face and comb, especially around the beak, and frequently discharge a thick, yellow matter. Isolate the sick birds at once, and anoint the sores with carbolated vaseline. Administer the iron, quinine and strychnine tonic, or a good poultry regulator, and simple, nourishing food.

Cholera.—See Enteritis. A bacterial disease caused by contaminated food or drinking water. Highly contagious, with a heavy mortality.

Colds.—See Roup.

Crop Bound.—A hard and swollen condition of the crop caused by an obstruction to the gullet from the crop, or by gorging large quantities of grain, which swell and ferment. This trouble is described elsewhere in a special chapter.

Diarrhea.—See White Diarrhea, described in a separate chapter.

Egg Eating.—This is a pernicious habit that is almost always the result of accidents, though the accidents are very often due to the carelessness or ignorance of the keeper. Cramped nests or an insufficient number of nests are generally responsible for broken eggs, the hens taste them, form a liking for them, and thenceforth eat them whenever an opportunity presents itself. The ringleaders should be caught and broken of the habit, even if they have to be killed for the table. If not, they are sure to set a bad example to the rest of the flock. This trouble will

(*Courtesy Atlantic Farm*)

Fig. 302.—Ducks require low fences, which make it possible for the attendant to walk from yard to yard without the bother of entrance gates.

spread through a flock much the same as Cannibalism or Feather-eating. As a precautionary measure, install your nests in accordance with the best practices, as described in the chapter on poultry house fittings.

Enteritis.—This disease is caused by irritant poisons or bacteria, which develop an inflammation of the mucous membrane of the intestines. The fowls have poor appetite, roughness of feathers, pale comb, and their excrement is of a greenish color, or bluish green. The trouble is almost always fatal if allowed to reach an advanced stage. Administer nitrate of bismuth,

31

keep the birds isolated, and give them olive oil for nourishment. Clean up the premises, look for the cause of the trouble, and disinfect all drinking vessels and other utensils. This disease is quite serious, and can create havoc with the strongest, healthiest flocks. It is one of the cholera-like diseases.

Favus or White Comb.—This is a contagious disease caused by growth of a fungus, the filaments or roots of which do not penetrate deeply into the skin, but remain very near the surface, consequently the general health of the victim is not seriously impaired in the early stages. The trouble usually breaks out on the comb first, then the wattles and earlobes and, finally, the neck and other parts of the body are affected.

When limited to the comb and wattles, the trouble responds very nicely to treatment, and may even disappear of its own account. If it has invaded the feathered portions of the body it is extremely obstinate, and in very severe cases the "hatchet and block" is the safest and most satisfactory remedy.

First, wash the affected parts with warm water containing a mild disinfectant, at the same time removing any scabs that can be rubbed off without bleeding, and then apply sulphur ointment or carbolated vaseline. Good results have been obtained by painting the spots with tincture of iodine. Some breeders recommend an ointment of red oxide of mercury 1 part and vaseline 8 parts.

Feather-eating.—See chapter on Crop Bound.

Gapes.—See chapter on Worms.

Indigestion.—See chapter on Crop Bound.

Leg Weakness.—See chapters on the Care of Chicks.

Lice.—See chapter on Parasites.

Limber Neck.—Partial loss of control of the muscles of the neck, and is generally caused by eating putrid animal matter. A fowl so afflicted cannot hold its head upright, but twists it around from side to side and staggers about drunkenly. The disease is not contagious, though very often it is stubborn to cure. A pill of asafetida night and morning for a couple of days is a good remedy, also, borax in water, a tablespoonful to a pint, pouring a large dose of the solution down the fowl's throat three or four times a day. Castor oil and turpentine and warm water are two other highly recommended remedies.

Mange (*Scabies*).—This is caused by mites which live at the base of the feathers, where they bite the skin and cause intense itching. It is quickly spread throughout the flock, and while the general health of the birds does not suffer greatly, still the trouble is discomforting, and if allowed to continue the birds will lose flesh and become unproductive. Moreover, as the mites spread the plumage is destroyed until the birds are almost naked.

Apply to the affected parts, and for some distance around them, an ointment made by mixing 1 part flowers of sulphur with 4 parts of vaseline or lard. Carbolated vaseline may be used, too, and if it is mixed at home, use 1 part carbolic acid to 50 parts of vaseline. It is sometimes beneficial to wash the irritated surfaces with a solution of creolin or some other disinfectant.

Mites.—See chapter on Parasites.

Pasting Up.—See chapter on White Diarrhea.

Parasites.—See special chapter devoted to these pests.

Roup.—This is probably the most dangerous, fatal and contagious disease with which the poultryman must contend, and it is certainly the most disagreeable to treat. It is a contagious catarrh, resembling the more malignant forms of influenza in the larger animals and in man; this and canker, which is a chronic form of roup, are generally the aftermath of such ailments as colds.

The first symptoms of roup are similar to those of a cold, except in the former there are more fever, dullness and prostration. There is sneezing, accompanied by a watery discharge from the eyes and nostrils. Later this discharge becomes thick and obstructs the breathing, and as the inflammation, which begins in the nasal passages, extends to the eyes and the spaces below the eyeballs, the fevered condition hardens the secretions into a cheesy matter, which accumulates in the tissues of the head, causing the eyes or other parts of the face to bulge. This cheesy matter has a very offensive smell, sometimes it obstructs the windpipe and the victim is suffocated. Other times the head swells twice its normal size, blinding both eyes, and the victim is a miserable-looking creature, indeed. When this stage is reached it is quite useless to attempt a cure.

The most common form of roup is an exaggerated cold, and nothing worse. It is caused by exposure for a prolonged period to those conditions which produce colds. Obviously, the first step is to rectify the conditions which foster the disease. Imme-

Fig. 303.—The ordinary household scales come in handy for the poultryman.

diately a case of roup is detected, it is a good plan to treat the entire flock with a roup preventative for about a week, or until one is assured the remaining birds have not been infected. This may be done through the drinking water. Sick birds should be isolated at once, and the houses whence they are removed, particularly the drinking fountains, thoroughly cleaned and disinfected.

Treatment for roup, if it is to be treated at all, must begin in the early stages. The affected membranes should be given applications of antiseptic and healing mixtures, either sprayed on, or by dipping the fowl's head in the solution. The following are simple remedies for this treatment: One ounce of permanganate of potash to three pints of water, or one and one-half ounces of boric acid and a half ounce of borate of soda to a quart of water, or one ounce of peroxide of hydrogen to three ounces of water, or a two per cent solution of carbolic acid.

There are several reliable roup remedies on the market which have given excellent results for a number of years; they should be used according to the directions which accompany them.

Chlorate of potash, alone, or mixed with sulphur, is recommended for dusting on the inflamed tissues caused by cankerous growths. Another way is to dissolve 1 part of chlorate of potash in 10 parts of glycerin, and swab it on the affected parts.

Scabies.—See Mange.

Scaly Leg.—This condition is caused by a parasite that lives under the scales of the shanks. The scabs or crusts that appear is the excrement thrown off by these mites. It can be cured by rubbing the shanks with an ointment containing a little sulphur or kerosene. Gas tar is excellent, also—a mixture of one-third carbolic acid and two-thirds glycerin. The trouble is harmful in that it is very discomforting to the fowls. It is easily spread by fowls coming in contact with the parasites on the perches. No careful poultryman will tolerate this condition, and there is no excuse for its existence, though it is frequently seen.

Sore Head.—See Mange or Favus.

Vent Gleet.—An inflammation of the cloaca, which causes frequent passages of a white, offensive discharge that collects on the skin and feathers around the vent. It is very difficult to cure, and such specimens are better off dead. Though not contagious, the trouble is transmissible, especially by the males, consequently such birds should be removed from the flock.

White Diarrhea.—See chapter on this subject.

Worms.—See chapter on this subject.

CHAPTER XXXV

CROP BOUND

Common Form of Indigestion.—Almost everyone who raises chickens in any numbers will be troubled at some time or other with an ailment known as *crop bound*. It is a form of indigestion, perhaps the commonest form of crop trouble, and is generally caused by improper feeding. The poultryman, however, is not always to blame, for the condition is very often brought about by the stupidity or gluttony of the fowls.

Easy to Detect.—As the name implies, *crop bound* is a compaction or hardening of the crop, and fortunately, it is easily discernible. Instead of the crop having a full, close appearance, in fact, scarcely noticeable in the well-proportioned bird, it is seen to hang down like a bag, and on closer inspection it will be found to be greatly enlarged, hard and heavy. The fowl thus afflicted is usually droopy and inactive, and frequently a bad-smelling liquid runs from the mouth. In an advanced stage of the ailment the fowl's comb will be purple in color, and the bird may gasp for breath.

Two Forms.—Generally speaking, there are two forms of crop bound: One is the result of a weakened or paralyzed condition of the crop muscle, and is usually observed in old fowls whose vitality has been impoverished by old age and improper care, or in chicks of low stamina. It is possible to relieve this condition by careful medical treatment, though a permanent cure is seldom effected. Therefore, unless the victim is considered very valuable for some particular purpose, it is generally more profitable to kill the sufferer, and to devote one's time and energy to correcting the conditions that brought about the trouble.

Clogging of the Crop.—The other form of crop bound is induced

486

by a clogging of the outlet of the crop by twisted grass or rough grain. It occurs most frequently among birds that are fed insufficient green food, and as a result of this craving they attempt to swallow pieces of hay, straw, tough blades of grass, cabbage ribs or some other bulky article. This obstructs the outlet of the crop and finally becomes so entangled and solidified with other food that the mass presses on the windpipe, or fermentation sets in and induces a form of poisoning.

Occasionally a ration contains too much middlings, or other sticky meal, fed either dry or moist, which, under certain conditions, bake together and clog the passageway. This food, al-

(Courtesy Kansas Experiment Station)

Fig. 304.—Brooder house. Note the covered platform in front of the building, under which the chickens emerge from the house.

though taken into the body, offers no nourishment until it is digested, consequently the bird continues to eat, which only distends the crop further. In a few days the fowl shows signs of weakness, and unless the obstruction is removed the bird dies.

If the trouble is discovered early, the treatment is comparatively simple and a cure is virtually assured. In the latter stages a surgical operation is necessary which, though simple in itself, is often accompanied by other complications that prove fatal.

Experienced, practical poultrymen make it their business to inspect their flocks very closely every day, especially at meal times and in the early morning. At these times it is easy to detect birds that are out of condition, no matter how trivial may be their ailment. If a fowl remains on the roost after day-light, or manifests little or no interest in food, it should be caught and examined. If the bird is crop bound the symptoms will be noticed as soon as the fowl is handled. The bird will be slow to avoid capture, its crop will be distended and hard, and in most cases a sour-smelling liquid will run from its mouth.

The treatment usually prescribed is an injection of castor oil or olive oil into the crop. If these are not available, melted lard will answer the purpose, or warm water, although the latter is not so active or effective in its action. A good plan is to begin by draining off any liquid in the crop, which may be accomplished by holding the bird head downward. Then inject the oil, about two tablespoonfuls, using a medicine dropper, small syringe or spoon. Hold the bird upright and gently knead and work the mass in the crop. After some minutes this operation will cause the injected liquid to mix with the solids, and when the mass is thoroughly broken up an effort should be made to remove it through the mouth.

Dislodging the Obstruction.—This is sometimes a matter of difficulty, particularly if the offending substance is long and fibrous, such as grass. If the massaging process is ineffectual in removing the contents of the crop through the mouth, and the case is not a serious one, it may be well to wait and see if the trouble will not pass off naturally. Kneading the crop some-times dislodges the obstruction in the outlet from the crop, and with the aid of the oil the mass will be assimilated.

When to Operate.—If, however, the crop is not materially reduced in six hours, there is but one remedy—an operation. It is a very simple one, requires no great skill, and if the fowl is in a vigorous condition it has a splendid chance to recover. Young chickens weighing under a pound are too small to undergo the surgical treatment, and had better be killed.

Instruments.—The task will be much easier if one person holds the bird while another performs the operation. The only instruments required are a sharp knife, lancet or scalpel, a pair of small scissors, a small spoon, preferably a mustard spoon—one that

Fig. 305.—Promising looking flock of ten-weeks-old Leghorn pullets.

(Courtesy DeVries Farm)

has a narrow bowl, and a needle threaded with white silk or surgical gut. Common sense dictates that the instruments should be absolutely clean, also the operator's hands, so as to prevent infection.

The first step is to trim the feathers from a space about one by two inches over the center of the crop, and to moisten and brush aside any other feathers that may be in the way. Clip the feathers with the scissors; do not pull them out. Wipe the bird's flesh with an antiseptic where the incision is to be made, and with the fingers of one hand draw the outer skin fairly tight; then make an incision with the point of the knife. Insert the point of the scissors and enlarge the cut until it is about an inch and a quarter long. Separate this outer skin by spreading the slit, and then make a similar incision in the crop wall.

Care should be taken to make the cut where there are fewest blood-vessels, and to avoid the largest ones altogether. A little blood will flow, and this should be removed with bits of absorbent cotton. The cut should not be made any larger than is necessary to gain access to the interior of the crop and to remove its contents with the small spoon. Sometimes the mass is so hard that it is difficult to remove it without first manipulating to break it up. It is usually very offensive, and considerable patience is necessary to perform the work thoroughly. The operation is practically painless, so that the operator need have no unnecessary qualms over the victim's comfort.

Washing the Crop.—After the contents are thoroughly removed, the crop should be washed out with a weak solution of boracic acid, permanganate of potash, or a similar non-poisonous disinfectant. To be sure that the fermented matter is entirely removed it is well to insert a finger in the orifice, otherwise the whole process may have to be done over again, or the operation will be unsuccessful. This done, the incision must be drawn together and sewed up. A bent needle is best, making the stitches about an eighth inch apart and tying them carefully. Sew the inner skin first, and then the outer skin, and tie each stitch separately.

Next swab off the wound with the antiseptic solution, and place the fowl in a clean, comfortable coop to rest. Food should be withheld for about twenty-four hours, and then only light feeds of an easily digested mash should be given at the regular feeding hours. In about a week's time the bird will have recovered sufficiently to be returned to the flock.

Some authorities advocate making the incision at the right of the neck and at the top of the crop, at the point where it is quite easy to see the contents of the crop, owing to the transparent nature of the flesh at this point. After the contents are removed the cut is allowed to heal naturally, without stitches, which is practicable because the opening is made in the top of the crop wall. In either method the subject must be kept isolated and on a very light diet. If permitted to join the flock too soon, the other birds will be attracted by the wound, and they will peck and aggravate it.

Mortality.—Many poultrymen consider it rather futile to operate for crop bound because the chances for recovery are discouragingly small. This is true in a sense, yet in most cases death is not the result of the operation, but because the complaint had reached an advanced stage. It is easy to understand that as soon as fermentation starts, poisons are formed, which are quickly absorbed by the victim's body, and which will eventually prove fatal. Or the crop may be so enlarged and create such pressure against the windpipe that breathing is made exceedingly difficult, and this is a great strain on the heart. As a general rule, if the fowl's comb has not turned a purplish color as the result of the trouble, it is well worth while trying to operate. If, however, the bird is already weakened, one had better use "the axe."

Feather Pulling.—One of the most distressing, troublesome and unmanageable habits of fowls is feather pulling, or feather eating, and it is this vice that frequently brings about a crop bound condition. Sometimes the feathers are merely plucked, for no apparent reason except the "joy" of plucking them, and at other times they are eaten as fast as they are removed. Quite naturally they are extremely indigestible; they are likely to form a mat inside the crop, and to obstruct the canal leading from the crop to the gizzard.

Causes.—The vice usually starts through fighting or accidents, or it may develop through lack of sufficient mineral and animal food—generally from insufficient animal food. It is also caused by idleness—close confinement or no opportunity for exercise. The vice spreads rapidly throughout an entire flock, unless the ringleaders are promptly caught and removed. Erroneous methods of feeding and management are largely responsible for this

(Courtesy Atlantic Farm)

Fig. 306.—Ducks can be raised without water, but not so successfully as with it.

trouble, so that the poultryman seldom has anyone to blame but himself, and the same general conditions are likely to encourage egg eating—another pernicious habit.

There is no medical treatment for feather eating, any more than the amount of animal food should be increased, and the fowls given as much liberty and exercise as possible. If the pens are small and the yards are destitute of green food, and there is no room in which to increase the range, the habit is sometimes controlled by changing the fowls to a different pen. This change in environment may arrest their attention long enough for the alteration in their diet to satisfy their peculiar craving for *blood*.

An old-fashioned remedy was to apply something very bitter to the plumage, such as oil of aloes, but in the writer's experience this practice was little more than a *faith cure;* the fowls continue to pluck the aloe-flavored feathers as though they considered this bitterness a relish. The nearest approach to a successful method for controlling the vice is to cut the tip of the lower part of the beak, which tends to prevent the fowl from getting a firm grip on the feather.

Occasionally feather pulling is developed by lice and mites, consequently the caretaker should investigate his fowls for these pests and treat them accordingly. The important measure adopted should be a well-balanced ration, one that contains skim milk, beef scrap, fish scrap, meat bone, vegetables or green feed, and frequently varied. A piece of fresh beef hung from a nail where the hens will have to jump for it slightly is one of the surest tricks for dissipating the feather pulling habit.

Failure to furnish the flock with a liberal supply of mineral substances is one of the contributing causes of crop bound and indigestion. Nature has not endowed birds with teeth as a means of masticating their food, but she has given them the equivalent in the gizzard. This is a tough, muscular organ, so situated that all food taken into the mouth must pass through it. When the food is received in the crop it remains there until soaked and acted upon by a secretion similar to that of the saliva in the mouth of animals. This partially digested food gradually leaves the crop and passes into the gizzard, where it is ground up, and thence it goes to the intestines, where, after being acted upon by other fluids, it passes on and the nutriment is absorbed.

Supply Grit.—We know that the gizzard is marvelously strong when provided with sharp grit, for it is the rotary action of these grindstones that crushes and masticates the solids. Hard, sharp substances are necessary, and without them the harder parts of the food are not digested. Husks and green food accumulate between the crop and the gizzard, and frequently cause a stoppage so that nothing but liquids can pass. In time this passage is completely obstructed, and the result is a sour or bound crop.

A person may live with defective teeth for years, or perhaps with none at all, yet we know that such persons seldom enjoy their food or good health. Surely, if the birds do not have the means of properly masticating their food, they can neither be healthy nor derive the greatest benefits from their food. In consequence they cannot be expected to give their master a good return for their food and care.

CHAPTER XXXVII

WORMS

Losses from Worms.—The question of worms in poultry is of far greater importance toward the success of the venture than most poultrymen realize. Well-built houses, carefully prepared food, close attention to sanitation, and good care generally are of little avail if the fowls are infested with worms. Where there are worms losses follow: if not actual death, at least there is a falling off in the egg yield. In any event the poultry keeper is not getting the proper returns from his feed and care, which is the equivalent to loss.

A postmortem examination at one of the State Egg Laying Contests suggested a careful examination of the dropping boards, which finally led to the conclusion that some of the pens were infested with intestinal worms. The flocks were given a vermifuge, followed by a purgative, which had the desired effect, and in a couple of weeks' time the change in the flocks was surprising. They were eating more, took on weight and their egg yield improved.

Widespread Trouble.—Numerous instances of anemia, liver trouble, indigestion, diarrhea, general physical debility and other complaints, due supposedly to lack of vigor in the breeding stock, have been traced to worms within the fowl's body. In fact, it has been said that of the strictly parasitic forms of life that affect poultry, worms play the leading rôle. I am not prepared to agree that worms are a greater menace to fowls than some other *varmints* of our acquaintance, such as mites and lice, but I do know that they work a great deal of loss and failure.

Kinds of Worms.—There are several varieties of worms, some of which take up their abode in the crop, stomach and intestines, not to mention the gape worm, which attaches itself to the windpipe and is made evident by frequent gaping, hence its name. Those that breed in the intestinal section are probably the most common and the most destructive.

There is something revolting about the idea of worms existing in the organs of a living creature; it is an unpleasant subject

(*Courtesy Purdue Experiment Station*)

Fig. 307.—Feed hoppers and water fountains should be located on a raised platform to prevent litter from being scratched into them.

to discuss. Nevertheless, since it is a foe, and a deadly one, we must take up arms against it, and to do so intelligently we must go into some detail.

Tapeworms.—It has been found that there are two principal kinds of intestinal worms, round worms and ribbon-shaped worms, commonly called tapeworms. The commonplace that tapeworms actually consume food is all wrong; they do not. If we examine them under a microscope, we find that they have no mouth or intestinal tract at all. They are a very low order of life, and attach themselves to the intestinal lining by means of a hook-like appendage. Free to come in contact with the digested · nutrients in the intestines, they absorb these elements, much as the intestines themselves absorb this food. Obviously the fowl is robbed of so much nutriment, and in due time it becomes poor and emaciated, depending upon the extent of the worms.

Under a strong glass we note that the worms consist of segments, each of which is a complete organism, if you can call it such. It absorbs its own food, develops its own eggs, and later separates itself from the other segments and finally is passed out to the soil. In each segment there are hundreds of tiny eggs which are scattered on the ground, among food and in the drinking water, only to be picked up by other birds, which are then contaminated. It has been discovered that flies devour these eggs, and that the eggs are hatched within the fly; and, of course, chickens eat flies, therefore they take over the incipient worms as well.

Numerous remedies are used to dislodge these parasites, and for best results they should be administered when the birds are fasting. The best way is to give the flock a light feeding at night and the following morning give them the vermifuge. Several hours later they are given a purgative, such as Epsom salts or castor oil, and the treatment is complete.

One of the most commonly recommended remedies and one of the easiest to administer is powdered pomegranate root bark. The dose is one teaspoonful for each fifty fowls given in a wet mash. Another good remedy is oil of wormseed (Jerusalem Oak). Mix a teaspoonful of the oil in a moist mash for every 12 fowls. In both treatments the purgative is given a few hours after the vermicide.

For individual treatment oil of turpentine is excellent, which may be mixed with an equal quantity of olive oil, and 20 to 30 drops of the mixture given at a dose. This is followed in a couple of hours with a tablespoonful of castor oil. Thymol is

32

Fig. 308.—Colony houses grouped for winter use as breeding pens.

also used, 1 grain to each fowl, or powdered areca nut, 30 to 45 grains; powdered male fern, 30 to 60 grains; and kamala, 30 to 40 grains for each fowl.

Clean Premises.—At the same time the flock is being treated medically, the premises must be treated by powerful disinfectants to destroy the worms and their eggs, otherwise the birds will only become re-infested. Be careful to drain off any stagnant water, and fill in any marshy places. Sunlight is one of the greatest insecticides, and the cheapest, therefore the soil should be exposed to it by thorough plowing and harrowing.

Slaked lime is highly recommended, and in extreme cases it may be well to spray the ground with a solution of carbolic acid. Birds that have died from worms should always be incinerated, or buried in quicklime, never allowed to decay on the surface of the ground.

The houses, especially the dropping boards and floors, should be thoroughly cleaned and disinfected, also all feeding troughs, hoppers and drinking fountains. Bear in mind, there is little use in combating worms in anything but a thorough, practical manner. To do the task half is wasted energy, for only strenuous efforts will rid the premises of these pests.

A heavy clay soil is much harder to rid of worms than a light, sandy loam. For that reason the latter soil is recommended for poultry. It is usually perfectly sanitary at all times.

Manifestations of Worms.—Enterprising poultrymen realize the danger of worms and keep on the lookout for manifestations of them. If a bird should die from any cause whatever it is examined, not only for the immediate cause of death, but for indications of worms. The intestines, stomach and crop are opened and their contents noted. Birds that are killed for the home table are also examined. If one fowl is troubled with worms it is pretty safe to assume that the remainder of the flock is afflicted also, in which case treatment is begun at once. Birds that are dull and listless, with pale combs and shrunken wattles, are likely victims of worms. *Post* one of them and see what the trouble is. It is better to lose one bird, and thereby determine the evil, than to risk losing half a flock later on.

Gape Worms.—Strictly speaking, gape worms do not come under the head of worms, as they are commonly understood, but under the term gapes, which is reckoned as a disease of chicks. Nevertheless, it is equally abominable, a kindred ill, hence its place in this chapter.

(*Courtesy Atlantic Farm*)

Fig. 309.—On commercial duck farms feeding is done almost exclusively by means of tramcars.

Gape worms exist at all seasons, though they are seldom observed as troublesome until the hatching months, when they affect young birds. Chicks are most susceptible from 10 days to 6 weeks old, since at this age they are not large enough nor strong enough to dislodge the worms from their throats. Vigorous birds and older stock are attacked by the worms, but they usually succeed in getting rid of them without the keeper's aid.

The worm which causes gapes is in reality two worms—male and female, and they are so firmly grown together that they cannot be separated without tearing the tissues. The female worm is the principal member; it is about a half-inch long, while the male is little more than one-fifth of an inch. The heads of both are attached to the mucous lining of the windpipe or trachea, which causes such an irritation that undue secretions collect and make breathing difficult. Sometimes so many worms collect in the trachea, and grow to such size, for their eggs develop while they are in this state, that breathing becomes impossible and the host, the afflicted chick, dies from suffocation.

Chicks affected with gape worms will be seen to cough and sneeze with labored effort in a vain attempt to dislodge the pests, which is very difficult to accomplish. Soon they commence to gape, extending the neck and opening the beak, indicating that they are having great trouble in breathing. Later, as they become weakened by their struggles against the parasites, their appetites fail and they grow dull and listless, their wings droop and they stand with half-closed eyes and head drawn back into the body feathers. In this condition they are apt to die from suffocation, or be trampled by their fellows.

In dealing with this complaint the poultryman should learn to rely more upon a preventative than a cure, because very young chicks are very difficult to treat individually and therefore expensive. Good results have been obtained by extracting the worms with a feather, twisted horsehair, or one of the patented extractors. These devices are forced down the victim's throat, either dry or moistened with turpentine, then twisted about vigorously in an effort to dislodge the worms, and removed.

Recently, good results have been reported from medicating drinking water, or by injecting 3 to 10 drops of a 5 per cent solution of salicylate of soda. The best method of prevention is to put the chicks on fresh ground, or soil that is known to be perfectly sanitary, and if any trouble is experienced, to treat the yards and premises with a strong disinfectant, the same as for other kinds of worms or parasites.

CHAPTER XXXVIII

WHITE DIARRHEA

Terror and Plague.—No term, perhaps, strikes greater terror to the poultryman than white diarrhea. It is synonymous with such words as plague, scourge, epidemic and pestilence. That it has earned this opprobrium is attested by the fact that thousands of chicks are lost annually by this infection, and also because of its resistance to any known treatment. It is successfully combated, of course, but by preventative measures rather than curative ones.

Exaggeration.—There is no gainsaying the malady exacts an enormous toll from the poultry raisers, yet I am inclined to think that much of the alarm is the result of sensational writers and highly imaginative persons, who find it more to the liking of their morbid minds to spread terrorism instead of optimism. Calamity is always more lurid than sublimity. These scares, like the alarm about cholera, small pox and infantile paralysis in the human species, are very much exaggerated and do more harm than good.

Investigations.—It is natural that white diarrhea should have been the object of a great deal of investigation; chemists and bacteriologists have struggled with its mysteries for many years. While, perhaps, they have not been particularly successful thus far in establishing a positive cure for the disease, they have at least succeeded in isolating the germ, learned how to detect it, studied its development and propagation, and the conditions under which it thrives best, and devised satisfactory methods of preventing its spread.

Causes.—We are told that white diarrhea is caused by at least four different kinds of infection, the most common of which is a bacillus called *bacterium pullorum*, which means in ordinary

502

terms, the bacteria or germs of the pullets. *Coccidiosis* and *aspergillus fungus* are two other forms of the disease, but so far as we are concerned these technical phrases are merely names, difficult to pronounce, harder to remember, and of no value to the unsophisticated mind.

All of these microbes infect the adult fowls and are generally communicated from them to the chicks directly or indirectly.

Fig. 310.—Ideal location for ducks and geese.

It is not necessary for the affected hen to have any external appearance of having white diarrhea or a diarrhea of any kind. In fact, the affected bird may be in the pink of condition, a good layer and a fine-looking fowl. By analyzing the eggs laid from an infected hen, we may find white diarrhea germs in very active form, or none at all. Apparently, some eggs are inoculated while others escape, though there is no method of determining this from external appearances

Symptoms.—In chicks the symptoms of white diarrhea are virtually unmistakable, the most prominent one being a more or less profuse diarrhea, the droppings consisting almost entirely of mucus from the intestinal tube and the white secretion of the kidneys. The white substance predominates, hence the name white diarrhea. It is caused by an irritation of the intestines, fever, and a rapid breaking down of the tissues of the kidneys.

Hens infected with this malady produce chicks which have the germs of the disease within them when they are hatched, and these chicks usually show manifestations of the trouble within the first few days of their life. From the experiments conducted it appears as though chicks are most susceptible to infection during the first twenty-four hours, and that after the fourth or fifth day they are practically immune.

Chicks which sicken of the disease later must have taken the microbes into their systems at an early age, which for some reason remained dormant for a time. Adult fowls are practically resistant to the germs, and do not show any symptoms though they may be inoculated with the disease and lay infected eggs. Thus incubators and brooders, as well as coops, become infected with the disease and preserve the contagion indefinitely, unless scrupulous measures are taken to destroy it. The ground is also impregnated with the contagion, and should be plowed under and sowed to plant life.

Pasting Up.—A large brood of chicks may be hatched from eggs subject to the germs of white diarrhea, and to all appearances they are hail and hearty when taken from the machine. But they soon commence to wilt. The first indication of something wrong is a disposition to huddle together and remain under the hover or under the hen, as the case might be. Apparently they suffer from chills. They are listless, stupid and sleepy, and take no interest in food or their environment. They stand still, heads drawn in, eyes closed, and chirp and peep almost constantly. Their wings droop or project slightly from the body, instead of being folded tightly against it, and the characteristic diarrhea soon appears. Usually the excreta is mucilaginous, adhering to the downy feathers about the vent, where it dries and cakes and continues to accumulate until it completely covers the opening and causes a stoppage. Unless relieved, this condition, known as *pasting up*, will bring about an early death. The mass should be removed as gently as possible, and the affected parts treated with vaseline or soothing ointment.

Spread of Contagion.—There is added mischief in these masses of excreta due to the chicks picking at them and thereby con-

(*Courtesy Purdue Experiment Station*)

Fig. 311.—Artificial pond constructed of concrete. Ducks can be raised without a swimming hole, though best results are secured with one. The eggs are likely to run more fertile.

tracting the disease. In this manner the germs are spread through an entire flock, unless the caretaker adopts prompt means of isolating the affected members.

Chronic Type.—Sometimes the disease is less severe, but of a more chronic type, and takes longer to run its course. The chicks thus afflicted waste away and gradually become weaker and more emaciated, until their legs are unable to support their bodies. They lean against walls or other objects for support, or squat down with outstretched wings until they die. As death approaches the breathing becomes labored, and at intervals the poor little creatures give utterance to a faint shrill cry, indicating that they are seized with paroxysms of intense pain. Most of these victims have the peculiar form of body described as *short back*, which results from the distension of the abdomen.

The most disastrous phase of this complaint is the heavy mortality. The losses vary from 50 to 80 per cent of the chicks

(*Courtesy Cornell Experiment Station*)

Fig. 312.—Well-proportioned poultry house. Front wall is of novelty siding, ends and rear wall are made of matched lumber covered with patent roofing, as is the roof. Note position of doors, curtain frames, windows and ventilator.

affected, and often it is impossible to raise any of them. Furthermore, it is questionable if it is advisable to try and raise any of them, in view of the probability that they may later become what is termed *bacillus carriers*. Because those that do not succumb still carry the germs in their bodies, lodged in the ova—the undeveloped eggs, which are ultimately laid, perhaps, and thus transmit the disease from one generation to the next.

No Absolute Cure.—For bacillary white diarrhea in young chicks there is no absolute cure so far as is known. Several so-called remedies have been extensively advertised, but most of these are in reality preventatives. Furthermore, the medical treatment of individual chicks is virtually impracticable, as it is too expensive, and flock treatment, once the chicks are afflicted,

(*Courtesy Atlantic Farm*)

Fig. 313.—White Pekin ducklings.

is of little consequence because the chicks cannot be induced to eat or drink in sufficient quantity to be of any avail.

The feeding of sour milk to young chicks as soon as they are taken from the incubator appears to be the most successful treatment toward controlling the disease. The purpose of the sour milk is to suppress any intestinal putrefaction which the bacillus may set up. In other words, the sour milk contains ferments or bacteria, which are calculated to counteract or offset the parasites of the white diarrhea.

Lacking the sour milk, it has been found that 15 grains of powdered catechu dissolved in a gallon of drinking water tends to prevent the development of the diarrhea. This treatment should be continued for about ten days, or until the danger period is past.

Preventative measures should begin, of course, with the eggs used for hatching. In the first place, no eggs should be used which are known to be laid by hens afflicted with white diarrhea or any other communicable disease at any time. If the eggs are purchased the buyer should insist upon some assurance as to the health of the parent stock, and if possible he should visit the farm and inspect the flock.

Before the eggs are set under hens or placed in the incubator they should be disinfected, which is easily accomplished by several methods. Wipe the shells with a soft cloth saturated with grain alcohol of 70 to 80 per cent strength. Or the eggs may be dipped in a weak solution of creolin and water. Both methods have been used effectively, and neither treatment has had any serious influence on the hatchability of the eggs.

If the hatching is done in an incubator it goes without saying that the interior of the machine, especially the egg trays and nursery drawers, should be carefully disinfected after each hatch. The same precautions should be adopted in regard to the brooder and any other coop or device that is used by the chicks. If the hatching is done by hens the broods should be put upon fresh ground, and frequently moved to fresh ground, so that if any cases develop the risk of contagion will be reduced to a minimum. It may be asserted positively that, though there is no positive cure for the disease, it may be eliminated, and if reasonable sanitary measures are practised as a part of the routine work, there is little reason to fear the ravages of white diarrhea.

Agglutination Test.—It is possible to detect the presence of white diarrhea in grown fowls by a blood test, called the *aggluti-nation* test, but owing to the delicate nature of the work it must be performed in a laboratory. Some of the Experiment Stations will perform this work for a nominal charge.

CHAPTER XXXIX

PARASITES AND PESTS

Fowls are Subject to Vermin.—In every enterprise there are certain realities and circumstances which may be classified as *grim realities*. To combat them seems to be a part of the philosophy of life. Therefore, to attempt a task without due allowance for this struggle is in the nature of folly, since it is quite likely to result in bitter disappointment, maybe failure.

Farming is no exception to the above. In fact, it is probably the most graphic example of a struggle against adverse circumstances. There is scarcely a tree, shrub, vine or plant that is without its natural enemy. For almost every stalk that sprouts there is some other form of life eager to feast upon it. Horticulture is a constant battle against blight, worms, beetles, weevils, moths, grubs and countless other insects. Successful animal husbandry necessitates a corresponding struggle against similar pests, for as such we have come to know these lower forms of life. And not the least of these are the parasites affecting poultry.

No matter how much we would like to think of our fowls as being nice and clean and free from anything so objectionable in name and nature as lice and mites, just as certain as dogs are likely to be bothered with fleas, and cattle and sheep are susceptible to ticks, poultry, especially chickens, are prone to become infested with vermin. It seems to be a part of the general scheme of things.

Be on the Lookout.—On the well-organized, progressive farm, where poultry is made a specialty, there is less trouble resulting from the ravages of vermin than on the general farm or backyard, where small flocks of fowls are kept principally as a side line. There are several reasons to account for this condition.

509

In the first place, the operator of large numbers of fowls, having considerable capital invested in his plant, is more likely to have a keener appreciation of the needs and requirements of his stock. Usually he makes a thorough study of the conditions affecting his birds, for he knows only too well if he is an experienced poultry man, that the conditions which affect his flocks adversely or beneficially are almost immediately reflected in his bank account.

Owners of small flocks are sometimes indifferent to improved methods, for no reason except they do not take the work seriously. Then, again, on large farms where the hatching is done in incubators and the chicks are brooded artificially, never coming in contact with hens, it is very much easier to keep vermin in check on the young stock. They are not so apt to become infested until they are fairly well grown, and not even

(*Courtesy Wisconsin Experiment Station*)

Fig. 314.—Painting the perches with crude oil or disinfectant to exterminate mites.

then, unless the buildings are seriously over-run with vermin. This is quite a factor, indeed, because vermin is particularly fatal to young chicks, and is responsible for all kinds of trouble.

Realizing the importance of safeguarding his flocks against

parasites, the owner of a well-organized poultry plant makes it his business to establish a regular sanitary schedule—a system of spraying and disinfecting, also whitewashing, which he adheres to quite as rigorously as feeding and watering. Not so frequently, of course, but just as systematically.

The point is—that it is equally important for the keeper of a small flock to exercise proportionate care. Because the flock is small, or because it is kept merely as a side line is no excuse for exemption, and no guarantee that the birds will not be troubled. However unpleasant the idea may be, you must make up your mind to the fact that wherever you keep fowls you are going to have vermin, unless you fight these pests, and fight them strenuously and continuously. There is an affinity between fowls and vermin. They must be fought the same as the farmer fights potato-bugs and cut-worms.

Kinds of Parasites.—The parasites that attack poultry are of two kinds, commonly known as lice and mites. There are several varieties of

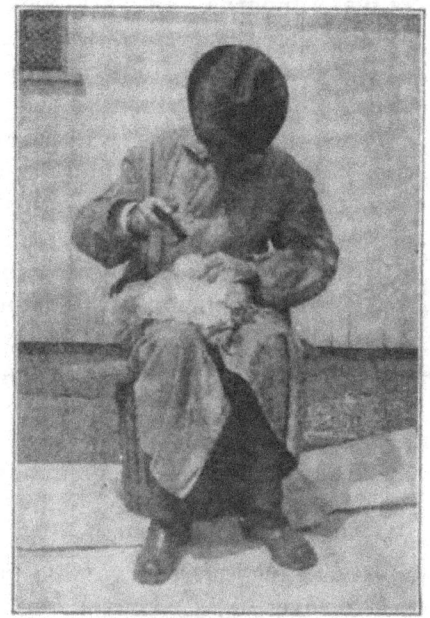

(*Courtesy U. S. Dep't Agriculture*)

Fig. 315.—Dusting a fowl with insect powder.

the former, but since they have the same general characteristics and are combated in the same way, we need not concern ourselves with a study of individual species. They live mainly on the secretions of the body, feathers and skin, and inhabit the fowls day and night. They are found largely on the head and neck, under the wings and about the vent. When allowed to multiply in great numbers, as they are prone to do, they will sometimes become so thick as to cause death. Sitting hens are especially annoyed by them.

The female lice deposit their eggs on the feathers of the fowls, cementing them to the quills near the skin. In a few days these eggs hatch, in a short time they breed, also, propagating more and more, so that it is possible for thousands to be multiplied in a couple of months. In warm weather conditions are particularly favorable for their reproduction, consequently with the approach of spring and throughout the summer months extra care must be taken to destroy them.

Mites.—If anything mites are more troublesome than lice. They abound in larger numbers, and aside from tormenting the fowls, they actually subsist on the blood of the birds. They are not usually found on the bodies of the fowls except when on the roost or in the nest. During the day mites inhabit cracks and crevices of the walls, roosts and other fittings. Many houses, unsuspected, actually swarm with them. When these pests have accumulated in such hordes that they are unable to get enough blood from the fowls during the night, they are found to remain on the birds during the day.

Potent Enemy.—Though infinitesimal in size and, therefore, almost insignificant individually, collectively mites represent an enemy with the potentiality of a fair-sized animal. A host of blood-sucking mites is capable of absorbing more blood, more vitality over night than the fowl can replace by the assimilation of large quantities of food during the day. In addition to this actual loss of blood, which is a terrific drain on the fowl's strength, the bird must endure the torment of being constantly pierced and chewed by these pests. Sitting hens are often so tormented that they are compelled to leave their nests in order to relieve themselves.

The mouth of the louse is formed for biting and chewing, and since they have a fondness for moisture, they can be poisoned by spreading a mercurial ointment where they are seen to congregate on the fowls. This treatment should be used on mature fowls only, since the mercury is likely to prove dangerous to chicks.

Lice breathe through spiracles or pores in their sides, hence they can be suffocated by stopping up these breathing tubes with a fine powder. The instinct for a hen to take a dust bath is nature's way to kill these pests. The most effective powder for this purpose is that which contains a drying and burning ingredient, or one giving off fumes.

(Courtesy Wisconsin Experiment Station)

Fig. 316.—Spraying the roosting compartments with an insecticide is part of the sanitary schedule on a well-regulated poultry farm.

A reliable home made powder can be mixed as follows: Add an ounce of 90 per cent carbolic acid to a peck of sifted coal ashes, mix thoroughly, and then add an equal quantity of tobacco stem dust.

Here is another recipe which has given satisfaction at the Maine and Cornell Experiment Stations for a number of years: Add one part crude carbolic acid to three parts gasoline, mix these together carefully, and while stirring add enough plaster-of-Paris to take up the moisture. When enough plaster has been

33

added, the resulting mixture should be a dry pinkish—brown powder, having a fairly strong odor of carbolic acid.

The dusting powder should be worked into the plumage of the fowl, the greater part of the application being in the fluff feathers, near the vent and under the wings. See Fig. 315. Its effect is noticeable almost immediately. The dusting should be repeated in about two weeks to catch the new brood of lice, which are in the form of eggs at the first dusting, and are not disturbed by it to any extent. Fumigation and spraying with a good disinfectant will destroy lice about the roosting compartments and cracks of the house. Once the hens are fairly rid of lice, it is no trouble to keep these parasites under control by a systematic spraying. See Fig. 316.

Destroying Mites.—Unlike lice, the young of the mites are not hatched on the fowls, but in the hiding places where the mites live. Consequently, to destroy mites the poultryman should confine his attack to their breeding places in the structure of the building, and not on the fowls.

Spraying or painting the perches, nests, drop boards, and all other interior fixtures, including the ceiling and walls, with a solution of crude oil or other powerful disinfectant, is the surest way to rid a house of mites. See Fig. 314. Enough of the liquid should be used to thoroughly saturate the surface of the wood, and to run freely into all cracks and openings. Frequently a poultry house is so dusty that unless large quantities of the spray are used, the dust will absorb the greater part of the liquid. To overcome this, it is well to first give the building a good brushing down with a stiff broom, and then follow with the spraying.

It is a mistake to think that because a house is idle for some time it will rid itself of mites, believing that the mites will starve where there are no fowls to feed upon. They will live for months without food, if necessary an entire winter, only to reappear as soon as a flock of chickens is placed in the house.

Whitewashing.—In the minds of many, *whitewashing* means slaking some lime in water and splashing this solution rather carelessly over a prescribed surface. The result is all too familiar

—whitewash so badly streaked as to be most unsightly, and at the slightest touch brushing off the woodwork. In fact, frequently it does not remain long enough to be of any real value.

To execute a good, permanent job—a job that will not only clean and disinfect the building, but improve its appearance as well—one must regard the task much the same as painting. First of all, the walls, sills, and all projections that have accumulated large quantities of dust should be carefully brushed clean with

(Courtesy U. S. Dep't Agriculture)

Fig. 317.—Spraying outfit for disinfecting and whitewashing.

a stiff broom; otherwise the whitewash will simply form a coating or deposit over the dust and will subsequently curl up and fall off, thus exposing the original dirt. Moreover, this dust is frequently the refuge and abiding place for disease germs and vermin and should be removed, not merely put out of sight temporarily. Whitewash is a good exterminator if allowed to come in contact with vermin, and will penetrate cracks and crevices which are in a horizontal position, such as those on dropping boards, in a thorough manner; but if sprayed on the dusty sides of a building, it is very apt to run off the dust like water from an oiled surface.

Spraying is much easier and quicker than applying the whitewash with a brush, and if the above cleaning precautions are taken, it is equally as effectual. But if one insists on following the careless method, by all means use a brush, and put some carbolic acid in the wash—for in so doing the woodwork is at least partially scrubbed with the solution.

Formulas.—The following are formulas for whitewash that will not rub off: Slake a half-bushel of good strong lime in boiling water, using just enough water to cover the lime and keep it from burning. When the slaking process is completed, add a little more water, and then strain the solution to remove any sediment of sand or foreign substance. Dissolve a peck of salt in warm water and add this to the lime solution; mix it thoroughly and allow it to stand for a couple of days. When ready for use, thin to the proper consistency and apply it hot.

Mixing Wash with Skimmed Milk.—Another well-known recipe: Slake the lime as before, and then add two pounds of sulphate of zinc and one pound of salt dissolved in water. The addition of a half-pound of whiting will improve the wash for outside use, and if skimmed milk is used instead of water, the results will compare favorably with oil paint.

Disinfectant.—The following is an inexpensive and efficient poultry house disinfectant for spraying purposes, and one that is very simple to make at home: Dissolve a pound of strong, hard soap or soap powder in a gallon of boiling water; immediately this is removed from the fire add a gallon of kerosene and one pint of crude carbolic acid, and churn and agitate the solution until the ingredients are thoroughly mixed. If the oil and water separate, it is because the soap was not sufficiently caustic. It is important that crude carbolic acid be used and not the refined product, for the crude acid—a dark brown, dirty-looking liquid—contains tar oil, which is to be desired. Add ten gallons of water to the above to make a stock solution, and when ready for use, mix this stock solution with an equal quantity of water. It is then in condition for spraying.

Fig. 318.—Interior of pigeon house showing feed hopper, roosts, nest boxes and different kinds of nest pans.

Coal ashes are an asset rather than a nuisance on the poultry farm, and may be used to splendid advantage. They are an excellent absorbent, and when sprinkled under the perches and mixed with the droppings, the volatile properties of the hen manure are carefully preserved. They differ from wood ashes in this respect, for wood ashes, though a good insecticide, contain considerable lime, which liberates the ammonia in the droppings and thus destroys much of their fertilizing value.

Finely screened coal ashes make the most effective sort of a dust bath for the hens. The fine dust penetrates the fowl's feathers, and coming in contact with lice serves to stop the breathing passages of these parasites, causing them to suffocate and die. Wood ashes are even better for this purpose, because the particles of dust are finer, but here again the lime is objectionable, since it tends to take the gloss off the plumage.

Coal ashes should be used freely on the floors of poultry buildings, for they penetrate cracks and crevices and assist in destroying mites and other vermin, dissipate noxious odors, and improve conditions generally. Still another advantage—large quantities of the cinders will be eaten by the birds as grit, and will contribute some of the mineral nutrients.

Road Dust.—Dust removed from a road during dry weather, and which is only an annoyance to travelers, will be found beneficial in the dust boxes also. Every poultry farm should have a supply on hand for winter use, for unless dirt floors are used, these artificial means of supplying the fowl's toilet requisites must be provided. A dust bath is quite as essential to the well being of poultry as is the *soap-and-water* variety to the human.

CHAPTER XL

DUCKS

Popularity of Duck Meat.—Time was when the duck was not considered sufficiently profitable to warrant the trouble in raising it. Its flesh was never keenly sought after by the masses, consequently it returned low prices and farmers generally declined to show any enthusiasm. In those days, however, ducks were raised without constraint in waterways and made to forage for their living, subsisting almost entirely upon fish and other water foods, which naturally imparted a strong fishy flavor to their flesh and made them undesired except by the few who might be partial to that kind of diet.

In the last twenty years breeders have awakened to the necessity for improving the flavor of the flesh by feeding grain almost exclusively, whereupon their efforts have been rewarded by a steadily increasing demand, until to-day duck raising has developed into a flourishing industry, and on some of the well-known plants, especially those on Long Island, a single farm will market 100,000 ducks a year.

Standard Varieties.—There are numerous standard varieties of ducks, among which are the following: Pekin, Muscovy, Indian Runner, Aylesbury (see Fig. 322), Rouen (see Fig. 323), Cayuga, Call and Swedish. Of these the Pekin, Muscovy and Indian Runner are the most widely bred, and probably the best suited to farm purposes.

Pekins.—None stands higher in popular esteem than the White Pekin, which was imported from China in the early seventies. See Fig. 319. It is valuable for raising on a large scale, and while naturally a very timid bird, it may be raised more easily, perhaps, than any other. It has a distinct type of its own, and differs

519

from all others in the shape and carriage of its body. They are large birds, sometimes attaining twenty pounds to the pair; yet their flesh, if properly nourished, is delicate and free from any taint, and they are considered among the best of table fowls. They mature early, are excellent layers, non-sitters, and require no particular skill in breeding.

Shape of Pekins.—The standard-bred Pekin has a long, finely formed head, neatly curved neck, and a bill of orange yellow,

Fig. 319.—Pekin ducks.

medium-sized, with no trace of any other color. The eyes are of a deep, leaden-blue color. The back is long and broad, the keel proportionately long and deep, the breast round, full and very prominent, and the carriage of the body elevated in front, sloping downward toward the rear. The wings are short, folded closely against the body, and are not capable of sustaining flight; hence a two-foot fence is sufficient to restrain the Pekin. The plumage is downy and of a faint creamy tint throughout, though recently there is a preference for an absolute white. The standard weight of the drake is 8 pounds, and of the duck 7 pounds.

Muscovy ducks, of which there are two varieties—White and Colored, the plumage of the latter being black and white—have several distinct peculiarities. See Fig. 320. They are sometimes called the Musk duck, owing to the odor of musk which pervades the skin, but which is not objectionable when the fowl is cooked. Also, their appearance is rather grotesque by reason of the long, crest-like feathers on the head, the sides of which and the face are

Fig. 320.—White Muscovy ducks.

covered with scarlet caruncles. This large, red face gives them a savage look, and to some minds it makes them quite hideous.

They are found in a wild state in the warmer regions of South America, but in Brazil they have been extensively domesticated and are highly prized for eating. They find great favor in Europe as well, especially in Germany, where they are raised in large numbers. They are the largest of all ducks, the standard weight of the drake being 10 pounds and of the duck 8 pounds.

The Muscovy is notorious for its pugnacious, quarrelsome nature, and for this reason it is frequently inadvisable to keep them on the farm with other poultry. The temper of the drake is violent, indeed; he will fight with another drake, inflicting serious harm, if possible; and his persecution of other poultry is unceasing and abominable.

Unlike Pekins, the Muscovy is a strong flier, and when fright-

Fig. 321.—Indian runner ducks.

ened, they will fly into trees or into high places of concealment where they remain for long periods of time. When bred, they must be kept in yards by themselves, care being taken to clip their wings to keep them from flying. The flesh of the Muscovy compares favorably with that of any other duck, if eaten young, but they cannot be credited with laying so many eggs as some of the other breeds.

The Indian Runner duck, which is said to have its origin in India—hence the forepart of its name—has rapidly come into wide popularity by reason of its prolific egg yield, and by reason of which it is aptly termed the Leghorn of the duck family. . See Fig. 321. The name *Runner* is appropriate, for they literally run, instead of waddle, as do other ducks, and sometimes present a very comical appearance.

Runner ducks are exceedingly active in their habits, and by reason of their willingness to forage are able to find a large part of their food. Though easily domesticated, they do not stand confinement well; they are non-sitters, are hardy and easy to raise, and while their flesh is of prime quality, their size is rather against them for market purposes, except as broilers. The standard weight of the drake is 4½ pounds and of the duck 4 pounds.

The body of the Runner somewhat resembles the shape of the Penguin; it is long, narrow and carried very erect, with little or no indication of keel. They are very good-looking, the plumage being light fawn or gray and white, which blends admirably with the orange-red of their shanks and the green of their bills. In the past five years White Indian Runners have become very popular, and are probably destined to be more extensively bred than the Fawn variety.

Water is not Necessary.—We naturally associate ducks with water—but as a matter of fact it is not at all necessary that they have access to it, except such as is given them for drinking purposes. If a pond or stream is available, it is well to allow the breeders the freedom of it, also the young ducks until they are about eight weeks old, when they should be penned and fattened for market. On the other hand, equally good results are being obtained by poultrymen who have no water on their premises; the only noticeable difference between these ducks and those having access to water is that the water ducks have somewhat prettier and cleaner plumage.

Duck raising is to be recommended for those who wish to make use of marshy land not suitable for chickens; yet this fact must not deceive one in the belief that damp, wet quarters are available for ducks. Houses for ducks are simple affairs, but they must be free from dampness and located so as to be assured of good drainage—preferably on sandy soil. The feathers of a duck are almost impenetrable to cold, but its feet are rather susceptible, much as a hen's comb and wattles. The duck likes to warm its feet, and to this end it is advisable to have a dry dirt floor in the duck house, covered with straw, hay, shavings or

Fig. 322.—Aylesbury ducks.

other litter. Some breeders advocate a board floor, but this is scarcely necessary unless it is impossible to maintain a comparatively dry earth floor.

The beginner should start modestly and develop his plant only as his experience increases and his capital warrants; for duck raising is an arduous task and requires an absolute knowledge of the business before success can be reached. Every phase of the work should be carefully studied with the idea of minimizing labor, and each department—the incubator cellar, brooder house, breeding pens, fattening pens, picking room and feed house—located where they are the most convenient and the most accessible. The task of feeding and watering ducks alone is no small one, and on large plants this factor receives very careful consideration. Artificial incubation and brooding, com-

Fig. 323.—Rouen ducks.

bined with improved machinery for handling and mixing the feeds, are largely responsible for the growth of the duck industry.

Inspire Confidence.—One of the essentials in breeding ducks is a quiet, even-tempered caretaker—the man or woman who will inspire confidence in the flocks and become, in a sense, a companion to them. Ducks are high-strung, excitable birds, skeptical toward strangers, and yet responsive to good treatment and regular attention. Furthermore, they are confirmed creatures of habit, and any serious change in their diet or management is likely to disturb their appetite and egg yield. For example, mature stock that has never been fed on corn will not eat it at first, and may never really learn to relish it, whereas if they are brought up on a part corn diet it is one of their best-liked grains.

Feeding and Watering.—Under natural conditions, the food of the duck is both animal and vegetable, consisting of fish, water insects, grasses and so forth, therefore when the birds are raised under a somewhat artificial environment this diet must be imitated to secure the most satisfactory results. Unlike the hen,

Fig. 324.—Ducks are heavy drinkers.

the duck has no crop, the food passing from the bill to the gizzard; in consequence the food must be largely of a soft character. Too much hard food does not agree with these birds—they will not thrive upon it, hence it is inadvisable.

While a proper selection of the food is highly important to secure rapid growth, the necessary ingredients are the simplest grains, usually wheat middlings, wheat bran, corn meal and low grade flour, to which should be added beef scraps and, where necessary, shredded alfalfa or other green food. The following is an excellent ration for breeders, whose food should not, of course, be so forcing as that intended for the market birds: Equal parts corn meal, wheat bran, green food; 5 per cent beef scraps, and 5 per cent coarse sand or grit. Mix with water to a moistened, crumbly state—not *sloppy*—and feed three times a day, the last feed to be given an hour before sundown.

Grit is as essential to ducks as it is to chickens, and should be kept before the birds at all times, in addition to being fed in the mash. The sand used in the mash supplies a certain amount of the necessary grinding material, and the duck will pick up more on range; yet this is not sufficient to fully satisfy the digestive demands. A supply of oyster shells should also be kept in a convenient place.

Heavy Drinkers.—As previously stated, water for bathing is not necessary for growing ducks, but it is most important that they have access to a liberal supply of fresh drinking water at all times. Ducks are heavy drinkers, and it has been said that the only neglect which will kill young ducks is failure to provide them with sufficient water, in a vessel deep enough for them to get their heads beneath the surface. Ducklings like to immerse their eyes—a habit which seems to be essential to their well-being. Obviously, the water should be kept as fresh as possible.

CHAPTER XLI

GEESE

Virtues of Geese.—It is pretty safe to say that we Americans do not raise enough geese, and do not fully appreciate their possibilities. Like the guinea, the goose is not taken seriously enough by the average farmer; yet there is no kind of livestock that can be fed with so little loss, and so little attention, and that requires such inexpensive equipment, as the goose. They are heir to very little sickness; in fact, a gosling one week old is virtually a grown bird, requiring little else but a stretch of pasture over which to roam and forage for itself.

Need of Grazing Land.—That goose raising is not practised so extensively as duck growing is probably due to the fact that geese require an abundance of grazing land, and will thrive best where there is a certain amount of water. They are the most persistent grazers of any kind of poultry, and though they prefer meadowland rich in plant life, which will sustain them in prime condition, they are, nevertheless, capable of adapting themselves to poor, waste land on which, perhaps, no other form of livestock could be supported. For this reason alone they should be considered by farmers, especially those who have tidewater farms, or low land bordering on rivers or ponds. A day's ride through the Eastern Shore section of Maryland will convince the most skeptical that there must be profit in geese, for nearly every farm will be seen to have its flock. Many times, when other crops have proved disastrous, the returns from the geese have been the mainstay of these farmers.

There are seven standard varieties of geese: Gray Toulouse, White Emden, Gray African, Brown Chinese, White Chinese, Wild or Canadian, and Colored Egyptian. Of these, the first two breeds are the most widely bred in this country, and are to be recommended for the average farm.

Toulouse geese (see Fig. 325) are named for a district in France where they are extensively bred. They are fairly good layers, and are well thought of as market birds, though their flesh is somewhat coarser and not so white as some of the other varieties. Their plumage being a dull gray in parts, merging

Fig. 325.—Toulouse geese.

into a lighter gray and then white on the underbody, they are not so valuable for their feathers as the pure white breeds.

The Emden geese (see Fig. 326) are probably the most desirable for all round purposes. They are rapid growers, good foragers, and are more suitable for the early markets. They originally came from Emden, hence their name, but they have been bred in this country for many years. Although not so prolific as the Toulouse and other breeds, they have other qualities which commend them as the most profitable, or at least the most

34

desirable. Their flesh is finer, whiter and of a better flavor; they have a wealth of pure white plumage which has a higher marketable value, and in disposition they are less pugnacious, more placid and more contented than the other breeds. In point of weight they are about the same as the Toulouse, according to the Standard, though as a general thing the Emden is bred somewhat smaller than the Toulouse.

Gray African geese are considered by many to be the most

Fig. 326.—Emden geese.

profitable, because of their great gain in weight in the least possible time. They are also very prolific, and many breeders cross them with the Emden for this purpose. According to the Standard they are the same weight as the Toulouse and Emden geese, yet they generally exceed the weights of these two. They have rather long necks, and their heads are conspicuous for the knobs which protrude from the base of the bill, the same as in the Chinese varieties. They also have a heavy dewlap under the throat, which is of a gray color. The knob of the African goose is black, and in the White Chinese it is orange colored.

The Chinese geese (see Fig. 327), of which there are two varieties, Brown and White, have never become extensively bred in this country, probably because of their small size. What they lack in size, however, they endeavor to make up in egg production, for they are the most prolific of all breeds of geese, averaging about sixty eggs a year. Their flesh has a superior

Fig. 327.—White China geese.

flavor and texture, and they are easy to fatten and manage. The standard weight of the adult gander is 12 pounds, and the adult goose 10 pounds, as against 20 pounds and 18 pounds for the other three varieties.

Gray wild geese, or Canadian geese, as they are also called, are about the same weight as the Chinese. Recently they seem to have come to the fore, and are prized very highly for table purposes. They are frequently crossed with African ganders, which has increased their weight as goslings. They are very hardy and easy to rear. The head and bill are black; neck black, shading to a light gray on the chest, and to a dark gray on the back. The plumage of the underparts of the body is white.

Colored Egyptian geese, sometimes called Nile geese, are the smallest of the goose family, also the most beautiful. The standard weight is 10 pounds for the adult gander and 8 pounds for the goose. They are not to be recommended for general farm use, and are bred almost exclusively for ornamental purposes. They are of a very quarrelsome nature, especially the males, who will frequently fight among themselves until dead.

Houses of the most simple construction are used for shelters for geese, little more than sheds, in fact, having nothing but a supply of straw or other material for litter on the floors. These may be of dirt or concrete; the latter is best to conserve the manure, which is in large quantities and a very valuable by-product. If ground phosphate rock is sprinkled over the manure at regular intervals, and fresh litter is added as required to keep the houses in a sanitary condition, there will be a surprising amount of the finest kind of fertilizer produced by even a moderate sized flock.

Fields that are worthless for cultivation may be turned into goose pastures, and those which have streams or unused springs are especially suitable. Unless too many birds are turned into a small area, which is then likely to become depleted of its plant life, the geese will gather the largest portion of their food, consisting of grasses, insects, and other animal and vegetable life. Or they may be made to work in the stubble of the grain fields, in place of hogs, for it is a simple matter to drive geese to distant pasture and home again at night.

Geese are much maligned, in that they are accused of destroying pasture for cattle and horses, which is true only if they are kept in large numbers in a comparatively small area. The same is true of sheep or almost any other form of livestock. Where there is a sufficiency of grass and other plant life, they may be left to graze in the same pasture with cattle and horses. Many poor pieces of land have been converted into good pasture lots by being stocked with geese for a few years.

They require drinking water in abundance, consequently, unless they have access to large bodies of water in which to swim, their drinking water should be supplied in fountains in which they can only get their bills to drink. Otherwise they will contaminate the water.

Although the goose is aquatic, and it must be admitted they seem to do better when given access to a body of water, especially in the summer, it is not absolutely essential for them to have a swimming place, any more than for ducks. They will keep themselves cleaner if a stream is available, and the chances are the fertility of their eggs will be greater. Then, too, the stream of water affords a large element of their food, which is of immense value in the cost of their upkeep. But it does not follow that they are not to be reared on farms without a watering place.

Age.—One of the most remarkable characteristics of the goose is its long life. Many have been known to attain the age of forty years, and have been handed down from father to son, as though they were a fixture on the farm. It is not at all uncommon for birds to live fifteen years, and as a general rule they will maintain their laying and hatching qualities throughout their life. Ganders are at their best as breeders at three years of age. The use of immature stock should be avoided as much as possible, especially for the renewal of breeding geese. To produce early goslings for market it is sometimes necessary to use eggs from young stock, as they usually lay earlier than the older birds, which is perfectly proper.

In selecting geese for breeders excessive size should not be sought at the expense of other important features, such as width of breast in proportion to length of body, depth of keel and shortness of leg. Care should be taken to avoid inbreeding, and to be sure of this it is sometimes necessary to procure ganders from a distant point. If so, the ganders should come from the same flock to insure their dwelling together amicably.

Mating.—As a rule three geese are mated to one gander, but in the case of very large specimens it is sometimes better to mate two geese to one gander. It is well to start this mating in the autumn, for geese are rather eccentric creatures, and require some time before they become accustomed to new surroundings and settle down to work in earnest. They also make strong companionships, and will pine and worry for weeks at the loss of a mate.

It is usually more economical to keep geese in one large flock, in preference to several small flocks, particularly if they have to be driven to pasture. When the ganders are admitted to the flock for the first time, and they are to be mated one to three, each gander will select his three wives to whom he will remain devoted for years. One of these three will probably receive the most attention, however, and will be his chief consort in their ramblings.

Laying.—Young geese usually commence laying in February, and the older ones in March, although if the weather is unusually mild they will start a month earlier. During the winter months, preparatory to this breeding season, the stock should not be allowed to become too fat, for an over-fat condition is not conducive to either productiveness or fertility. So long as there is pasture for them, they require very little grain.

When pasture is not to be had, and they are fed a grain diet, the greater part of this ration should be soft food, such as bran, middlings, corn meal and so on, with a little beef scrap. At least ten per cent of the bulk of this food should be green stuffs of some kind, either parings, cooked vegetables, or steamed clover or alfalfa. It should be moistened by skimmed milk or water. Grit and oyster shells are kept before them at all times the same as for chickens. Most breeders feed the soft food in the early morning, and a light feeding of cracked corn at night.

Broodiness.—The goose usually lays an egg every other day, until from ten to fifteen eggs have been laid, when she will become broody. As soon as this inclination presents itself, the goose should be removed from the nest and her maternal instincts broken up, whereupon she will join the flock and again commence laying. After she lays the second clutch and becomes broody, she should be discouraged again, and made to complete the third laying. The second and third clutches are not apt to be so large as the first, being one to three eggs less.

Nests.—Geese make their own nests from straw on the floor of their houses, if they are encouraged to do so, otherwise they may lay outdoors or in remote spots where the eggs are likely to spoil. By the time the goose has completed laying a clutch of eggs, she will have lined the nest with a thick covering of down plucked from her breast, which makes a nice warm place for the goslings to hatch. From ten to twelve eggs is the correct number to place under a goose, and care should be taken that the broody one is not too warlike in her attitude. If such is the case, she is very apt to crush her eggs, especially during the last few days of hatching, when the shells become more or less fragile. For this reason many goose breeders prefer to have the hatching done by hens, giving the hens about four eggs each. It is well to give the first eggs laid by the geese to hens, or they may be hatched in incubators with good success.

Hatching.—From twenty-eight to thirty days are required to incubate goose eggs, and they require a great deal of moisture, much the same as duck eggs. They should be sprinkled at frequent intervals, and given plenty of time to cool after the first week.

When the hatching is done by geese, the little goslings should be carefully removed from under the goose as they are hatched, allowing but one to remain to reassure the mother; otherwise the great weight of the goose is apt to crush them, or she will trample them. The goslings may be kept in a warm box, and when the hatch is completed and they are sturdy enough to walk about, which is usually on the second day, and at which time the yolks have been absorbed, they may be given back to their mother. The goose and her brood should be housed in a sheltered spot, and the mother confined for the first week, at the same time giving the goslings their freedom, which prevents the old goose from taking her charge too far afield and exhausting them.

When a gosling is a week old it is usually reckoned as a grown bird, for, barring accidents, it is a very hardy creature and will make rapid growth. If given good pasture they require but one feeding daily after they are two weeks old, but they should be returned to the security of the goose house every night.

Fig. 328.—Muscovy ducks are sometimes regarded as geese.

Turkey hens make good mothers for geese, because they can cover so many eggs; but they should be confined with their broods for the first week. When goose eggs are placed under chicken hens, the caretaker should make it a point to turn the eggs daily, for they are too heavy for the hen to do this.

The prices obtained for geese in the large city markets run from fourteen to twenty cents per pound live weight, which quickly mounts up when we consider their great weight. In addition to this revenue, there is a nice profit to be made from the feathers. A prime goose will average about one pound of feathers a year, and feathers of good quality will bring from forty-five to sixty cents a pound. If the down is separated from the feathers, it will bring about a dollar a pound. The feathers should be plucked when there is no blood in the ends of the quills.

CHAPTER XLII

TURKEYS

Turkey Hearsay.—There is a widespread impression in some localities that turkeys are exceedingly difficult to raise, and that due to a heavy mortality among young turkeys the chances for profit are very precarious. Much of this hearsay is nonsense, gossip—pure and simple, or let us call it turkey tradition, mysticism. Like other traditions or prejudices, these notions are hard to eradicate. The beginner with turkeys should disabuse his mind of these notions, discard them utterly, since they contribute nothing to the industry but fear and worriment.

Susceptible to Exposure.—It is true that young turkeys are delicate and that they are susceptible to exposure, to cold and dampness, but they are nothing like as frail as one might suppose, judging from the popular idea. For that matter chicks are delicate creatures, too, and unable to endure exposure. The young of all fowls require a great deal of care for the first few weeks; it is a part of the business of growing livestock of any kind. The point is to master the details in the most practical, labor-saving manner. With proper care a good proportion of the poults can be raised, and when the holiday season comes round a handsome profit has accrued.

Standard Varieties.—No doubt it will surprise those who have but a casual knowledge of turkeys to learn that there are seven different varieties of domestic turkeys in the United States, each with certain points of excellence, and ranging in color from white to black. Many of us think of turkeys as being of a single breed, commonly known as the Bronze, or Mammoth Bronze; a few are familiar with the White Holland and Narragansett varieties; whereas only those who have made a study of these fowls know about the Bourbon Reds, Slate turkeys, Black turkeys and Buff turkeys. The Buffs and Slates have always been rare, and to-day the Blacks are seldom bred. The Bronze is the most popular, next comes the Narragansett, and then the White Holland, though in recent years the Bourbon Red has grown into considerable favor and may be entitled to third position.

The exact origin of the domestic turkey will probably never be satisfactorily settled, for ornithologists are greatly at variance on this subject. The most accepted view of the matter is that all the turkeys of the world have descended in some way from the three forms of wild turkeys, the North American, the Mexican, and the Honduras, or Ocellated turkey. There seems to be no question concerning the transportation of these birds from America to Spain about the year 1520, and that they were subsequently shipped to England, in 1524, where they soon became very popular and were extensively bred. Many improvements were made among the English breeders, but it remained for American fanciers to develop the present standard varieties.

The color of the North American wild turkey is much the same as the Bronze. It is black, wonderfully shaded with bronze, the breast plumage being dark bronze, illuminated with a lustrous copper or gold color. The name Bronze is derived from this beautiful metallic sheen.

Mexican Turkey.—The wild turkey of the southern part of the continent, known as the Mexican turkey, is shorter in shank than the North American species. The color is much the same, except for the white markings on the tips of the feathers, which is considered to be responsible for the color of the domestic variety known as the Narragansett. From the meager records available, the Mexican turkey was the first variety to be taken to Europe by the Spaniards.

The Ocellated turkey, indigenous to Honduras and other Central American countries, is considered to be the most beautiful in color, and may be compared to the Impeyan pheasant. The ground color of the plumage is a bronze-green, banded with bars of gold, blue and red, or a lustrous black. The head and neck are devoid of feathers, and unlike the other wild varieties, it has no breast tuft. Unfortunately, this breed will not thrive in northern climates; it seems to be too sensitive to cold.

The domestic Bronze turkey is too well known to require any particular description. See Fig. 329. It is being raised almost to the exclusion of all other varieties, and holds the post of honor for size and market requirements. Hens of this species

Fig. 329.—Bronze turkeys.

run from sixteen to twenty pounds or more, and while the Standard of Perfection calls for a weight of thirty-six pounds for gobblers, they are often brought to much higher figures—even exceeding fifty pounds.

The Narragansett is next in size to the Bronze, and back in the days when Rhode Island was the leading turkey state of the Union, this variety was one of the most widely bred throughout New England. The ground color of the plumage is black, with markings of white and black which imparts a grayish cast to the entire surface. The female is lighter in color throughout than the male. The standard weight for hens is eighteen pounds, and for cocks thirty pounds, though they are grown almost as heavy as the Bronzes. If anything, the Narragansett is more suitable for market purposes than the Bronze, inasmuch as it has fuller, plumper breast, and will mature slightly earlier. Furthermore, they seem to bear confinement better.

The Bourbon Reds have attained great popularity in the West, though still rarely bred in the Eastern states, and rank very high as a market bird. They are hardy, mature rapidly, have excellent quality of flesh, and their weights are about equal to the Narragansett. The plumage is a chestnut color, which is made strikingly beautiful by brownish red markings and pure white tail and wing feathers.

White Holland turkeys are now quite widely known; at first they were small and delicate and not so desirable. See Fig. 330. They are beautiful birds, with snowy white plumage and pink bills and shanks, and are considered *sports* from other turkeys. In recent years the breed has been improved in size and vigor by the infusion of blood from the white *sports* of Bronze and Narragansett varieties. Just why the name Holland attaches to this species is not definitely understood. They may have originated in Holland or been brought to this country by Hollanders, but it is certain that they were not natural to the Netherlands. They have been known to exist in England for over a hundred years, and are sometimes referred to as "Austrian Whites."

Dress Well for Market.—It is thought that the Whites are more difficult to raise than the darker varieties, though they mature rapidly, attaining market size in five to eight months. They dress splendidly for market, as with all white poultry, the pin feathers show less than in darker birds, and their feathers command higher prices than those of the colored breeds. The standard weight for hens is eighteen pounds, and for cocks twenty-eight pounds. At recent exhibitions I have seen toms that weighed thirty-five pounds, but this is unusual.

The Black turkey is much the same as the English Norfolk turkey, and is very desirable for table purposes. The young are quite hardy when produced by strong, non-related stock, and when it is necessary to confine turkeys upon a more or less restricted area, the average farmer will do well to select this variety. They are not so large as the other breeds, except the

Fig. 330.—White Holland turkeys.

Giant Blacks, which closely resemble the Bronzes in everything but plumage.

The Slate turkey, sometimes called the Blue turkey, and the Buff turkey, range from ten to twenty-five pounds, according to age and sex, and may be raised to advantage in almost any locality. These varieties have been neglected for some reason or other, in spite of the fact that they possess qualities that are the equal of the more popular varieties. Lately, fanciers have devoted considerable attention to the Buffs, which are truly beautiful specimens.

Profitable.—When we consider that from the time turkeys are six weeks old until winter sets in, they will obtain the greater part of their sustenance from the fields and woods over which they roam, and this assures their keep at virtually no expense to the grower, the question naturally arises: Why is the farmer not more enthusiastic about growing them? And what is responsible for the losses that we hear so much about?

These losses really do exist, of this there can be no doubt, and with the industry in its present stage it is likely to be accompanied by grave uncertainties. Turkey raising demands the best efforts of which we are capable—and then some. It is no secret, however, that many of the failures are attributable to gross carelessness or ignorance, or both. The whole question of deriving a profit from turkeys resolves itself into the ability of the grower to rear the poults, the young turkeys. If the same intelligent care in selecting the breeders is applied to turkeys as we unquestionably devote to the scientific breeding of cattle, sheep, hogs and horses, then we are in a fair way to achieve success.

Deterioration through inbreeding is the greatest foe of the turkey industry, and it has been brought about by the heedlessness of hundreds of farmers who have declined to consider the necessity of infusing new blood into their flocks. For generations turkey growers in many of our eastern states have depended upon their neighbors for the service of male birds, giving no thought to the inevitable consequences, until in some localities it is difficult to find any unrelated stock. This total disregard of the fundamental laws of nature has in some sections reduced the condition of turkeys almost to a state of imbecility, and so undermined the vitality of the birds as to make it difficult to rear a tenth of the number of poults hatched.

Avoid Inbreeding as You Would a Plague.—New blood is of vital importance. It is better to send a thousand miles for a new male than to run the chances of inbreeding. Whenever possible the tom should be a yearling, and the hens not less than two years old. The hens from good stock will cost about five dollars each, and the gobbler from six to ten dollars. A less expensive way to start may be made with eggs purchased from reliable breeders. When this is done, it is better to secure settings from different localities, and the poults carefully marked when hatched so that they can be properly crossed another season.

Time to Start.—The fall and early winter is the best time to make a start with turkeys, for at these seasons there is a greater number of birds from which to make a selection, and they are generally offered at better prices. Furthermore, stock bought at this time will become accustomed to each other and to their new quarters before the breeding season commences, consequently better results are likely to be secured.

Parent Stock.—Every precaution should be taken to obtain strong, vigorous stock. Do not imagine that size is the main point of excellence. A medium-sized gobbler weighing about twenty-five pounds will usually render more satisfactory results than an over-heavy specimen. In all fowls, remember that size is largely influenced by the female, and the color and distinguishing characteristics by the male. The hens should be well matured, weighing not less than fourteen pounds, intelligent and tame, as distinguished from wild and unduly excitable birds, and of pronounced constitutional vigor.

A safe rule for mating is to have a tom for every four to six hens. Good fertility is reported from matings of a male to every twelve females, but I am inclined to think this is unusual. On farms where the flocks are yarded it is customary to keep two cocks for every eight or ten hens, and to alternate the males about twice a week, keeping one penned aloof, while the other is with the flock. When turkeys are given unlimited range, which is the most successful method of raising them, they naturally divide into flocks.

Management.—It is said that the real secret of success in rearing turkeys is exercise. They must have an abundance of food, and to maintain the necessary health to assimilate large

Fig. 331.—Shelters such as this provide shade and protection against sudden storms.

35

545

quantities of food, they must have an abundance of exercise. The idea is entirely logical. Turkeys are large birds, semi-wild by nature, possessed of a roving disposition, and fully capable of taking care of themselves. To confine them is to impose a feeling of constraint and worriment, over which they never cease to fret. To do well turkeys must have range. Only a few should be attempted in a confined space, and even then they will require painstaking care.

Feeds.—Much of the so-called *bad luck* in turkey raising—infertility, soft-shelled eggs and impaired vigor—is due to improper feeding. Avoid having the breeding stock too fat. If they have become so during the winter season, endeavor to reduce them to medium flesh before the mating season. Oats is one of the best feeds during the breeding months, with an occasional feeding of wheat, corn, barley and ground bone. Grit, oyster shells and charcoal should be kept within easy reach of the birds at all times, also a plentiful supply of fresh drinking water.

Roosts.—Turkeys do better when they can roost in the open. Only in storms do they seek protection, and not always then. Fences and trees are preferable to tight houses, for the turkey must have unrestricted ventilation. It is most unwise to compel them to roost with other poultry. If found necessary to house them, which is recommended in extremely cold climates, their quarters should be roomy and perfectly ventilated. The ideal shelter consists of an open-front shed or house, which is sufficient to protect them from heavy storms and from enemies, such as dogs. Foxes take a heavy toll of turkeys annually, especially if there is much brush or wooded areas nearby.

With the approach of cold weather, when insect food and greens become scarce, an increased grain diet must be provided for the growing turkeys. Do not give the flock large quantities at first, but work up the supply gradually, until they are having all they will eat up clean. Wheat and corn is about the best ration. Keep them growing and fattening as fast as possible, so that by the time Thanksgiving week arrives, and prices are usually the best, they will be in prime condition for marketing.

To grow the best is more expensive than to grow the poorer grades, but the profits to be gained are almost double.

After the turkeys are ready for market quite as much care should be given to the killing, dressing and shipping, not to forget grading, as to the growing. If these conditions cannot be obtained, it is better to sell the birds alive to someone who makes a business of handling such stock.

Kill Nothing but Well-fattened Stock.—It never pays to send poor stock to market. Skinny, gawky, crooked-breasted carcasses are undesirable. Keep the stock away from food or water for at least twelve hours before killing, preferably for twenty-four hours. The food tract must be emptied, otherwise there is danger of discoloration or spoiling. Full crops and full entrails may increase the weight slightly, but they discount the price so heavily, there is nothing to be gained, only disappointment. Crops distended with food are sour, sometimes tainting the flesh, but in any event they are uninviting to the careful buyer.

There are several methods of killing, but the most popular way is to suspend the fowl by the shanks, head down, and cut or stick it in the roof of the mouth with a sharp, narrow-bladed knife. This severs the arteries, causing a hemorrhage, and at the same time pierces the brain, causing insensibility. The flow of blood should be copious, for poorly bled fowls are likely to be purplish-tinted.

Dislocation.—Another method is to dislocate the neck by a sudden twist and jerk. The disjointed part of the neck is then pulled away, so as to form an open space into which the blood may settle. Dislocation is claimed by some to be the only sanitary, up-to-date method of killing, since there is no opening by which air can get into the body. It is used more for chickens than for turkeys, and requires considerable practice to do it well. Then there is the old-fashioned method of beheading with an ax, which should never be used, except on birds intended for home use, and even then it is a very poor mode.

Dry-picking is the only way to pluck poultry for a fancy market. As soon as the bird is stuck, and while the blood is still flowing, commence to remove the feathers, taking great care not to break the skin or tear the flesh. Purple abrasions, often noticeable on plucked fowls, are due to bruises and rough handling. Avoid these. Nothing detracts so much from the appearance of dressed poultry as careless workmanship in the picking. It will also mean a reduction in the selling price. Remove the pin feathers with a blunt knife.

In dressing turkeys a small ruff of neck feathers and the wing tips are undisturbed; this is a conventionality found in most markets. To complete the dressing, cleanse the mouth and head of any blood, and wash the feet thoroughly. Never remove the head, feet or entrails. Some years ago it was the practice to remove the viscera, but modern efficiency has found that the undrawn carcass, from which all animal heat has been expelled, is the most sanitary.

Cooling.—When the fowl is plucked hang it in a cool place, head down, until the heat is entirely gone from the body; it is then ready for packing. Poultry should not be allowed to freeze for it will spoil the appearance.

Packing and Shipping.—Barrels are generally preferable to boxes for shipping poultry, and they are easily obtained at a small cost. Line the package with manila paper,—do not use soiled or printed paper,—and pack as tightly as possible to avoid shifting about in transit. In warm weather use ice, or if the consignment is billed for a long distance. Head the barrel securely, and mark its contents plainly on the head to whom it is shipped, and the name of the shipper. Never ship mixed lots of poultry in the same package if it can be avoided. Graded shipments invite good treatment on the part of the merchants, facilitate sales, and are rewarded by larger returns. In short, it pays handsomely to take a few extra pains.

Turkey Nests.—If left to follow her own inclinations the hen turkey will select some secluded place for her nest, probably under a pile of logs, in the brush or in the lee of a stone wall; but, unless the flock is particularly wild, the hens can be induced to adopt more suitable laying quarters, than which nothing is more practical than barrels laid on their sides and blocked to keep them from rolling, and lined on the bottom with sod covered with straw or hay for nesting material. Sugar barrels answer the purpose nicely; they are easy to handle, and not only make an excellent protection from cold winds and rains, but hot days as well. Later, when the brood is hatched a board may be

Fig. 332.—Combination chicken and squab farm in New Jersey.

nailed to the lower end so that the little turkeys cannot start to roam at too tender an age.

Turkeys usually commence laying about the first of April, and the earlier the eggs can be hatched after the middle of May the longer the period of growth before the first holiday demand. Six months at least are required to bring them to a profitable marketable size, yet if their environment and feeding are correct this can be readily done.

The hen will lay from 15 to 20 eggs before becoming broody, and if one wishes to obtain a second clutch of eggs, it is not difficult to break her of broodiness, whereupon she will soon start laying again. The first clutch of eggs may either be sold or placed under chicken hens to be hatched. The eggs should be gathered as promptly as possible and stored in a clean, cool place. They should not be kept too long, for their vitality depreciates rapidly. The poult issuing from the egg that is set within a few days of its being laid is noticeably sturdy.

As a rule 18 eggs are sufficient for a turkey hen to cover properly, and 10 eggs for the chicken hen to manage. If more than these are placed under the birds there is danger of the eggs being chilled at times, and poor hatches will result. Before setting the hen dust her thoroughly with a lice-expelling powder, also the nest and nesting material. This will prevent trouble from lice or mites, either of which may cause the hen to desert her nest. When the hatch is ready to come off the hen and nest should be dusted for the second time. Nothing is more fatal to turkey health than parasites.

Eternal vigilance should be the watchword. When the little ones droop search for vermin, and be satisfied with only the closest scrutiny. Vermin is responsible for some of the greatest losses. How the turkeys become infested is sometimes an enigma—from coops, from other poultry, from sparrows—despite the best care these pests will put in an appearance, and a vigorous campaign against them is necessary. Poults so afflicted will sicken and die as though stricken with some wasting disease. In fact, the poor little things are frequently dosed internally with medicine, when their dire need is an insect powder.

From 27 to 29 days are required to hatch turkey eggs, depending upon their freshness, weather conditions and the devotion of the hen in charge. During this time she should not be disturbed in any way. If frightened or driven from the nest the hen is apt to abandon the eggs entirely. Some hens will rear two broods in a season, and while the late-hatched brood will not mature early enough to meet the demand of roasters, nor are they so desirable as breeders, they may be grown as broiler poults and as such sold to advantage.

Poults Unlike Chicks.—Many poultrymen have met with disaster with turkeys because they tried to apply the same care to the poults that they were accustomed to giving little chicks. This is a great mistake, for on vital points the turk and the chick are widely different. Poults have a ravenous appetite, without the chicks' capacity for digesting and assimilating large quantities of food within a short space of time. They require a certain amount of exercise as an aid to digestion, yet too much

(*Courtesy U. S. Dep't Agriculture*)

Fig. 333.—Typical pigeon house and fly.

running around wearies them to the point of complete exhaustion. To guard against this, especially if chicken hens are being used for mothers, the hens should be confined for the greater part of the day for the first week, or until the poults have gained sufficient strength to be taken far afield. Chicks seldom tire in this way, and will trudge along unceasingly from dawn to dark without ill effects. Still another point: soured food or partially decayed food, which a chicken might eat with safety, works havoc with the turk's digestion.

Diet.—Many of the difficulties with turkeys spring from the attempt to make them conform to an unnatural diet and management. The turkey raiser should always bear in mind that poults are seed-eating chicks, not slop eaters. In the wild state their food consisted of the bugs, worms, seed sand other tidbits which they hunted for in the woods and fields. In this habitat there was no overfeeding of unnatural, concentrated foods, likely to impair health and produce bowel troubles.

Like quail and other wild birds, turkeys subsisted by their own efforts as foragers, and they were strong and vigorous, whereas in a domestic or even semi-domestic state, they are too often forced to eat unnatural foods with the idea of forcing them to make an unnaturally rapid growth. Some breeders feed bread and milk as soon as the poults will eat, while others feed dry bread; some adhere strictly to a grain diet, while others feed anything that happens to come to hand. The writer has seen farmers' wives feed an exclusive diet of corn meal and skim milk curds, believing that they were taking special pains with their turkeys. Either the corn meal or the curds alone was enough to kill the brood, and about the only thing that saved them was the combination. The evil of the corn meal helped to offset the evil effects of the curds.

Foods.—As a general practice dry foods are safer than moist ones, though milk is very beneficial when fed judiciously. Stale bread soaked in milk, with the milk pressed from the bread before feeding, is a good food for the first few days, and it should be sprinkled with a fine grit. The addition of a little hard-boiled egg is good. Fine oatmeal or finely cracked wheat and corn and a little granulated beef scrap are excellent rations. Bread baked from corn meal, middlings, bran and ground oats may be used, and after a few days add a little meat scrap. A small quantity of lean beef cooked and chopped into fine bits is a strengthening ration, but care must be taken that the meat is sweet. Nothing will start bowel troubles quicker than tainted beef. Raw meat and green bone should be avoided for this reason.

Feed the poults frequently, giving them but a little at a time, and be particularly sparing with the concentrated foods, such as grains in the hulls—millet, kaffir corn and so on. Too much hard-boiled egg or milk curds will congest the bowels. Do not forget to provide succulence; green stuff should constitute at least a half of the fare. Keep an abundance of clean water before the poults at all times, also a plentiful supply of sharp grit. Charcoal should be mixed with the food or fed separately; it aids digestion and guards against fermentation in the crop and gizzard.

Attention Required.—It should also be remembered that turkey hens are not apt to be so attentive to their young as chicken hens, especially in the matter of feeding; moreover, the turks are not so apt as chicks in learning how to eat. The chick commences to peck at objects almost as soon as it is hatched, but not so with poults. They seem particularly unintelligent little creatures in this respect, and the idea of looking on the ground or floor for food never seems to occur to them. Indeed, sometimes they run about with their heads in the air crying for food, until they weaken and die from exhaustion. The turkey grower must be on the lookout for this, and if necessary teach the poults by hand feeding. It involves considerable time and trouble, of course, but the advantages gained in giving the little turks a good start will more than compensate for one's pains.

By all means keep the brood dry—dampness is fatal. Do not allow the hen to take her brood into the tall grass early in the morning when the dew is on it, or trudge about on rainy days. Keep her penned up in an airy, roomy coop until conditions are favorable. It is also imprudent to expose the brood to intense heat, for they will wilt under it as though suffering with sunstroke. Shade of some kind must be provided in hot weather so that they can escape the direct rays of the sun.

Moderation in all things is the secret to success. Avoid dampness and filth, guard against vermin, do not overfeed and do not overcrowd; these are the essential features.

CHAPTER XLIII

GUINEA FOWLS

Nature of Guineas.—Generally speaking, until recent years the guinea has merely been tolerated on the farm, and seldom regarded as profitable. Semi-wild by nature, noisy, flighty and unmanageable, they exhausted the patience of the farmer and were too troublesome for serious consideration. Their chief virtue, it seemed, was their well-known habit of setting up a discordant chorus at the slightest provocation. If a hawk appeared, or an animal or person approached the barnyard, these alarmists immediately burst into a raucous denunciation. Woe betide those who trespass on lands where guineas abound; nothing seems to escape their notice.

Left to their own devices guineas will skirt the edges of civilized poultrydom, multiplying in distant fields and hedge rows, but rarely reproducing more than their own number. Though a hen will often hatch a large brood, it is seldom that she manages to raise more than two or three chicks, and often not that many. For some obscure reason the mother guinea does not seem to realize that her little ones are frail creatures, unable to withstand extremes of heat and cold, moisture and long tramps afield. Their habits with their young seem to be about as senseless as their noise, which probably accounts in a large measure for their culture having been neglected.

Snows and stormy weather sometimes drive the guineas to the barnyard for food and shelter, but as a rule they are very independent. Therefore, having cost the farmer nothing, he was satisfied to sell them in an indifferent market for forty or fifty cents a pair, or to tolerate them for the sake of an occasional Sunday dinner for the family. No one will ever become rich raising guineas, and it is hardly likely that any one will find it feasible to raise them as a commercial proposition, that is—to make a living by keeping them as a specialty; but they can be kept as a profitable side line.

The delicately wild flavor of the guinea's flesh and its tenderly plump, dark breast commend it to the epicure. The appetites of hotel and restaurant patrons are keen for game birds, but owing to the increasing scarcity of these delicacies, those who cater to epicurean tastes have had to seek substitutes for quail and

(Courtesy Purdue Experiment Station)

Fig. 334.—Open-front poultry house. Windows at top permit sunlight to flood the rear of the building, where it is most needed.

pheasant that were hitherto plentiful. In consequence, much of the so-called game listed on menus is—guinea. Whether the epicure eats it under its own name, or deludes his palate with the thought of wild fowl, he must admit that the guinea rivals quail. Hence more young guineas are eaten now than ever before, and the demand is steadily growing. The truth of this is reflected in the prices received from dealers; instead of forty or fifty cents a pair, guineas now bring about a dollar a pair and more. And their eggs are considered quite a delicacy.

Varieties.—There are two popular varieties of guineas—Pearl and White, the only difference being in their color. The Pearl variety should be bluish-gray in color, each feather marked with white spots resembling pearls, hence its name, but it must be free from any solid white feathers in any part of its plumage. The White variety should be a pure white in plumage, with orange or yellowish-white bill and legs.

In the Pearl variety the bill and legs are brown. Some specimens of this variety have white breasts, or breasts of a lighter gray color than the back and other plumage, which denotes a cross between the Pearl and White, hence they are mongrels. For size, egg production and other characteristics both varieties are equally desirable, although the Pearl Guinea is probably the most common. The flesh of both is dark, but that of the White, or of the White crossed with Pearl, is a shade lighter.

Except in size, a newly hatched Pearl Guinea is the replica of a baby partridge; markings, colorings and contour are identical. Lately, there has come to be another variety known as the Dove Guinea, but it is scarcely popular enough to be recognized.

Sex.—The amateur has great difficulty in telling males from females; at a casual glance they are indistinguishable. The most accurate method of distinguishing the sex is by the cry— the hen has the preponderance of vocabulary. The well-known *potrack*, *buckwheat* or *too quick* is uttered by the female only, while the *che* or *tck* is typical of both male and female. Moreover, the female seldom screeches like the cock. Those who are well acquainted with the fowl will observe that the male has a larger spike on his head, and that the ear-lobes are also larger than those of the hen, and that the lobes generally curl in a sort of semi-circle toward the beak. It will also be noted that the cocks usually hold their heads higher than the hens.

The male selects his mate and his devotion is steadfast. While the hen attends to the duties of maternity, the male remains close by, ready at the least sign of danger to utter his shrill cry of alarm. When the brood appears, he shares the

responsibility of food and shelter, and should misfortune over-
take the hen, the cock assumes her duties in a thoroughly competent manner.

Nests are often located by the observance of guineas feeding
solitarily, since this is a pretty sure sign of the male bird, and
that the hen is laying nearby. They will make their nests in
remote, out-of-the-way places, under hedges, bushes, brush-
heaps or wheat shocks, and if their nests are disturbed they will
move to another place. Laying begins about the middle of April,

(Courtesy Purdue Experiment Station)

Fig. 335.—Another view of house shown in Fig. 334, taken in midwinter.
Fowls are in splendid condition, proving that they require an abundance of
fresh air, providing the house is free from drafts and dampness.

and if broodiness is discouraged it will continue throughout the
summer.

Rearing Young.—Owing to the heavy mortality among broods
reared by guinea hens all attempts to raise a large flock by
natural methods will be discouragingly slow. On the other hand,
to raise the guinea chicks by artificial brooding methods is
equally difficult. When placed in a hover, they either remain
there all the time, or come out and cannot find their way back.
Unlike young chickens, they are particularly stupid in learning

how to eat or drink without the aid of a hen, consequently they soon pine away and die. They do not seem to have the imitative ways of other young fowls, and in many respects they resemble young turkeys.

Experiments have been made by placing newly hatched chickens with the young guineas, with the hope that the guineas would learn how to eat and drink from their companions and become hover broken; but not so. The guineas stood around until they were weakened by cold and hunger, while the chickens thrived. Apparently, the only other way to raise guinea chicks is under chicken hens, and this may be done very successfully.

Hatching with Chicken Hens.—The eggs should be given to reliable sitting hens, such as Plymouth Rocks or Wyandottes— about sixteen eggs to each hen, or the eggs may be started in an incubator and later given to the hens. From twenty-six to twenty-eight days are required to hatch guinea eggs, and true to their heritage of fear, as soon as they leave their shells the little fellows slink into corners of the nest away from the prying eyes of the attendant. Naturally, their foster-mother's call is a foreign language to them, which they find difficult to understand at first, and until they get to know the meaning of *cluck*, and the hen becomes accustomed to their peculiarities, they must not be allowed to roam.

Care should be taken at the beginning to see that the hen accepts her responsibility kindly, for sometimes chicken hens are antagonistic to young that is not their own, and will kill the guineas. When the hen has proved her dependability she may be given twenty chicks; she can easily take care of this number; and if she is a very large hen, twenty-five chicks are not too many.

The hen and her brood must be confined in a coop for the first few days, after which they may be given the freedom of a small yard. Later, after the attendant has observed that the chicks respond to their foster-mother's guidance, they may be allowed complete freedom with the hen. Do not confine the hen within a slatted coop and allow the chicks to run abroad, as is

(Courtesy Purdue Experiment Station)

Fig. 336.—Colony house for growing stock on edge of corn field. Range of this sort cannot be excelled.

the custom with young chickens, for unlike young chickens, the guinea chicks will not always return to the hen in the coop.

Furthermore, it is important to keep the chicks off the wet grass and out of the rain for the first month, for like young turkeys, dampness is fatal. Failure to appreciate this fact is responsible for the heavy losses among broods reared by guinea hens. They have been known to start out in the morning with large healthy broods, and return at night without a single chick —the entire flock having perished along the route from exposure and exhaustion.

Guinea chicks are ravenous little creatures, and for the first week they should be fed five or six times a day. If allowed to become too hungry they will over-eat, and digestive troubles may result. They thrive on bread crumbs and rolled oats, mixed with hard-boiled egg, or on fine ground chick-food. Sour or fresh milk may be substituted for the egg. Owing to its insectivorous nature, the guinea requires a large proportion of animal food, also green food, and they must have plenty of water to drink, grit and charcoal. As soon as they are large enough to be given free range, which should be done at the earliest possible moment, for guineas do not thrive well in confinement, they will forage for the greater part of their keep. At such times they may be fed a coarser chick-feed, wheat, and later, corn.

Maturity.—Broods hatch from May to September, and in three or four months they will reach the marketable weight of one and a half to two and a half pounds. They rarely suffer from any of the countless diseases that poultrydom is heir to, and no houses are essential for their comfort, except during the brooding season. The hens are prolific layers, and as the chicks reach a marketable size at an early age, they yield a quick return for their feed and care. The one serious problem is in raising the chicks for the first month, after which success is assured.

Guineas should be killed by *sticking*—severing the blood-vessels on the inside of the throat, so that no cuts are visible, and after the birds are dead and have finished bleeding, all blood clots and smears should be carefully washed off, that they may present an attractive appearance when marketed. Guineas are almost always shipped without removing any of the feathers, which is another factor in favor of the producer. Ninety cents a pair is probably the average price in the large Eastern markets, and at this figure there is a nice profit for the producer.

Distrustful Nature.—In no other fowl does the instinct of distrust seem such a conspicuous characteristic as in the guinea, and this timorousness is responsible for its hitherto limited list of friends; yet for all this senseless hysteria and shyness, they can be made to yield a certain amount of confidence to the attendant who treats them kindly and feeds them regularly, and in so doing there is both profit and interest in their culture. To frighten or treat them roughly is to alienate them beyond hope, and their propensity for flying makes them exceedingly difficult to capture; in which case they will have to be shot. In this wild state they are likely to be pugnacious, and to frighten and drive off other poultry. It is also found that the wilder they are the noiser they will be; hence it behooves those who raise guineas, for their own peace of mind, to treat them as gently and sympathetically as possible.

The guinea may shriek hysterically at a shadow, and it is subject to peculiar *nightmares*—seeing ghosts and goblins; nevertheless they are excellent guards, and for this reason alone they are a valuable adjunct to the barnyard.

In buying mature guineas, the poultryman should confine them for about two weeks, so that they may become accustomed to their new home; otherwise, on giving them their freedom, they are very apt to take flight and never come back. The best plan is to purchase eggs from a reliable breeder and to raise one's own stock.

36

CHAPTER XLIV

PIGEONS

Fad.—Some years ago there was a big boom in the squab industry; it became quite a fad and received a great deal of publicity. Many failed, and considerable money was lost before people awakened to the fact that the profits in the business had been greatly over-estimated, and that the care required by the birds, the necessary skill, was greatly underestimated. Furthermore, there was not the demand for their products that beginners were led to expect. High prices were not sustained.

Failures.—It is quite likely that more money has been lost on plants erected for the production of squabs on a large scale, than in any other branch of the poultry business. Stories were told of the great success of a few breeders; they were plausible, and the figures were so seductive that many unfortunate men and women were led to invest all their savings in ventures they were in no way equipped to operate. Failures became so numerous that the business was viewed askance, as a sort of joke, and detracted much from its real credit. If one spoke of being in the squab business he was apt to be regarded with suspicion. Most of this feeling has passed, along with the boom spirit, and left in its wake a great deal of knowledge of practical value.

Profits.—There is profit to be made from the breeding of pigeons, just as there is a profit to be had from chickens or from ducks, but we would not advise the amateur to expect to make a fortune or even a good living at producing squabs, unless he is trained and equipped to operate a fairly large plant. From my observation, only large lofts return substantial profits, and most of the successful pigeon farms make a business of selling breeding stock, and are not devoted primarily to the production of squabs for market.

562

My advice to the beginner would be to start with a few pairs of birds, and not attempt to engage in the business on a commercial scale unless experiments with a few pigeons clearly indicate a worthwhile profit and success. Fifteen pairs can be handled nicely in a back lot, and will help the beginner to splendid working knowledge, furnish squabs for home use, and add to the future mating-pen. Do not plunge into the enterprise under any circumstances, especially if you have never had actual experience with livestock, and fail to appreciate that careful

Fig. 337.—Homer pigeon.

attention to details and the most exacting personal supervision are required.

Side Line.—On general farms, where a flock of pigeons may obtain the greater part of their living from the fields, they will return a nice revenue, at very little expense or trouble to their keeper. The one difficulty with a flock at large, the pigeons may be a nuisance to neighbors, or losses may occur by shooting and by cats and hawks. Many pigeons are kept as a side issue on general farms in the Middle West, but they are mostly of common origin and, therefore, not worth much as squab-producers. Common pigeons are not so prolific, and they produce small squabs of poorer quality. In consequence, the average value of pigeons in the Middle West and in the South is only from fifteen to twenty-five cents apiece. Compared with the prices received by reliable specialty breeders, the foregoing figures are insignificant. If a little more attention was paid to these general farm flocks, and a fresh supply of properly bred stock was introduced, they could easily be converted into a profitable side line.

The squab is a young pigeon just before it leaves the nest, and is considered quite a delicacy. It makes a delicate food for invalids, and is used to replace the supply of game, notably the quail. There is a fairly uniform demand for squabs in the large cities, and they bring from two dollars to six dollars per dozen, depending on quality and the season of the year.

Weights of squabs run from six to eighteen pounds to the dozen, with nine pounds as a good average. Four to six weeks, depending upon the variety, the stamina of the flock, and the care they receive, is the time required to bring squabs to marketable size. When the down disappears from the head and they are fully feathered under the wings, these are indications of the correct time for killing. At this stage they are plumpest and heaviest. If allowed to pass this period their fat decreases, and the flesh that was once tender becomes hardened. They soon learn the use of their wings and are likely to leave the nest.

There is a feeling that squabs must be kept stuffed with food from hatching time until they are ready for market. It is marvelous the quantities of food they can consume without ill effects from lack of exercise. They just sit contentedly in the nest, while the parent birds feed patiently and faithfully.

Varieties.—There are a great many varieties of pigeons, and though all will give squabs, only a few breeds are used extensively for squab culture. Of these the Homer was generally considered the most desirable variety. See Fig. 337. It is the popular standard breed, hardy, prolific and of fair size. Lately it is being supplanted by the Carneaux. See Fig. 338. Several other varieties, such as the King, Dragoon, Runt (see Fig. 339) and Maltese, which are larger than the Homer, are used in crossing to increase the size of the squabs. Crosses produce the largest squabs, and the practice seems to prevail. See Figs. 340 and 341. Too much inbreeding results in degenerates and undersized stock.

The Homer derives its name from the fact that it will usually find its way home from distant points, even when taken under cover for hundreds of miles. Records of the flights of racing

Fig. 338.—Carneaux pigeon.

pigeons are wonderfully interesting. This characteristic makes it essential to confine Homers very carefully if they are purchased. They are bred very largely for racing as well as squabs, and come in a variety of colors.

Strain.—As with all forms of livestock, good breeding birds are one of the chief essentials to success; in fact, they are the foundation of profit or loss. The beginner should secure his stock from reliable breeders, and wherever possible from those who will guarantee their products. The age and sex of pigeons are hard to determine by casual observation, and when one considers the delay and loss likely to occur from unmated birds, or from stock so old as to be past their period of prolificacy, it becomes apparent that much depends upon the word of the seller.

Unmated birds will cause a lot of trouble in a loft. Unlike poultry, pigeons are monogamous; they mate in pairs, and usually remain devoted for life. Obviously, the experienced breeder sees to it that his birds are properly mated, either

Fig. 339.—White runt, female pigeon.

naturally or by special methods. The presence of unmated males means fighting and jealousy.

Unscrupulous dealers sell anything as mated birds. Some claim that so long as there is a male and a female they are relieved of any responsibility for further devotion. An authority on squab raising has this to say about mated pigeons: "A mated pair of pigeons consists of a male and female that have built a nest, laid eggs and hatched a pair of squabs which are fit for market in four weeks from the time of hatching. The only safe way in buying breeders is to get a written guarantee that they are mated, and a list showing the pairs. The purchaser who buys birds thus represented has a right to expect that they have actually been mated, and will prove it by going to work and rearing squabs."

Production Age.—Pigeons are most productive between the age of two and six years. The larger varieties will breed at the age of eight months, smaller breeds at six months. They are sometimes serviceable as old as ten years, but this is the exception. A good rule is dispose of old breeders at regular intervals.

Fig. 340.—Runt cross pigeon.

If one does not wish to buy mated birds, a good plan is to secure young stock, about eight weeks old, and mate them at the proper age. Squabs intended for breeders should be leg-banded before they are old enough to leave the nest, and a record kept of their breeding. Otherwise it is difficult, and a matter of chance, to prevent inbreeding. Later, when the sex is definitely determined, the males are banded on one leg, usually the right leg, and the females on the left leg, to distinguish the **sex** of the birds in the pens.

The mating-pen is a separate compartment through which new pairs are added to the regular lofts. In it the young pigeons, males and females, are placed, also the doubtful pairs. Here the young birds reach maturity, and after an interesting courtship they choose partners. This is usually indicated by the male

Fig. 341.—Runt cross pigeon.

driving and pecking at the female. If properly mated, the pair will start to erect their nest, and they will be found together at night, whereas unmated members of the pen generally remain alone.

Discerning Sex.—It takes a very keen observer and one intimately versed in pigeon ways to discern the sexes before pairing, and even after the courtship has started experienced breeders are sometimes deceived in their selection of the male bird. The male is apt to be larger and more active in the love-making, and his voice is more guttural and his expression more masculine— more determined.

It is customary for the first squabs to be reared in the mating-pen, after which the parent birds are permanently leg-banded and numbered and removed to their permanent quarters on the farm. This is the natural method of mating birds.

The forced method of mating consists in confining a male and female in a mating coop, a cage about three feet long and twelve to fifteen inches high and the same in depth, with a wire partition in the center which can be removed or hinged back as desired. The hen is placed in one side and the cock in the other, where they can watch and study each other at close range for a week or ten days, and become enamoured of each other's charms. The partition is then removed, and if they take to each other's society and the mating is successful, they are taken from the mating coop and given their freedom in one of the regular pens. This method is used successfully, and is of practical benefit where special matings are desired. For example, some matings produce undesirable qualities in the squabs, in which case it becomes necessary to cull the flock and remate along other lines.

Quarters.—Pigeons are accommodating creatures; they will adapt themselves to almost every condition, from the eaves of the barn to the nests of a well-appointed loft. They do best, of course, in quarters that are fairly roomy, dry, well-ventilated and sunshiny. Almost any style of building can be converted into a satisfactory pigeon loft with very few modifications. To avoid dampness the location should be well drained. A southern or southeastern exposure is best, and the same general principles that apply to hen houses also apply to pigeon lofts. The walls and roof should be tightly constructed to prevent leaks and drafts, and above everything else the house must be proof against rats. These pests are notorious thieves in a pigeon loft, killing hundreds of young birds and destroying the eggs.

Arrangement of House.—It is customary to divide the house into pens holding from twenty to seventy-five pairs, and to have a narrow passageway or alley in the rear of the building connecting with the pens, feed-house and other conveniences. Two to three square feet of floor space per pair is sufficient room if a number of pairs are kept in the pen. When pigeons are confined, which is customary on the large squab plants, outdoor flyways or covered yards are necessary. These are generally located on the south side of the building, and are made eight or ten feet high, twenty to thirty feet deep, and extend across the widths of the pens. They should be covered on top and sides with inch mesh netting to keep out the sparrows, which will otherwise come in swarms and eat much of the food.

Alighting Boards.—A six-inch board or shelf should be placed along the two ends and possibly one side of the flyway, for the pigeons to alight upon, but it is not considered advisable to erect roosts across the center of the flyway. The pigeons are apt to strike against them and be injured. A few holes are cut in the front of the house at a convenient height, say, about five feet from the ground, for the pigeons to enter and leave the building. These need only be about four inches high and three inches wide, and three or four to a pen are sufficient. Lighting boards, six inches wide, similar to the perches in the flyways, should be placed in convenient relation to these holes, on the inside and outside of the house.

The other interior fixtures are very simple, and they should be made as easy to clean as possible. Two nest boxes are provided for each pair, in recognition of the fact that they often run two families at one time. They frequently start to lay eggs in the second nest, while ministering to the needs of a pair of squabs in the first nest. It is a good plan to have a few extra nests.

Each nest should be not less than twelve inches square. They may be built in tiers, but not made to extend above the level of the eye, else it will be difficult to clean and inspect them. Undersized nests offer cramped quarters and are a source of trouble to birds and attendant alike.

Cleanliness is the slogan in pigeon raising, the same as in keeping any form of livestock. An abundant supply of drinking water must be kept before the birds at all times, and it must be pure and fresh. All food must be placed before the birds in a wholesome condition, and they must have grit, oyster shells and charcoal, also salt. Fountains in which the pigeons cannot bathe are best for drinking vessels, while a galvanized iron pan three or four inches deep and about twenty inches in diameter makes a good bath pan. Baths are usually furnished about three times a week, and except at these times the pans are removed.

Feeding.—Many varieties of grains are used in feeding pigeons. A good grain ration may be made from the following: equal parts by weight of cracked corn (sifted), hard red wheat, kafir corn and Canada peas, with a small quantity, perhaps ten per cent, of hemp and millet seed added during the molting season. Canada peas are expensive, but for best results they seem to be indispensable, especially during the breeding season. They seem to take the place of green feed. Other grains which may be added to the ration are peanuts, which are used to some extent in place of Canada peas, hulled oats, Egyptian corn, barley, cowpeas, and milo maize. In addition to these a small amount of stale bread, rape, rice, vetch and sunflower seed may be fed for variety. Lentils are sometimes fed as a tonic to breeding birds during the molting period.

The grain may be fed on the floor of the pen, in troughs or kept before the birds in hoppers. It is not well to feed the grain outdoors on the ground for fear that it may become moldy and sour.

The young of pigeons are fed by the parent birds on a thick, creamy mixture, called pigeon milk, which is secreted in the crops of the pigeons. The squabs are usually fed shortly after the grown birds have eaten, consequently great care should be exercised not to disturb the breeders at this time. In fact, all the work about the pens should be accomplished in a quiet, orderly manner, and in as few visits as possible, for pigeons are easily alarmed. It is poor judgment to enter the lofts at night, especially with a light. The hen pigeons are likely to be frightened from their nests, and in the darkness fail to find them again, consequently chilled eggs and fewer squabs will result.

Squabs intended for market should be caught in the morning before they are fed by their parents, so that their crops will be empty. They are killed in the same manner as poultry, by cutting the arteries in the back of the roof of the mouth and piercing the brain, then bled, after which the feathers and down are plucked clean, with the exception of the head. To pluck eight squabs an hour is good work, though there are some experts who claim a record of fourteen.

Packing.—Squabs should be cooled the same as other poultry, either by plunging them in cold water, or by hanging them in a cool place. If the crops contain any food, it may be advisable to cut them open and clean it out. When the birds are thoroughly chilled they are carefully graded as to size and color, and packed for shipment in buckets with perforated bottoms. Pack them with their breasts up, in layers, with paraffin paper between the layers, and a generous supply of cracked ice throughout the entire package.

Naturally, the production of squabs from each pair of breeders varies widely, much the same as the egg yield will vary on a chicken farm. They are known to yield ten or twelve pairs a year. This is exceptional; a fair average would be seven pairs. They sell at the highest prices during cold weather, for pigeons do not breed so freely in the winter months.

INDEX

37